"For anyone interested in the meaning and value of religion in the lives of ordinary people in Sub-Saharan Africa and southern Asia, this book is an extraordinarily important and valuable one. Rakodi lucidly weaves together a deeply captivating range of insights of how religion intersects with the everyday lives and practices of people in the most religious areas of the world. This book will be immensely helpful for undergraduate and graduate-level instruction as well as researchers in the field, and it demonstrates that while religious change remains arguably the most profound and far-reaching influence on social change in Sub-Saharan Africa and southern Asia, it is perhaps the least empirically researched and understood. It also shows that to understand other forms of social change such as economic and political change, lived religion is *sine qua non*: no one who thinks about these processes of change can ignore this book!"

Asonzeh Ukah, Associate Professor and Director, Research Institute for Christianity and Society in Africa [RICSA], the University of Cape Town, South Africa

"This book proposes and elaborates some of the basic knowledge that should be an integral part of the analytic framework for much development work – what we call, often too loosely, 'religious literacy'. Such a framework, Rakodi argues, includes a basic understanding of beliefs and practices, organization, and ways in which religions relate to society more broadly, and her analysis underscores the complexity and dynamism of each of these categories. The vital roles that religious teachings and expectations play in understandings of wide-ranging norms are linked to the ways in which they shape world views. In short, this thought-provoking work introduces a vital and demanding topic. The lessons it offers have relevance far beyond the development policy field."

Katherine Marshall, Senior Fellow, Berkley Center for Religion, Peace, and World Affairs, Georgetown University, USA, and Executive Director, World Faiths Development Dialogue

"This book addresses a key gap in religious literacy for all those interested in the nexus between religion and society in Sub-Saharan Africa and southern Asia and beyond. Further, it provides a nuanced, informed and wide-ranging coverage of the links between religion and human development, both conceptually and empirically. A must-read for students of development, religious and area studies."

Mariz Tadros, Professor, Coalition for Religious Equality and Inclusive Development, Institute of Development Studies, University of Sussex, UK

Religion and Society in Sub-Saharan Africa and Southern Asia

This book analyses how religion is entangled in people's lives in Sub-Saharan Africa and southern Asia. It provides an introduction to the teachings, practices and values promoted by the main religious traditions in these regions and an overview of the evidence on what religion means to people in terms of their beliefs and religious practices and how it influences their values, attitudes and day-to-day relationships with others, especially their families.

Over the course of the book Carole Rakodi explores similarities and differences between and within religious traditions and identifies some of the key factors that influence and explain the roles played by religion in people's personal lives and social relationships. A separate companion volume will go on to focus on the social and political roles and relationships of religious groups and organisations.

This book will be of great interest to academics and students working in a range of disciplines, especially sociology, religious studies and development studies but also anthropology, geography and area studies.

Carole Rakodi is Emeritus Professor in the International Development Department, School of Government and Society, at the University of Birmingham, UK.

Routledge Research in Religion and Development

The *Routledge Research in Religion and Development* series focuses on the diverse ways in which religious values, teachings and practices interact with international development.

While religious traditions and faith-based movements have long served as forces for social innovation, it has only been within the last ten years that researchers have begun to seriously explore the religious dimensions of international development. However, recognising and analysing the role of religion in the development domain is vital for a nuanced understanding of this field. This interdisciplinary series examines the intersection between these two areas, focusing on a range of contexts and religious traditions.

Religions and Development in Asia
Sacred Places as Development Spaces
Matthew Clarke and Anna Halahoff

Between Humanitarianism and Evangelism in Faith-based Organisations
A Case from the African Migration Route
May Ngo

Christianity's Role in United States Global Health and Development Policy
To Transfer the Empire of the World
John Blevins

Religion and Society in Sub-Saharan Africa and Southern Asia
Carole Rakodi

Religion and Society in Sub-Saharan Africa and Southern Asia

Carole Rakodi

Routledge
Taylor & Francis Group
LONDON AND NEW YORK

First published 2019 by Routledge

2 Park Square, Milton Park, Abingdon, Oxon, OX14 4RN
605 Third Avenue, New York, NY 10017

Routledge is an imprint of the Taylor & Francis Group, an informa business

First issued in paperback 2020

British Library Cataloguing-in-Publication Data
A catalogue record for this book is available from the British Library

Library of Congress Cataloging-in-Publication Data
Names: Rakodi, Carole, author.
Title: Religion and society in sub-Saharan Africa and Southern Asia / Carole Rakodi.
Description: 1 [edition]. | New York : Routledge, 2019. | Series: Routledge research in religion and development | Includes index.
Identifiers: LCCN 2018055753 (print) | LCCN 2019008359 (ebook) | ISBN 9781315543314 (Master) | ISBN 9781138685307 (hardback) | ISBN 9781315543314 (ebook)
Subjects: LCSH: Africa, Sub-Saharan—Religion. | South Asia—Religion.
Classification: LCC BL2462.5 (ebook) | LCC BL2462.5 .R35 2019 (print) | DDC 200.967—dc23
LC record available at https://lccn.loc.gov/2018055753

ISBN: 978-1-138-68530-7 (hbk)
ISBN: 978-0-367-73110-6 (pbk)

Typeset in Goudy
by Apex CoVantage, LLC

Contents

Figures

Tables

Boxes

Preface

The idea for this book germinated while I was directing a large international research programme on religions and development (2005–2011) based in the International Development Department, School of Government and Society, University of Birmingham. In the early 2000s, development studies was, for a variety of reasons, largely religion blind. The study of religion and society had been pursued by religious studies, was central to much anthropology, and was a concern of various sub-disciplines within sociology, political science and economics, although these focused mainly on the United States and Europe and paid little attention to developing countries.

The research programme was, therefore, exploratory. It sought to identify what was known about the role of religion in developing societies and its influence on the achievement of development goals, to develop a framework for analysing the links between religion and development, and to undertake a series of research projects to investigate aspects of the links between religion and development, selected because of their obvious salience and/or their relevance to the research funder, the British government's Department for International Development. Researchers in the UK worked with colleagues in four countries of interest to DFID, selected because of their different religious traditions and also because they had experienced conflict between or within those traditions, with adverse effects on the achievement of development goals. Despite the challenges of assembling research teams from scratch, forging mutual understanding between team members with different disciplinary and religious backgrounds, developing suitable conceptual tools and methodological approaches, the difficulties of working in contexts characterised by religious rivalry, the sensitivity of religion as a research topic and the pressures of academic life, pathbreaking research was undertaken in Nigeria, Tanzania, India and Pakistan, in the course of which several world religions were considered – Christianity, Islam, Hinduism, Buddhism, Sikhism and African traditional religions.

Some of this research has been published, initially as a series of working papers and subsequently in a variety of other outlets. The programme provided some of the illustrations used by Emma Tomalin in her introductory text, published in 2013, which sought to provide a guide to different approaches to understanding the relationships between religions and development studies, policy and practice and was aimed primarily at students of religious studies and development studies in the global North. Subsequently, two collections of research on the theme of religions and global development drew together chapters on different religions, areas of the world, development themes and debates, issues in development policy and case studies of practice (Clarke (ed), 2013; Tomalin (ed), 2015). A few of the papers in each drew on research undertaken during the RaD programme, as did the papers in a special issue of *Development in Practice* (Rakodi (ed), 2012, later reproduced as a single volume, 2014). However, most of the outputs were written by researchers involved in one or two individual research projects, and time ran out before the intended synthesis could be produced.

As is widely recognised, 'development' refers to a particular international discourse and 'industry' in which the policy and research agenda is heavily influenced by Northern governments and funding agencies, as reflected in the focus of much recent research and the contents of those volumes, which seek to provide basic religious literacy, improved understanding of the social contexts in which students and practitioners of development work, and insights into the relationships between religion and development concerns, policies and practice. Development issues and debates are important, but they do not necessarily reflect the concerns and aspirations of many people in developing countries. Here, therefore, the preoccupations and ways of living of Southern actors (citizens, religious and secular organisations, governments) are not assumed to be identical to those of mainstream development discourse and organisations, and this book and its companion volume seek to examine wider aspects of the links between religion, societies and states. They will draw on research undertaken during the RaD programme, which provides a rich source of original material, but also on other published work in a range of countries in Sub-Saharan Africa and southern Asia.

Any analysis of social phenomena brings an intellectual, cultural and in this case religious baggage to the task. The first step is to acknowledge that baggage – something that many researchers, policy makers and practitioners fail to do. Responses include developing, as far as possible, literacy in unfamiliar cultural and religious traditions and disciplines other than one's 'home' discipline and treating people with different priorities and perspectives with respect. Of course, it is impossible to escape one's intellectual, cultural and political heritage, and this is particularly difficult for adherents to one of the religious traditions which believes its claims to truth are unique and superior, conveying an obligation to promote those truths to others.

However, the best of social and development research is highly reflexive, and in that tradition, I will briefly set out my own position. First, because development studies is a multidisciplinary field, most researchers in this area make a conscious effort to develop some competence in disciplines other than their own, albeit in less depth than many working within a single discipline might consider satisfactory. In my own case, the issues which have intrigued me over the years have led me to develop an awareness not only of my own discipline of geography but also of sociology, economics, political science and, more recently, anthropology and religious studies. One benefit of this less discipline-bound perspective on the world is, hopefully, an ability to spot aspects of social life that have been neglected by or invisible to others.

Second, I acknowledge that my religious and cultural heritage is Christian and British. I can never set this aside completely. However, membership of a religious group (Religious Society of Friends, Quakers) that adheres to the tenet that 'there is that of God in every person' engenders mutual respect and a commitment to equality. In addition, my own religious stance is universalist. In my view, religion is a human response to a perceived or experienced spiritual dimension of life. As a human response, there is always an inescapably human element within it: the historical and contemporary contexts of ideas, cultural assumptions and practical circumstances shape people's concepts and images of God/the ultimate reality/the divine/the sacred. How people respond to the divine or ultimate reality that both transcends and, in the case of some traditions, is believed to be immanent within us and the world depends on where and when they are born. Each of the religious traditions can therefore be seen as constituting "different human responses to the ultimate transcendent reality to which they all, in their different ways, bear witness" (Hick, 1999, p. 77).

This book has had a long gestation period, as I reflected on the findings (and the strengths and weaknesses) of the RaD research, edited and produced various outputs and extended my reading in ways prevented by the exigencies of managing a large international research programme. In the course of that programme and the subsequent process of analysis and reflection, I have accumulated many debts, most of which cannot be individually acknowledged here. I am grateful to all those researchers based in various universities in the UK and overseas who participated in the research, especially Gurharpal Singh, the Deputy Director of the programme, members of the programme's advisory group, our link advisers at DFID, and those who coordinated the research teams in Nigeria, Tanzania, India and Pakistan. I am particularly grateful to Nida Kirmani and Emma Tomalin, research associates based in Birmingham and Leeds, respectively, and to the research support team in Birmingham.

References

Hick, J. (1999) *The Fifth Dimension: An Exploration of the Spiritual Realm*, Oxford: One World.

Clarke, M. (ed) (2013) *Handbook of Research on Religion and Development*, Cheltenham: Edward Elgar.

Rakodi, C. (ed) (2014) *Religion, Religious Organisations and Development: Scrutinising Religious Perceptions and Organisations*, London: Routledge.

Tomalin, E. (2013) *Religions and Development*, Abingdon: Routledge.

Tomalin, E. (ed) (2015) *The Routledge Handbook of Religions and Global Development*, London: Routledge.

Glossary

Buddhism (Pali/Sanskrit)

arahan/arhat	One who is worthy but not yet a fully enlightened person (Buddha)
bhikku/bhikkuni	Ordained monk/nun
bodhisattva	Individual who progresses his/her own development towards full Buddhahood but delays entry into *nirvana* to help others achieve enlightenment
dukkha/duhkha	Human suffering
kamma/karma	Action and its results
nibbana/nirvana	Enlightenment
panna	Wisdom
paritta/pirit	Recitation of verses recommended by the Buddha to ward off misfortune or danger, heal, bring peace. Also blessings by monks on auspicious or inauspicious occasions (e.g. a new temple or house, a funeral, anniversary of a death, respectively).
samsara	The cycle of rebirths
suttas/sutras	Pali texts, the oldest scriptures
tanha/trsna	Insatiable craving and desire for things external to the self

Christianity

AD/BC	Anno Domini (the year of the lord)/BC Before Christ (according to the Gregorian calendar), equivalent to CE the Common Era/BCE Before the Common Era
baptism	Sacrament or rite of admission and adoption, usually involving the use of water
Eucharist/holy communion	Sacrament commemorating Jesus' sharing of meal of bread or wine with his disciples before his death; also Lord's Supper
faith	Belief, confidence, trust

glossolalia	'Speaking in tongues', the vocalising of speech-like syllables that lack any readily understood meaning
sacrament	A rite which is believed to be a visible symbol of God's grace, or an ordinance or commendable practice
salvation	Deliverance of the human soul from sin and its consequences, believed to have been made possible by the life, death and resurrection of Jesus

Hinduism

atman	Individual self or soul
darsan/darshan	Seeing, the mutual gaze of a deity (often believed to be in an idol or image) and devotee
dharma	Duty, role
guru	Person with knowledge, wisdom and authority who guides others
jati	Basic social units that govern marriages, social networks, rituals and food taboos
moksha/moksa	Spiritual freedom/liberation
puja	Act of devotion to a personal deity
rita	Cosmic order
samsara	Cycle of death and reincarnation
shivam	Grace, compassion
swami	Spiritually accomplished and learned devotee, normally celibate
varna	Class, of which there are believed to be four in society

Islam

AH/BH	In the year of the hijra (AD/CE622), the year Muhammad and his followers migrated from Mecca to Medina, a 12-month lunar calendar that lags behind the solar calendar by about ten days/Gregorian year. Each month starts when the moon is sighted, the entire cycle repeating every 33 lunar years. BH before the hijra
alim/imam	Leader of a mosque, especially in prayer
Allah	God
dargah	Shrine of a Sufi saint
da'wah	Preaching Islam, proselytising
deen	A believer's faith and practices, a moral and social system
fatwa (pl. fatawa)	Authoritative scholar's interpretation of or opinion on an aspect of *shari'a* (traditionally issued by a mufti or qualified jurist but also other scholars)
fiqh	Jurisprudence, the rules of derivation used to apply Islamic law to situations not directly covered by the source materials (and the methodology for development of such rules)

hadith (pl. ahadith)	A report of a saying or action of the Prophet. The *ahadith* were compiled between one and three centuries after his death and categorised into reliable, less reliable and weak
hajj/hajji	Pilgrimage to Mecca/Makkah/one who has performed the pilgrimage
ihsan	Responsibility to attain excellence/perfection in worship
ijma	The consensus of scholars/a community
ijtihad	Critical reasoning
iman/eeman	Personal faith
jinn	A spirit
khangah	Sufi building, hostel (*dargah* in South Asia)
marabout	Leader of a group of disciples (*daara*) in Sufi Islam
pir	Spiritual leader, teacher, saint
qiyas	Analogical reasoning
Qur'an	Lit. recitation, the word of God as believed to have been revealed to the Prophet Muhammad
salah/salat	Ritual prayers
sawm	Fast, especially during the month of Ramadan
shahadah	Profession of faith, a declaration of belief in the oneness of God and acceptance of Muhammad as God's prophet
shari'a	Islamic law (the rules that govern Muslim behaviour in all realms of life, derived from the Qur'an and *ahadith*). Sometimes a distinction is drawn between *shari'a* as a whole and the body of Islamic law developed (and mutually recognised) by the four main legal schools by about the ninth century, which may be referred to as classical *shari'a* or the Shari'a.
shaykh (F. shaykah)	Scholar, spiritual leader, senior teacher/cleric, originally an Arabic honorific, used of royalty. Also founder of a Sufi *tariqa*.
sunnah	Trodden path, the custom and practice of the Islamic community, based on the teachings, sayings and actions of the Prophet Muhammad, his tacit permissions or approvals and reports about his companions
taqwa	God-consciousness
tariqa	Brotherhood, order
tawhid	The oneness/uniqueness of God
ulama (pl. ulema)	Religious scholar, guardian, interpreter and transmitter of religious knowledge, Islamic doctrine and law (in Sunni Islam)
ummah	Community of believers, the global Muslim community
zikr	Reiteration of God's name or Qur'anic verses

Part I

The religion–society nexus

Introduction

The chapters in the first part of the book provide a starting point for the consideration of interactions between religious beliefs, practices and values and societies in Sub-Saharan Africa and southern Asia that follows. The analysis has two primary aims: to provide readers unfamiliar with some or all of the main religious traditions in these regions with basic religious literacy in each and to review empirical evidence emerging from recent research in a number of social science disciplines on the links between religion and societies. This rationale and the analytical questions to which it gives rise are outlined in the first chapter, which also provides background information on trends in religious affiliation and religiosity in these regions, including the apparent 'resurgence' in religion – in practice the emergence of revivalist movements rather than increased religious affiliation and participation.

The second chapter in this part sets out a framework for the analysis that follows in Parts II–IV. It explains the overall perspective, that of 'lived religion', and briefly discusses the three key concepts used throughout: religion, society and culture. These have been constructed by scholars from different disciplines, overlap with each other and continue to be contested, so there is no consensus on their definition. However, I hope that because I am setting out my conceptualisation at the outset, readers will be provided with an understanding of my approach to the analysis that follows. I find the metaphor of a tapestry more useful than the notion of a theory that purports to explain the complex, changeable and varied links between different aspects of religion and the characteristics of people's lives and societies. The analytical framework thus suggests that to identify and understand these links, it is necessary to consider three dimensions of the social field, analyse the links of interest at a variety of levels and examine four basic domains of religion (discourse, practice, community and institutions). Finally, to help the reader assess the empirical evidence identified, the chapter concludes with a discussion of some of the methodological issues that arise when considering lived religion.

1 The entanglements of religion in everyday life

Why religion and society?

Religion is embedded in societies and intertwined with many aspects of people's lives: often, it is not experienced or regarded as a separate social phenomenon or sphere of social life. Historically, religious ideas have provided ways of explaining the world, conveying value and meaning, defining norms and ethics for personal and collective life and promoting cooperation or conflict. Religion cannot, therefore, be fully understood without considering whether and how it plays these roles and influences individual lives and social relationships, while society cannot be fully understood without considering the ways in which religious beliefs, practices, values, institutions and organisations influence individuals and social institutions at every level. Yet the relationships between religion, societies and states in contemporary developing countries are complex and poorly understood. Aspects of these relationships have, in recent years, been analysed and intensely debated, but the coverage of both issues and countries is uneven, and the outputs of relevant research are scattered.

To date, most analyses of the links between religion and society have relied heavily on the nineteenth- and early-twentieth-century writings of the 'founding fathers' of sociological research (Durkheim, Weber, etc.), many of whom had an international outlook. However, from about the middle of the twentieth century, research and commentary on the links between religion and society, especially in sociology and political science, originated in and concentrated on North America and Europe. The study of different religious traditions also dates largely from the nineteenth century and has included research in Sub-Saharan African (SSA) and southern Asian countries, mainly by anthropologists, and, more recently, in political science and development studies. However, the sociology of religion has continued to be concerned mainly with the United States and Europe, and development studies, policy and practice have neglected religion and the roles it plays in society. In the last few decades, however, the amount of research in a number of disciplines and a variety of geographical contexts has increased. This includes a few comparative studies, and there is a growing number of edited collections, but to date there has been no overview of the available research on the links between religion and society in Sub-Saharan Africa and southern Asia.

Religious views and practices are reflected not only in individual lives and social relationships but also in movements for social change, social action by religious organisations and politics. Our current understanding of these relationships will be explored in two linked volumes, with a view to identifying the key factors that influence and explain the varied roles religion plays in a range of Sub-Saharan African and southern Asian countries. Social arrangements, cultural characteristics and religious traditions in the contexts covered vary greatly. The book draws together recent empirical evidence from a variety of disciplines to provide reviews of the state of the art. This volume will provide those interested in the links between religion and society with basic literacy in the main religious traditions and an understanding of these varied contexts. It reviews what is known about how some important dimensions of religion are entangled in the everyday lives of individuals, families and social groups. Its multidisciplinary approach and wide coverage of important religious traditions, two important subcontinents in the global South and key features of contemporary societies will be of interest to postgraduate students and researchers in a variety of social science disciplines, including the sociology of religion, development studies, religious studies, anthropology and geography.

Here, I conceive of religion, culture and society as dimensions of social life. All are constructed concepts, which have, over time and in different places, been contested and reconstructed. I hold that while they should not be essentialised, they nevertheless have analytical value. The book explores how religion is intertwined in the social lives of (ordinary) people in contemporary SSA and southern Asia, considering what it means to people and the influences it has on their values, with respect to ethics, morality and right social ordering, and on their social practices. It examines how the teachings and experiences in various religious traditions are interpreted and how they influence religious beliefs and practices. It also explores how these beliefs and practices affect some of the social and religious institutions with which people are engaged, particularly their families and other immediate social relationships.

The analysis recognises the great variety between and within religious traditions, because of their different origins, as well as the changing historical, cultural and political contexts in which they have evolved, and so presents material on the major faith traditions in the regions under study: Islam, Christianity, Buddhism, Hinduism and folk religion.[1] The beliefs, practices and values advanced

1 I adopt the term 'folk religion' in preference to the alternative terms. 'Popular religion' seems inappropriate in countries where the vast majority affiliate themselves with Islam, Buddhism or another religion. 'Indigenous religion' also seems inappropriate when imported religions have now been established for many centuries and can hardly be portrayed as 'non-indigenous'. The term 'traditional religion' also has unsatisfactory connotations, because while contemporary beliefs and practices may be rooted in the past, they have often been adapted to meet the demands of modern life, new practices have emerged, and new groups have been initiated by prophets or charismatic leaders. Here, I refer to 'folk religions' to denote the local roots and distinctive characteristics of such religio-cultural traditions, or 'folk religion' when recognising their many similarities.

in the formal teachings and guidance of the religions under study are outlined for readers unfamiliar with these traditions (or with those of traditions other than their own), but it focuses less on official interpretations of the texts and sources than on how adherents understand the principles, practices and ethics of the religion with which they are affiliated. All religions emerge and are practised within social and political contexts, and so it also seeks to identify domestic and international influences on people's interpretations, including socio-economic, ethnic and gender differentiation within societies, economic change, globalisation and the mass media. It then explores ways in which the beliefs, religious practices and values identified play out in people's attitudes and lives, especially their relations with those with whom they interact on a day-to-day basis. The second volume will focus on the links between religion and social organisations, economic arrangements and trends and politics and governance. It will build on this analysis of religious beliefs, practices and values and how they influence the institutions (formal and informal rules) that underpin social behaviour and relationships but will shift the focus to organisational aspects of religions and societies, especially with respect to social and political action.

Four general analytical questions are, therefore, addressed in this book:

- How do people experience and understand religion? What religious beliefs do they hold, and how do these inform their understanding of the world and their own place in it?
- In what religious practices do people engage, and what do these reveal about their understanding of religion?
- How does religion inform people's values, ethics and morals?
- How do people's beliefs and values inform their views and behaviour with respect to key dimensions of their own lives and right social ordering at various scales?

To prepare the ground for the consideration of these questions in subsequent chapters, in the remainder of this chapter some of the contributions and shortcomings of the relevant disciplinary literatures will be briefly reviewed, the sources of material for this volume identified and some basic information will be provided on the religious context in SSA and southern Asia.

Understanding the links between religion and society in Sub-Saharan Africa and southern Asia: recent disciplinary literature and sources

As noted above, although the roles of religion in societies in colonial countries, as well as Europe and North America, were a central concern of many of the 'founding fathers' of the social sciences, subsequent generations of social scientists either did not consider religion or limited their analysis to European and North American contexts. For many, especially in Europe, such neglect seemed justified by the prevailing view of religion as a hindrance to 'progress' and 'modernity' and the expectation of further secularisation. In addition, despite the persistence and pervasiveness of religious beliefs and organisations in most parts of Africa and Asia, the sensitivity of religious issues in many newly independent countries and the lack of contemporary theoretical frameworks and analytical vocabularies to discuss them exacerbated the neglect of religion in the social sciences.

The sociology of religion is based on the conviction that "Relations between religion and society are fundamental to the nature of religion and . . . intrinsic to the nature of society" (Capps, 2005, p. 8461). It sets out to describe and explain religion in society, with a particular focus on how social factors affect religious attitudes and behaviours and vice versa. As noted above, the founders (Durkheim, Weber and others) did not confine their analyses to the West or their interest in religion to Christianity. They continue to be the departure and reference point for most contemporary sociologists of religion, although their enduring influence on sociology as a whole and the sociology of religion in particular has been, according to some, unhealthy (Voas, 2015). Meanwhile, the study of religion became isolated from many central theoretical conversations in sociology (see, for example, Beckford, 2003; Turner, 2014; Winchester, 2016), cross-cultural analysis in the field of the sociology of religion was, as recently as the mid-2000s, largely absent (Riesebrodt and Konieczny, 2005), and empirical research focused on North America and Europe. Even when sociologists did take an interest in different religious traditions and adopted an international perspective, some acknowledge that the dominant theoretical perspectives and main themes have been heavily influenced by this focus, emphasising, for example, secularisation, individualism and choice, personal spirituality and particular types of new religious movements (see, for example, Beckford, 2003; Capps, 2005; Turner, 2010, 2014; Voas, 2015).

Anthropology uses ethnographic methods to undertake small-scale in-depth studies of particular contexts and population groups, documenting the dynamics of human interactions, social organisation and social change. Examinations of how religious beliefs inform the shape of everyday life, culture and identity have always been central to the discipline and have made an important contribution to our understanding of the links between religion, culture and societies. Ethnographies of religious communities, congregations and religious events abound, while themes like spirit possession and witchcraft, healing practices, myths and oral traditions, rituals and rites of passage, gender, identity, the nature and effects of violence and the impact of religious and socio-economic change enable some broader comparisons. However, the amount of anthropological research on particular contexts varies, coverage of religious groups and organisations and their impact on societies is uneven, and collections tend to bring together case studies which have been designed and carried out independently. In addition, the colonial antecedents of anthropology resulted in it being side-lined in many newly independent countries and, for many years, from mainstream development debates.

Overlapping with these fields of study is that of religious studies, which sought, not entirely successfully, to distinguish itself from insider disciplines (especially theology) and to understand what religion means to believers rather than explaining its origins or functions. Theoretically and methodologically interdisciplinary, it is interested in the content and nature of religious beliefs and their interpretations and how these impact on social difference and social change. Research in this field may be religion or context specific, but it is also more likely than anthropology to engage in comparative discussions of issues such as the impact of patriarchal religion on women or the links between religion, politics and violence. The volume edited by Woodhead et al. (2009), which provides analyses of eight religious traditions as well as discussing a variety of themes relevant to analysing the links between religion and society internationally, is a good example.

Some of the social scientific focus on North America and Europe has changed in recent years, in response to new trends and concerns. First, the rise of political Islam has led to a dramatic increase in studies of its characteristics and implications for individual countries and international relations. Second, concern with human rights has fuelled discussion of the compatibility of certain religious beliefs and practices with human (including women's) rights. Third, state failure and downsizing has renewed interest in the roles religious organisations play in service delivery. Fourth, growth in the scale and influence of piety movements, especially charismatic Christianity in the US and beyond, has led Europeans to recognise that social secularisation is exceptional rather than universal.

Work produced within the disciplines mentioned above and these recent changes have rubbed off on development studies, which historically neglected religion, but has, in recent years, produced an increasing volume of work (for example, Clarke, 2011, 2013; Deneulin and Rakodi, 2011; Rakodi, 2014; Tomalin, 2013, 2015). The association of this work with the development agenda is manifest in its concerns, orientations and biases. First, it often fails to adequately consider and define the concept of 'religion', leading to a "default conceptualisation [of it] . . . as substantivist, essentialised, ahistorical and universal" and 'development' as a secular domain, set apart from 'religion', which conceals the historical specificity and value-laden nature of mainstream development thinking and practice (Fountain, 2013). Second, it tends to regard religious organisations in either a positive or a negative light, recognising, for example, their contribution to service delivery while condemning their patriarchal structures and attitudes. Third, it is concerned with development policy issues more than deepening understanding of the political and cultural contexts in which policies are devised and implemented and external agencies function. Fourth, it produces research that reflects the concerns and theoretical orientations of northern researchers and research funders – indeed, development studies may be perceived as a 'critical partner' of the aid industry. Finally, much of the work has a normative orientation, generally promoting the positive potential of religion and religious organisations in achieving development goals (Clarke, 2011; Deneulin and Bano, 2009; Marshall and Keough, 2004; Ter Haar, 2011; Tyndale, 2006), although sometimes lamenting its potentially divisive role and social conservatism (De Kadt, 2011). While these orientations do not necessarily mean that the analysis is biased or uncritical, on the whole writing in this field refers to a particular international discourse and 'industry' in which the agenda is heavily influenced by Northern governments and funding agencies.

Overall, therefore, much of the relevant social science research has emerged from an academic community largely based in the global North, where the research capacity of universities is more highly developed, researchers can obtain funding and most publishers of books and journals are based (Connell, 2007). As a result, it often reflects the concerns and theoretical orientations of Northern researchers and research funders, and the main market for research outputs continues to be the student communities and university libraries of Northern universities. This is reflected in the widespread use of English and the remit and content of research outputs, journals and textbooks (see, for example, Tomalin, 2013, p. 5). In contrast, much of the social scientific work that has emerged from countries such as India or Nigeria has taken the prevalence and importance of religion as self-evident, steered clear of studying it because of the sensitivity of the issues, and/or focused on issues determined by the agendas and funding of Northern partners, including researchers and aid agencies. As a result, recent research is unlikely to fully reflect the preoccupations, concerns and aspirations of many in developing countries. For the reasons given above, this volume does not seek to provide a systematic analysis and critique of development thinking, policy and practice. Instead, its central concern with the links between religion and society is intended to provide a better understanding of the social, cultural and religious contexts in which researchers, policy makers and practitioners work. It endeavours to be sensitive to the issues and concerns raised above, drawing on work by local researchers where it is available, adopting a critical approach to the available studies and providing a synthesis that will be useful to not only North American and European but also African and Asian students, researchers and other readers.

To achieve this, it will draw primarily on studies of one or more religious traditions, particular countries and contexts and selected issues in a range of countries in Sub-Saharan Africa and southern Asia to throw light on key characteristics, links and trends, explore the similarities and differences between national and sub-national settings and religious traditions, identify the factors that need to be taken into account when seeking to understand and explain the connections between religion and social trends and suggest fruitful avenues for further research. In particular, it will draw on work undertaken between 2005 and 2011 during a large international research programme on religions and development, which the author directed. During this programme, a series of research projects considered a number of religious traditions (Christianity, Islam, Hinduism, Buddhism, Sikhism and African traditional religion)[2] and several dimensions of society and poli-

2 This volume will not consider Sikhism or Confucianism. Although the former is a significant and distinctive religious tradition and was an important focus of work within the Religions and Development Research Programme, it is almost entirely confined to India, and here space is restricted. Confucianism has been influential in some Southeast Asian countries but is the dominant religious tradition only further north, particularly in China. For a definition and conceptualisation of African traditional religion, see Pew Forum, 2010, p. 6. The debate over whether African traditional religion can be treated as a single category is longstanding and intense. By definition, 'indigenous

tics in four countries (India, Pakistan, Nigeria and Tanzania). With the exception of Pakistan, these countries were selected because of the significant presence of two or more religious traditions. The inclusion of a Muslim-majority country was a deliberate response to the difficulties experienced by many Western agencies when working in Islamic contexts, largely because of the (often implicit) assumptions associated with their Christian religious and cultural roots. The opportunity to work in four countries and study six faith traditions provided a rich source of original material, although the research undertaken was necessarily selective in terms of geographical location, the research questions addressed, the focus of individual research projects and the exploration of variations within faith traditions. Case studies of particular issues, groups and organisations in individual countries and some comparative analyses have been published in a variety of outlets, but no overall synthesis of the research has been produced.

In addition to this published and unpublished material, the book will draw on the work of others writing on these and other Sub-Saharan African[3] and southern Asian countries[4] in order to deepen and extend the analysis. Much of the relevant work has emerged from the disciplines of sociology, anthropology and religious studies, as well as theology and Islamic studies. Many of the relevant studies focus on particular social phenomena in a specific cultural context and adopt an ethnographic approach. Quantitative analyses are scarce, but data from large-scale surveys will be included when available to situate the more detailed studies within a wider landscape.

The religious context: trends in affiliation and religiosity

Debates about the interactions between religion and society need to be informed by how important religion is perceived to be, which is usually assessed in terms of

religions' have distinctive characteristics. However, not only do these traditions in Africa seem to have a number of features in common, the term 'African traditional religion' is commonly used in African scholarship, and so here it will be used despite the debate about whether individual traditional religions can be seen as a single religion. However, case studies of particular religio-cultural traditions will be used for illustrative purposes where available.

3 Muslim-majority North African countries, which have closer religious, political and cultural links with the Middle East than the rest of Africa, are not considered. After centuries of cross-Saharan trade, countries lying to the south of the desert continue to display religious, cultural and political influences originating in the Muslim world. Inevitably, any boundary is arbitrary. Sudan (91% Muslim) is not included here, and there is little or no reference to Somalia, Eritrea, Chad, Niger, Mauritania, the Western Sahara or South Sudan, not least because there is little research on these often unstable or war-torn countries.

4 In addition to India and Pakistan, the Religions and Development programme built on research related to the theme of lived religion in Bangladesh. Relevant research has also been undertaken in other South Asian countries and a number of countries in Southeast Asia, where the main religious influences are similar, including Indonesia, Malaysia, the Philippines, Thailand and Vietnam. Afghanistan, which has stronger cultural links with the Middle East than with South Asia, is excluded, and only limited reference is made to Nepal, Myanmar, Bhutan, Cambodia, East Timor, Brunei or Laos, where little relevant research seems to be available.

trends in religious adherence and religiosity at both the global and local scales. In the mid-twentieth century, the main perceived trend in religious affiliation, belief and practice was secularisation. However, by the 1980s, it had become clear that secularisation was not a worldwide trend and that even within Europe and North America, the nature and direction of change varied considerably. Gradually, secularisation was displaced as the dominant narrative in favour of 'religious resurgence' or 'revival', although some analysts commented that, as religion had never ceased to be important in many societies in Africa, Asia and Latin America, it was as important to recognise and understand its persistence as to become excited or alarmed about its resurgence.

The identification of trends at the global, subcontinental and national levels is strongly influenced by the availability and quality of data. Religious adherence and importance is generally assessed in terms of adherents' own claims to 'belong to' a particular tradition or group, an individual's response to questions about whether or not s/he considers her/himself a religious person, whether or not s/he believes in 'God' or regards her/himself as a 'convinced atheist' and self-assessment of how important 'God' and/or participation in religious practices are in his or her own life. Supposedly objective, it is clear that the choice of questions and the data generated are influenced by the assumptions made by those who design and analyse the results of international surveys, especially their conception of 'religion'.[5] Nevertheless, with care, these sources can be used to provide a broad overview of religious affiliation and religiosity in SSA and southern Asia.

Recently, the Pew Research Center, as part of a major project to assess the future of world religions, produced an estimate of the religious composition of the world's population in 2010. This draws on data from 2,500+ censuses, surveys and population registers as a basis for projecting trends to 2050, based on assumptions about age, fertility, mortality, migration and religious switching (Pew Research Center, 2015). The analysis shows that the religious profile of the world is rapidly changing, driven primarily by differences in fertility rates and the size of youth populations among the world's major religions. Over the next four decades, it is estimated that Christianity will remain the largest faith tradition but that Islam will grow more rapidly, so if current trends continue, by 2050 the number of Muslims (30%) will nearly equal the number of Christians

5 Several organisations have compiled data or carried out international surveys, using a variety of methods. These include the Center for the Study of Global Christianity in the US, which maintains the World Christian and World Religion databases, www.worldchristiandatabase.org/wcd and www.worldreligiondatabase.org/wrd_default.asp, and their printed predecessor; www.adherents.com; the Pew Research Center http://assets.pewresearch.org; the World Values Surveys www.worldvaluessurvey.org, and Gallup International www.gallup-international.com. Even when data gathering exercises are repeated, the data are not necessarily comparable, since different definitions may be adopted in successive rounds, different sampling procedures adopted, different questions used and different approaches adopted to combining data from different sources. For an assessment of the reliability of five alternative data sources, see Hsu et al. (2008).

(31%), because, while the world population and the number of Christians are expected to increase by 35%, the number of Muslims is expected to increase by 73%. This is related to the relatively youthful age structure of the global Muslim population (in 2010 34% were under 15 and only 7% 60+, compared to 27% and 14% for Christians and 30% and 8% for Hindus). This age structure is associated with high fertility rates (3.1 for Muslims globally compared to an average of 2.5 to 2.7 for Christians, 2.4 for Hindus and less for other faith traditions) (Pew Research Center, 2015, p. 7, 10). The global Buddhist population is expected to be about the same size as it was in 2010 (7%) because of low fertility rates and ageing populations in, for example, China, Japan and Thailand. Although the numbers of atheists, agnostics and others who do not affiliate themselves with any religion are expected to continue to increase in Europe and North America, they will make up a declining share of the world population (13% in 2050).

These trends are influenced not only by the fertility rate and age distribution of each religious group but also where each is concentrated today – religions with many adherents in the global South, where birth rates remain relatively high and infant mortality rates are falling, will grow relatively rapidly. They are also affected by patterns of religious switching, which are complex and differ between groups. Information from surveys in 70+ countries shows that in some places it is fairly common for adults to leave their childhood religion, while in others it is rare, legally cumbersome and even illegal. Christianity is expected to experience the largest net losses from switching (followed by Buddhism), with most joining the ranks of the unaffiliated, while Islam (especially in Africa) and folk and other religions are expected to experience modest net gains. The trends are also influenced by patterns of international migration.

The best available data on historical trends are for Christianity, partly because of the organisational and doctrinal character of this faith tradition, in which adherents are typically expected to become 'members' of a specific denomination, and partly because of the data available. In 1910 about two-thirds of all Christians lived in Europe, but this continent's share had dropped to 26% by 2010, alongside an increase in the share of the Americas from just over a quarter to over a third, the Asia-Pacific region from 5% to 13% and, most dramatically, Sub-Saharan Africa from 1% to nearly a quarter. Today, although only a quarter of the total population of the global South are Christians, they account for about 61% of all Christians, reflecting the enormous population growth in this region. A quarter of all Christians live in the US, Brazil and Mexico, while five of the ten countries with the largest Christian populations are in Africa (Nigeria, the Democratic Republic of Congo, Ethiopia) and Asia (the Philippines and China) (Pew Research Center, 2011).

Much of Sub-Saharan Africa's current pattern of religious affiliation can be attributed to its history of trade, mission and colonialism, with the patterns in different parts of the subcontinent reflecting their geographical location, precolonial and colonial trading links (including the effects of the trade in slaves in both West and East Africa), the proselytising strategies of different faith traditions and denominations and the religious composition of the colonial powers. These are reflected in the trends shown in Figure 1.1, which shows the estimated increase in the proportion of Christians and Muslims between 1900 and 2010 (9%–57% and 14%–29%, respectively) and the dramatic decline in the proportion claiming African traditional religion as their primary religious affiliation (from 76% in 1900 to 13% in 2010) (Pew Forum, 2010, p. i).

In 2010, the Pew Research Center's revised estimates put 63% of people as Christian, 30% as Muslim, 3.3% as adherents of African traditional religions and 3.2% as unaffiliated (Pew Research Center, 2011). Because of continued high fertility rates, it is expected that the numbers of Christians will double by 2050 and that they will remain the largest group but that their share of the population will decline to 59%, while the Muslim share will increase from 30% to 35% because of this population's higher rate of natural increase (Pew Research Center, 2015). As will be discussed below, self-stated religious affiliation is by no means straightforward in the African context: today Christianity and Islam tend to be identified as 'religions' and the beliefs and practices associated with traditional religions as 'culture', so many hesitate to claim that their primary religious affiliation is with so-called traditional religion. Of the 500 million+ Christians in SSA in 2010, 34% were Catholic, 57% Protestant (including African Independent Churches), 8% Orthodox (mostly in Ethiopia) and 1% other denominations (Pew Research Center, 2011, p. 55).

The data assembled by the Pew Research Center for its attempt to predict future trends in religious affiliation can be drawn on to estimate the broad religious composition in 2010 of 38 Sub-Saharan African and 14 southern Asian

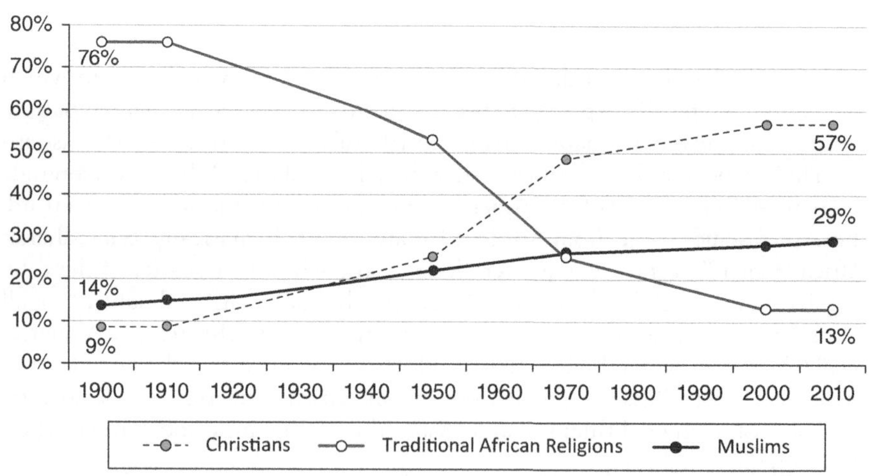

Figure 1.1 Growth of Islam and Christianity in Sub-Saharan Africa (1900–2010)
Source: Pew Forum, 2010, p. i, based on the World Religion database.

countries to provide contextual information for the subsequent discussions and case studies in this book. They are presented in Table 1.1.

Here, it is worth drawing attention to trends in the religious composition of Nigeria, the sub-continent's most populous country, which had 80 million+ Christians in 2010. In 1953, according to the last census before independence, 21% of the country's population was Christian, 45% Muslim and 33% other religions (including ATRs). By 1963, at independence, the latter had declined and the share of Christians had increased to about half the population, including nearly 60 million Protestants, about 20 million Catholics and a quarter of a million others (Pew Research Center, 2011, p. 55). Nigeria has not had a reliable census since 1963, but the Pew Research Center estimates the current share of Christians and Muslims to be approximately equal, at 49% each, while the share of those acknowledging their religious affiliation to be traditional religion has continued to shrink to just over 1%. Stonawski et al. (2016) note that if the fertility gap between Christians and Muslims continues to widen, in the next couple of decades the latter may be in a majority, with significant political implications.

A relatively small proportion of the population of southern Asian countries are Christians, who are a majority only in the Philippines, which has the fifth-largest Christian population in the world (c. 87 million) and the third largest Catholic population (76 million), behind Brazil and Mexico. Following the arrival of Roman Catholic priests and missionaries in the sixteen and seventeenth centuries, Spain conquered the country, remaining in control until the Americans took over colonial rule in 1898. Today, 92% of the population is estimated to be Christian (81% Catholic, 11% Protestant).[6] Following missionary endeavours and colonial conquest, there are Christian communities in most other southern Asian countries. Although they are minorities, the large populations of some of these countries means that the Christian communities in them are large in numerical terms (31 million in India, 23 million in Indonesia).

Within Islam (and other faith traditions), there is no exact equivalent of denominations or 'membership' within Christianity, and what data are available on the size of various traditions or subdivisions within the religion are scarcer and more speculative.[7] As noted above, the relatively youthful age profile of the global Muslim population is associated with high fertility and rapid growth, including in the southern Asian countries considered in this volume, where the

6 Although see Goh (2005), who notes that after only modest growth during the twentieth century, the Protestant churches, especially newer independent groups, have grown in recent years to, he estimates, nearly a third of the population.

7 To help address this perceived gap, the Pew Research Center conducted national sample surveys of Muslims in 39 countries in 2008/9 and 2011/12. For a comparative analysis of Christianity and Islam in SSA, it carried out surveys in 15 countries with substantial Muslim populations in 2008/9 (Pew Forum, 2010). Niger was added in 2011/12, in addition to 23 Muslim-majority countries in North Africa, Asia and Eastern and Southern Europe, including three in Southeast Asia and three in South Asia (Pew Research Center, 2012).

Table 1.1 The religious composition of Sub-Saharan African and southern Asian countries (2010).

	Buddhists %	*Christians* %	*Folk religions* %	*Hindus* %	*Jews* %	*Muslims* %	*Other religions* %	*Unaffiliated* %
Sub-Saharan Africa	**-**	**63.1**	**3.1**	**0.1**	**-**	**29.5**	**1.2**	**3.0**
Angola	<1.0	90.5	4.2	<1.0	<1.0	<1.0	<1.0	5.1
Benin	<1.0	53.0	18.1	<1.0	<1.0	23.8	<1.0	5.0
Botswana	<1.0	72.1	6.0	<1.0	<1.0	<1.0	<1.0	20.6
Burkina Faso	<1.0	22.5	15.4	<1.0	<1.0	61.6	<1.0	<1.0
Burundi	<1.0	91.5	5.7	<1.0	<1.0	2.8	<1.0	<1.0
Cameroon	<1.0	70.3	3.3	<1.0	<1.0	18.3	2.7	5.3
CAR	<1.0	89.5	1.0	<1.0	<1.0	8.5	<1.0	1.0
Chad	<1.0	40.6	1.4	<1.0	<1.0	55.3	<1.0	2.5
Rep of the Congo	<1.0	85.9	2.8	<1.0	<1.0	1.2	1.1	9.0
DRC	<1.0	95.8	<1.0	<1.0	<1.0	1.5	<1.0	1.8
Ivory Coast	<1.0	44.1	10.2	<1.0	<1.0	37.5	<1.0	8.0
Eritrea	<1.0	62.9	<1.0	<1.0	<1.0	36.6	<1.0	<1.0
Ethiopia	<1.0	62.8	2.6	<1.0	<1.0	34.6	<1.0	<1.0
Ghana	<1.0	74.9	4.9	<1.0	<1.0	15.8	<1.0	4.2
Guinea	<1.0	10.9	2.7	<1.0	<1.0	84.4	<1.0	1.8
Guinea Bissau	<1.0	19.7	30.9	<1.0	<1.0	45.1	<1.0	4.3
Kenya	<1.0	84.8	1.7	<1.0	<1.0	9.7	1.2	2.5
Lesotho	<1.0	96.8	<1.0	<1.0	<1.0	<1.0	<1.0	3.1
Liberia	<1.0	85.9	<1.0	<1.0	<1.0	12.0	<1.0	1.4
Madagascar	<1.0	85.3	4.5	<1.0	<1.0	3.0	<1.0	6.9
Malawi	<1.0	82.7	1.7	<1.0	<1.0	13.0	<1.0	2.5
Mali	<1.0	2.4	2.7	<1.0	<1.0	94.4	<1.0	<1.0
Mozambique	<1.0	56.7	7.4	<1.0	<1.0	18.0	<1.0	17.9
Namibia	<1.0	97.5	<1.0	<1.0	<1.0	<1.0	<1.0	1.9
Niger	<1.0	<1.0	<1.0	<1.0	<1.0	98.4	<1.0	<1.0
Nigeria	<1.0	49.3	1.4	<1.0	<1.0	48.8	<1.0	<1.0

Rwanda	<1.0	93.4	<1.0	<1.0	<1.0	1.8	<1.0	3.6
Senegal	<1.0	3.6	<1.0	<1.0	<1.0	96.4	<1.0	<1.0
Sierra Leone	<1.0	20.9	<1.0	<1.0	<1.0	78.0	<1.0	<1.0
Somalia	<1.0	<1.0	<1.0	<1.0	<1.0	>99.0	<1.0	<1.0
South Africa	<1.0	81.2	<1.0	1.1	<1.0	1.7	<1.0	14.9
Swaziland	<1.0	88.1	1.0	<1.0	<1.0	<1.0	<1.0	10.1
South Sudan	<1.0	60.5	32.9	<1.0	<1.0	6.2	<1.0	<1.0
Tanzania	<1.0	61.4	1.8	<1.0	<1.0	35.2	<1.0	1.4
Togo	<1.0	43.7	35.6	<1.0	<1.0	14.0	<1.0	6.2
Uganda	<1.0	86.7	<1.0	<1.0	<1.0	11.5	<1.0	<1.0
Zambia	<1.0	97.6	<1.0	<1.0	<1.0	<1.0	<1.0	<1.0
Zimbabwe	<1.0	87.0	3.8	<1.0	<1.0	<1.0	<1.0	7.9
Southern Asia	**7.5**	**7.1**	**2.4**	**45.1**	-	**35.3**	**1.3**	**1.3**
Bangladesh	<1.0	<1.0	<1.0	8.5	<1.0	90.4	<1.0	<1.0
Cambodia	96.9	<1.0	<1.0	<1.0	<1.0	2.0	<1.0	<1.0
India	<1.0	2.5	<1.0	79.5	<1.0	14.4	2.3	<1.0
Indonesia	<1.0	9.9	<1.0	1.7	<1.0	87.2	<1.0	<1.0
Laos	66.0	1.5	30.7	<1.0	<1.0	<1.0	<1.0	<1.0
Malaysia	17.7	9.4	2.3	6.0	<1.0	63.7	<1.0	<1.0
Myanmar/Burma	80.1	7.8	5.8	1.7	<1.0	4.0	<1.0	<1.0
Nepal	10.3	<1.0	3.7	80.7	<1.0	4.6	<1.0	<1.0
Pakistan	<1.0	1.6	<1.0	1.9	<1.0	96.4	<1.0	<1.0
Philippines	<1.0	92.6	1.5	<1.0	<1.0	5.5	<1.0	<1.0
Singapore	32.2	17.7	2.4	6.5	<1.0	16.1	8.5	16.5
Sri Lanka	69.3	7.3	<1.0	13.6	<1.0	9.8	<1.0	<1.0
Thailand	93.2	<1.0	<1.0	<1.0	<1.0	5.5	<1.0	<1.0
Vietnam	16.4	8.2	45.3	<1.0	<1.0	<1.0	<1.0	29.6

CAR – Central African Republic, DRC – Democratic Republic of the Congo

Source: Pew Research Center, 2015, pp. 234–5.

absolute and relative size of the Muslim population is increasing, including in Muslim-majority Indonesia (87% in 2010), Pakistan (96%), Bangladesh (90%) and Malaysia (64%). Most of the Muslims interviewed in the Pew Research Center's surveys self-identified as Sunnis or Shias (estimated at 87–90% and 10–13% of the global Muslim population, respectively). Sectarian identities are most salient in the Middle East/North Africa or in countries where there is a sizable Shia minority (e.g. Pakistan 6%, where tensions between Sunnis and Shias and controversy over whether Shias are 'real' Muslims is rife) (Pew Research Center, 2012, p. 21). However, in other places, many Muslims responded that they consider themselves to be 'just a Muslim': in SSA 48% of Muslim respondents identified themselves as Sunnis and 23% as just a Muslim, especially in Mali (55%) and Cameroon (40%), while in Indonesia 26% identified themselves as Sunni and 56% just a Muslim (Pew Research Center, 2012, pp. 20, 28). Smaller sects include Ahmadiyyas, who are often not recognised as Muslims by other Muslims but have substantial minorities of affiliates in several African countries such as Ghana, Tanzania, Cameroon and Liberia.

Another overlapping affiliation is with Sufism, "a mystical movement within Islam that emphasizes the possibility of gaining direct knowledge of God through euphoric worship and other practices" (Pew Research Center, 2012, p. 31). Views differ on whether Sufism is 'true' Islam – it is more widely accepted as such in SSA and South Asia than elsewhere in the Muslim world. The proportion of Muslims who claim to belong to a Sufi order/brotherhood/*tariqa* is largest in SSA, where a quarter or more claim such an affiliation in 11 of the 15 countries included in the surveys, varying from 92% in Senegal to between a third and a half in other West African countries such as Cameroon, Niger, Liberia, Ghana and Nigeria (Pew Research Center, 2012, p. 31). The oldest and second-most-widespread order is the Qadiriyya brotherhood, which originated in twelfth-century Iran. Today, the Tijaniyya order, which emerged in Sunni Islam from the Qadiriyya order in the late eighteenth century, is the largest, while the Muridiyya order, founded in the late nineteenth century, is prevalent in Senegal (34% of Muslims) and the Chistiyya order in Bangladesh (12% of Muslims) (Pew Research Center, 2012, pp. 23–32).

Hindus are concentrated in South Asia, where the only countries with a Hindu majority in 2010 are located – India (80%) and Nepal (81%) (see Table 1.2). Although Hindus are expected to remain the majority in India, the country's overall fertility rate is decreasing, and so by 2050, it is expected that it will also have the world's largest Muslim population, exceeding that of Indonesia. Despite international migration, it is anticipated that 99% of Hindus will continue to live in southern Asia.

While in the past there were many Buddhists in India, today Buddhism there has almost disappeared (0.8% of the population in 2001, although this still represents 8 million people; see Table 1.2). Buddhists are a majority in Cambodia (97%), Laos (66%), Burma/Myanmar (80%), Sri Lanka (69%) and Thailand (93%), although about half the world's Buddhists live in China (Pew Research Center, 2015).

Table 1.2 India's changing religious composition

	1961		*1981*		*2001*	
	No.	%	No.	%	No.	%
Hindu	367m	83.4	550m	82.6	828m	80.5
Muslim	47m	10.7	76m	11.4	138m	13.4
Christian	11m	2.4	16m	2.4	24m	2.3
Sikh	8m	1.8	13m	2.0	19m	1.9
Buddhist	3m	0.7	5m	0.7	8m	0.8
Jain	2m	0.5	3m	0.5	4m	0.4
Other & NS	1m	0.3	3m	0.4	8m	0.7
Total	**439m**	**100.0**	**665m**	**100.0**	**1,029m**	**100.0**

Source: O.P. Sharma and Carl Haub, "Change Comes Slowly for Religious Diversity in India" (Washington, DC: Population Reference Bureau, 2009). All rights reserved. Reproduced by permission. Based on censuses.

The category of folk or traditional religions refers to those associated with particular groups of people, usually defined in terms of ethnicity or tribe. Some (both analysts and practitioners) suggest that such beliefs and practices are cultural rather than religious and are suspicious of attempts to conceptualise them in the same way as Christianity or Islam. Nevertheless, some international surveys include them as a category. The ten countries with the largest number of people who specify a traditional religion as their primary religious affiliation include China (294 million in 2010 – 73% of the world's folk religion adherents), India (5.9 million), South Sudan (3.3 million), Burkina Faso (2.5 million) and Nigeria (2.3 million) (Pew Research Center, 2015, p. 117). Relatively few of those who claim such an affiliation are young, and so their fertility rates are lower than other groups. As a result of both low rates of natural increase and religious switching, their share of the world population is expected to continue to shrink.

Affiliation may reflect a conscious choice, a tradition into which a person is born or a category to which individuals are assigned by government. It does not indicate how important religious beliefs and practices are to adherents. As noted above, international surveys attempt to capture this using the concept of 'religiosity', based, variously, on an individual's self-assessment of how important 'God' and/or religious practices are in his or her own life, whether or not s/he believes in God and whether s/he considers him/herself to be a religious person. Additional questions may refer to other beliefs and frequency of participation in religious practices. A decline in the prevalence of religious beliefs and practices is associated with the process of secularisation, alongside differentiation between religion and other institutions associated with the political, economic and scientific spheres and a relegation of religion from the public to the private sphere and to the remit of religious organisations, the functions of which may nevertheless be regulated by the state.

Driven mainly by trends identified in Europe, the perception that societies were secularising was supported most clearly by evidence from that region, although even there, the details vary between countries and the evidence is sometimes ambiguous (Bréchon, 2007). By the mid-twentieth century, it had become clear to those analysing parts of the world other than Europe and North America and religions other than Christianity that, while institutional differentiation seemed to be occurring, secularisation was taking a variety of forms, reflecting both distinct cultural pathways and different histories of interaction with the West (Burchardt et al., 2013). The concept of secularisation and its manifestations have been widely discussed (see, for example, Bremmer, 2008; Casanova, 1994, 2008; Demerath, 2007; Martin, 2005; Tomalin, 2013). No attempt will be made to summarise this literature here. What emerges is that how secularisation is framed, whether and how it occurs and how entrenched it becomes has to be examined within different cultural framings of the religion-secular distinction that are not only particular to different regions, countries and localities but are also often disputed (see, for example, Berger, 1999; Kaviraj, 2010). In addition, many of those who earlier believed that secularisation would occur worldwide have had to acknowledge that they were mistaken. Some, in their more recent writings, have tended to swing dramatically in the other direction, asserting that instead, a 'resurgence' of religion or 'de-secularisation' of the world is occurring (Demerath, 2007; Martin, 2010; Thomas, 2005).

In the regions of concern to this book, the vast majority of people consider themselves to be an adherent of a religion. According to a Gallup poll in 2016 in 69 countries, of which 11 are relevant (5 in SSA and 6 in southern Asia), only in two did less than 95% of the population consider themselves to be adherents of a religion (in South Africa 6% and in Vietnam, one of the few countries in these regions still adjusting to the end of communist rule, 34% considered themselves to be agnostic/atheist) (WIN/Gallup International, 2017).

However, self-stated adherence may reflect cultural identity, social expectations or government fiat rather than the importance of religion in people's lives. For example, it does not necessarily imply that religious beliefs are the sole (or even an important) basis for the cosmological lens through which people understand the world and their place in it or that they will hold a set of beliefs prescribed by or associated with only one religion, sect or denomination. As we will see, many people's belief systems are highly syncretic. Nor does it imply that they will engage deeply with religion or regularly participate in religious practices. The concept of 'religiosity' or 'religiousness' tries to capture the importance of religion to its adherents.

As noted above, the questions in international surveys are phrased differently, varying between organisations and successive surveys. The longest comparable datasets are available from five waves of the World Values Surveys, which asked respondents how important religion is in their lives, although they are only available for six countries of interest here and in only two of those for all five successive surveys (see Table 1.3).[8] In these six countries, over the last 20 years, the

8 In the WVS, samples of between 1,000 and 4,000 in each country are fairly similar in successive waves, with the exception of India, where a sample of 3,920 in the 2010–14 wave was double the

Table 1.3 Trends in the proportion of people who say that religion is a very important part of their lives (1989–93 to 2010–14)

	1989–93	*1994–8*	*1999–2004*	*2005–9*	*2010–14*
SSA					
Nigeria	85	91	93		90
South Africa	65	68	70	70	56
Zimbabwe			78		84
Southern Asia					
India	49	48	56	50	67
Pakistan		80	82		90
Philippines		78	87		86

Source: www.worldvaluessurvey.org/WVSOnline.jsp. Accessed 10/8/2017.

proportion of people who consider religion to be very important in their lives was consistently high in Nigeria and the Philippines. The proportion apparently increased in India, Pakistan and Zimbabwe but decreased in South Africa, which is not a typical Sub-Saharan African country.

Considering religion to be 'very important in one's life' is not equivalent to being a 'member' of a church or religious denomination, as confirmed by other data from the WVS, which show that in some countries (e.g. South Africa), the proportion of people claiming to belong to a religious denomination exceeds the proportion who consider religion to be very important in their lives, while the opposite is true in others (e.g. the Philippines). Moreover, in many countries with dominant religions that do not have a conception of 'membership' of a religious organisation (e.g. Islam, Hinduism, Buddhism), critics would suggest that the question is inappropriate, even meaningless (e.g. India, Malaysia, Pakistan, Thailand).

The question used by Gallup International in 2005 and 2012 asked whether a respondent considers him/herself to be a religious person, not a religious person or a 'convinced atheist'.[9] In the 39 countries in which polls were carried out in both 2005 and 2012, on average 77% of respondents described themselves as 'a religious person' in 2005 and 68% in 2012 (Gallup International, 2012, p. 10). Of these countries, nine are in the regions of interest here (see Table 1.4). In a third of these, the proportion who consider themselves religious changed by fewer than five percentage points. More marked trends included an apparent increase in Malaysia and Pakistan and a decrease in India and especially South Africa and Vietnam.[10] The proportion who consider themselves to be 'religious' is generally

size of previous samples. Both sample size and the means of selecting a sample may change in successive surveys.

9 Between November 2011 and January 2012, Gallup International polled 50,000+ men and women from 57 countries on five continents, mostly face-to-face (33,890) but also by telephone (11 countries, 7,661) or online (11 countries, 10,376). The question asked in the 2005 and 2012 surveys differed from those asked in earlier surveys.

10 These figures are largely borne out by figures from the Pew Forum's surveys carried out at roughly the same time (Pew Forum, 2010).

Table 1.4 Trends in the proportion of people who consider themselves to be religious (2005–2012)

	Respondent considers him/herself to be a religious person			*Respondent is a convinced atheist*		
	2005	*2012*	*% change*	*2005*	*2012*	*% change*
Overall average (32 countries)	77	68	−9	4	7	+3
SSA						
Ghana	96	96	0	0	0	0
Nigeria	94	93	−1	1	1	0
Kenya	89	88	−1	0	2	+2
Cameroon	86	82	−4	5	3	−2
South Africa [a]	83	64	−19	1	4	+3
Southern Asia						
India	87	81	− 6	4	3	− 1
Pakistan	78	84	+6	1	2	+1
Malaysia	77	81	+4	4	0	−4
Vietnam	53	30	−23	1	0	−1

[a] The sample was only 200. However, figures from other sources confirm that South Africa is atypical. NB The global *weighted* average, which reflects "the share of the covered adult population of the surveyed countries" (p. 8), shows that in 2012, 59% of respondents considered themselves religious and 13% claimed to be convinced atheists, pp. 15–16.

Sources: Gallup International (2012) *Global Index of Religiosity and Atheism*, figures selected from pp. 12, 13. www.Gallup-international.com.

slightly smaller than those who acknowledge adherence to a religion, but the data confirm that religiousness continues to be widespread in most typical African and southern Asian countries. Although the time period is limited, there is little evidence of either rampant secularisation or increased religiousness, except perhaps in the Muslim-majority countries of Pakistan and Malaysia. In addition, a majority of respondents in a 2011 survey in Sri Lanka perceived their society as becoming more religious – 63% believed people to be more religious than five years prior to the survey (70% of Buddhists, the majority religious group, 53% of Muslims, 39% of Hindus and 44% of Catholics) (Asia Foundation, 2012).

Various hypotheses about the relationship between socio-economic characteristics and religiousness have been advanced, three of which will be briefly considered here. First, it has been suggested that widespread and intense religiosity is associated with poverty and insecurity. There does indeed appear to be a relationship between religiosity and national wealth (see Table 1.5). However, like other organisations, Gallup International does not provide any commentary on these findings. Norris and Inglehart (2004), in contrast, on the basis of further analysis of World Values Survey results, attribute the higher levels of religiosity in poorer countries to the insecurity and vulnerability of people's daily lives. Crabtree and Pelham (2009) are not convinced by this argument, noting instead

Table 1.5 Religiosity and wealth/poverty

Country per capita income $	*'Religion is an important part of my daily life' (2009) %*
0–2,000	95
2,001–5,000	92
5,001–12,500	82
12,501–25,000	70
25,000+	47

Source: www.gallup.com/poll/142727/religiosity-highest-world-poorest-nations.aspx Accessed 9 Aug 2017 (global median 84%).

that in the lowest income group in 32 countries studied in 2008, religious people were more likely to say that they had experienced enjoyment the previous day and less likely to report that they had experienced a range of negative emotions, leading the authors to conclude that, although the results varied by country, the positive social and psychological effects of religion appear to be strongest in the poorest countries.

Second, based on data from countries in the global North, it has been suggested that religion is linked to knowledge. Gallup International findings appear to substantiate this, with a larger proportion of the least well educated in the 57 countries where surveys were carried out describing themselves as 'a religious person' than of those with higher levels of education (68% of those with primary school education or less, 62% of those with secondary education, and 52% of those with higher education) (Gallup International, 2012, p. 20). However, breakdowns for individual countries were not published, so it is not possible to assess the apparent link between educational level and religiousness in countries where the vast majority of people are religious.

Third, data on European and North American countries suggest that religiosity is differentiated by gender, with women more likely to be religious than men. A gender breakdown is available for 11 relevant countries in the 2010–14 wave of the WVS surveys (see Table 1.6). The proportion of men and women in these countries who consider themselves to be religious is generally quite similar, varying by four percentage points or less except in the Philippines, Singapore, South Africa and Zimbabwe, where the proportion of women who consider themselves to be religious is higher than the proportion of men who do so. These indicative results are borne out by data from the Pew Research Center's national surveys in 84 countries between 2008 and 2014. In 46 of the 84 countries, men and women were equally likely to say that religion is very important in their lives, although in 36 other countries, women were more likely than men to say religion is very important to them personally, often by large margins. In the countries relevant here, the proportion of women who said religion is important to them was 18 percentage points higher in South Africa, 14 in Vietnam and 7 in Kenya and Rwanda. Only in Mozambique was religion significantly more important to men (by four percentage points). Among Christians, religion is more important to women than men in many countries, although in 29, including most of the African countries covered, there is no significant gender gap and nor is there such a gap among Hindus in India or Buddhists in Malaysia. The Pew Research Center concludes that, as might be expected, where religious affiliation is nearly universal and self-professed religiousness widespread, differences between men and women are small (Pew Research Center, 2016).

The figures from different sources and repeat surveys support the idea that most people in Sub-Saharan African and southern Asian countries acknowledge affiliation with a religious tradition and also regard themselves as religious. Although evidence is scarce, there do not appear to be clear differentials related to wealth, education or gender. The evidence on religious revival is ambivalent – while it does appear that religiousness has increased in some countries, it appears to have decreased or remained unchanged in others, although the results may also have been influenced by such artefacts of surveys as differences in sampling, the phrasing of questions and individuals' responses. Voas (2015) cautions that perhaps a quarter of responses to any question on religion are unreliable because people see 'religion' in different ways, the question may be ambiguous, and test–retest reliability assessments show that a large proportion of respondents answer differently in successive surveys.

Trends in religious adherence and vitality are, of course, not just a matter of numerical growth or decline. Rather, they reflect the dynamism, scope and force of religious groups or movements, which may be manifest in, for example, the emergence of new ways of believing and belonging (Clarke, 2009). All the religious traditions have been characterised by successive revivalist movements. These are

Table 1.6 Proportion of men and women who consider themselves to be religious

Proportion of respondents who consider themselves to be a religious person			
	Total	*Men*	*Women*
Pakistan	100	99	100
Ghana	97	95	99
Nigeria	96	96	96
Rwanda	96	95	96
India	92	91	93
Zimbabwe	91	88	93
Philippines	81	78	93
South Africa	80	73	86
Singapore	60	54	65
Malaysia	54	55	52
Thailand	33	32	35

Note that these figures are not necessarily consistent with the Gallup results. For example, 81% of respondents in Gallup's survey in Malaysia consider themselves to be a religious person (Gallup International, 2012).

Source: www.worldvaluessurvey.rg/WVSOnline.jsp. Accessed 10/8/2017.

often responses to the perceived shortcomings of the authority/ies regarded as responsible for regulating relations between societies and the powers governing the cosmos or aspects of a religious tradition or group. They have sometimes emerged in the face of (a fear of) catastrophe and the coming to prominence of a prophet who renews the promise of an earlier tradition that promised salvation from oppression, anxiety or frustration. They have been associated with rivalry between new and older religious traditions, such as mission Christianity and folk religion. Even more commonly, they seek to renew an existing religion or denomination that seems to be moribund or unresponsive to wider social change. They may succeed or result in conflict and schism.

Significant recent trends include so-called fundamentalism in Christianity and other religious traditions, the growth of Pentecostalism in Christianity, a new wave of reform movements in Islam, including political Islam and pietist movements, and the emergence of nationalist religious ideologies and movements in other religions, including Hinduism and Buddhism. Some see religious revival as a response to the kind of world to which modernity has brought us, which is characterised by inequality, environmental deterioration and social stresses, as well as the end of a certain kind of faith in progress and the belief that science and technology can solve all contemporary problems. Rather than a perception that modernity has 'failed', they reflect a desire to re-evaluate how religion and modernity are related (Casanova, 2008; Rakodi, 2015). Earlier movements were closely linked to historical developments in Europe and the US, and twentieth-century movements were often responses to colonialism and struggles for independence. However, the increasing prosperity and international influence of the US and, later, the massive increase in oil-based wealth of Middle Eastern countries, mean that late-twentieth- and early twenty-first-century trends can only be understood in the context of contemporary economic and cultural globalisation (see, for example, Held and McGrew, 2000; Murray and Overton, 2014).

Some of the revivalist trends have been extensively documented and discussed. Contemporary fundamentalism, it is agreed, can be traced to a movement to return to the 'fundamentals' of Christianity in the US in the early decades of the twentieth century (Duschinsky, 2012; Munson, 2006; Shupe, 2009; Turner, 2010). At about the same time, and also in the southern US, Pentecostalism emerged. It shares many of the concerns, values and features of fundamentalism, although the distinct Protestant and independent churches that were founded in response have a less biblical and more experiential orientation, emphasising ecstatic experience of the Spirit and the practice of spiritual gifts. As the growth of Pentecostalism threatened the appeal, number of members and influence of the mainline Christian denominations, revivalist movements also emerged within later (often called charismatics, from the Greek *charismata*, meaning gifts) (Anderson, 2013; Pew Forum, 2006, Pew Research Center, 2011; Martin, 2001). However, revivalist movements emerged at about the same time in other geographical locations, for example southern Africa, where they led to the establishment of large numbers of African Independent (or Instituted) Churches (Adogame, 2010). Today it is estimated that at least a quarter of Christians worldwide are Pentecostals or charismatics, concentrated numerically in SSA, Latin America and the eastern fringe of Asia (Goh, 2005; Pew Research Center, 2011; Anderson, 2013).

As in Christianity, there have also been movements that seek to return to and reinforce the fundamentals of Islam, which had been weakened by the long decline and eventual collapse of the Mughal Empire in the mid-nineteenth and the Ottoman Empire in the early twentieth centuries, as well as European colonial expansion. Bruce (2000) suggests that Islam is susceptible to fundamentalist thought because of the status of the Qur'an as the literal and inerrant word of God, and the Muslim belief that the domains of religion and the world cannot be divided without compromising one's beliefs, so that religion cannot be relegated to the private sphere (see also Marranci, 2010). In this interpretation, it is thought that Muslims have been weakened by their failure to obey divine laws and ensure that religion pervades all aspects of life, implying the need to revert to a form of Islam which sets out to reclaim what it regards as traditional moral and political values, in reaction to aspects of modernisation. Thus on the one hand, it rejects both Sufi traditions within Islam and, on the other, the reason-based worldview of modernity (attacked as Westernisation). Muslim fundamentalists are, however, not traditionalists, because they seek to employ selected (especially scientific and technological) achievements of modernity in pursuit of an Islamic world order (Shupe, 2009; Tibi, 2005).

In Islam, fundamentalism can be traced to a number of elite-driven revival and reform movements, generally launched by eminent religious teachers, writers and ideologues (Tibi, 2005). In the eighteenth century, the most important was Muhammed Ibn Abd al-Wahhab, whose criticisms of the rule of Ottoman Turks in parts of Asia led him to see innovation as the enemy and to seek to lead Muslims back to the original 'pure' form of Islam. Wahhabism is associated with a particular Sunni legal school and the contemporary Saudi Arabian state. Salafists have a similar agenda. Named after the Companions of the Prophet and the pious Muslims of the first three generations, they seek to return to an unmediated Islam, based on the original sources. Salafism is often linked with Wahhabism, not least because Saudi Arabian money, intellectual influence and spiritual resources support Salafi communities around the world. However, the two movements were and are neither identical nor monolithic.

In the nineteenth century, religious reform movements tended to be a response to colonialism, mission and Westernisation (Asani, 2006). For example, Muslims as well as Hindus in India developed a heightened sense of their religious identity and tradition, leading to a desire to restore their religion to its imagined original vitality and purity and revitalise Indian society. In 1867, the Dar ul' Ulum madrasa was established at Deoband in northern India. This and other movements (e.g. Tablighi Jama'at, Ahl-e Hadith, the Barelwis) sought to purify Islam of Hindu accretions. Since the 1920s, their national and international influence has grown (Ruffle, 2013; Waines, 2009). In the twentieth century, Hassan Al-Banna and Sayyid Qutb in Egypt, Sayyid Abu Al-Ala Maududi in Pakistan and Ayatollah Ruholla Khomeini in Iran were the most prominent ideologues driving the next wave of revivalism (Zaidi, 2006). Hassan Al-Banna founded the

Ikwan al Muslimin (Muslim Brotherhood) in Egypt in 1928. Following rapid early growth, it was banned in 1948 because of its advocacy of violence, but its offshoots include Hamas, Al-Qaeda and Jama'at-e-Islami (Tibi, 2008). Qutb, who founded the latter, was influenced by Maududi's thought. In 1941, the latter in turn established Jama'at-e-Islami in Pakistan, which had increased influence following the military coup of 1977 (Waines, 2009).

It is clear from this list of influential thinkers that fundamentalist Islam has been associated with both the Sunni and Shia sects. For both, fundamentalist views refer to their own sect rather than Islam as a single religion. They disagree on early Muslim history, their choice of exemplars and the correct performance of religious rituals, in some instances leading to conflict and violence, including persecution of the Shia minority in Pakistan since the 1980s. However, both Sunni and Shia revivalists insist that Islam provides specific prescriptions for all aspects of life, insisting that true justice is only possible when religious law replaces secular law and religious practices are enforced. Islamic fundamentalism is therefore often equated with so-called 'political Islam', in which religion should, it is believed, be fused with a state order to institute an 'Islamic system', in which religious institutions are restored to the centre of politics and decision making. This ideology and programme has more appeal to many Muslims than the secular, imported political order associated with the elite in many countries, which is regarded as having failed to deliver on its promises (Shupe, 2009; Tibi, 2005). The scale and influence of political Islam has increased in recent decades, especially in Muslim-majority countries, including Pakistan and Indonesia. However, not only does it have many manifestations, not all of which believe that violence is a legitimate way of instituting an Islamic world order, focusing attention on it distracts from other aspects of Muslim revivalism. These include modernist attempts to reinterpret elements of the teachings to relate them to changing economic, social and political contexts (Tibi, 2005, 2008) and piety movements that focus on moral renewal (Schultz, 2011). In addition, most attention has focused on Islamic revivalism in the Middle East, North Africa and the Indian subcontinent, diverting attention from parallel trends elsewhere, especially in West Africa.

In some other contemporary religious revival movements, religion is closely linked with cultural and political nationalism, for example Hindu nationalism in India and Sinhala nationalism in Sri Lanka. While popular Hinduism continues to be widely practised, the former employs the ideology of Hindutva (or 'Hinduness') to equate India with an essentialised Hindu religion and culture. By emphasising selected aspects of Hindu religious practices, this promotes a collective identity intended to provide the cultural foundation for a strong Hindu 'nation' oriented towards Hindu concerns, especially vis-à-vis Muslim demands. Thus religio-cultural revival and political ambitions are closely intertwined (see, for example, Berglund, 2004; Corbridge and Harriss, 2000; Fuller, 2004; Ram-Prasad, 2003; Sheth, 2010). Similarly, in Sri Lanka and Myanmar, Buddhism is used to legitimise the political marginalisation of religious and ethnic minorities, while in other places, such as Nigeria, religious differences are closely linked to ethnic competition over State and natural resources. For the most part, these essentially political movements will not be discussed in this volume.

The influence of the recent trends and movements briefly introduced above on contemporary beliefs, religious practices, values and attitudes in SSA and southern Asia, as revealed by the available quantitative and qualitative evidence, will emerge in subsequent chapters. It is impossible to generalise about the nature of their influence, since this involves interactions between international trends and local characteristics. Nor can we be precise about the extent of their influence or predict their outcomes. Care should be taken not to exaggerate their scale and importance at the cost of acknowledging and explaining the persistence of longstanding religious organisations and practices. However, analysts' attention is often captured by the new, unfamiliar, extreme and inspiring or alarming rather than the ordinary and enduring, and so much recent research has focused on them. As a result, despite my efforts to focus the analysis on the ongoing and everyday, their manifestations and effects will figure prominently in the discussions of religious beliefs, practices and values that follow.

References

Adogame, A. (2010) Pentecostal and charismatic movements in a global perspective, in Turner, B.S. (ed) *The New Blackwell Companion to the Sociology of Religion*, Oxford: Wiley-Blackwell, 498–518.

Anderson, A. (2013) *To the Ends of the Earth: Pentecostalism and the Transformation of World Christianity*, Oxford: Oxford University Press.

Asani, A.S. (2006) Religious movements with Islamic origins, in Juergensmeyer, M. (ed) *The Oxford Handbook of Global Religions*, Oxford: Oxford University Press, 515–23.

Asia Foundation (2012) *National Values Survey, Sri Lanka 2011*, Colombo: The Asia Foundation.

Beckford, J.A. (2003) *Social Theory and Religion*, Cambridge: Cambridge University Press.

Berger, P.L. (1999) The desecularization of the world: A global overview, in Berger, P.L. (ed) *The Desecularization of the World: Resurgent Religion in World Politics*, Grand Rapids, MI: Eerdmans Publishing Co, 1–18.

Berglund, H. (2004) Religion and nationalism: Politics of BJP, *Economic and Political Weekly*, March 6, 1064–70.

Bréchon, P. (2007) Cross-national comparisons of individual religiosity, in Beckford, J.A. and Demerath, N.J. III (eds) *The Sage Handbook of the Sociology of Religion*, London: Sage, 463–89.

Bremmer, J.N. (2008) Secularization: Notes toward a genealogy, in de Vries, H. (ed) *Religion: Beyond a Concept*, New York: Fordham University Press, 432–7.

Bruce, S. (2000) *Fundamentalism*, Cambridge: Polity Press.

Burchardt, M., Wohlrab-Sahr, M. and Wegert, U. (2013) 'Multiple secularities': Postcolonial variations and guiding ideas in India and South Africa, *International Sociology*, 28, 6, 612–28.

Capps, W.H. (2005) Society and religion (first edition), in Jones, L. (ed) *Encylopedia of Religion, Second Edition*, Detroit: Palgrave Macmillan, Vol. 12, 8461–70.

Casanova, J. (1994) *Public Religions in the Modern World*, Chicago: University of Chicago Press.

Casanova, J. (2008) Public religions revisited, in de Vries, H. (ed) *Religion: Beyond a Concept*, New York: Fordham University Press, 101–19.

Clarke, M. (2011) *Development and Religion: Theology and Practice*, Cheltenham: Edward Elgar.

Clarke, M. (ed) (2013) *Handbook of Research on Development and Religion*, Cheltenham: Edward Elgar.

Clarke, P.B. (2009) Introduction: Towards a more organic understanding of religion within a global framework, in Clarke, P.B. (ed) *The Oxford Handbook of the Sociology of Religion*, Oxford: Oxford University Press, online ed.

Connell, R. (2007) *Southern Theory: The Global Dynamics of Knowledge in Social Science*, Cambridge: Polity Press.

Corbridge, S. and Harriss, J. (2000) *Reinventing India: Liberalization, Hindu Nationalism and Popular Democracy*, Cambridge: Polity Press.

Crabtree, S. and Pelham, B. (2009) Religion provides emotional boost to the world's poor, www.gallup.com/pol/116449/Religion-Provides-Emotional-Boost-World-Poor.aspx?g_source=&g_medium=&g_campaign=tiles

De Kadt, E. (2011) Should God play a role in development? *Journal of International Development*, 21, 6, 781–6.

Demerath, N.J. III (2007) Secularization and sacralisation deconstructed and reconstructed, in Beckford, J.A. and Demerath, N.J. III (eds) *The Sage Handbook of the Sociology of Religion*, London: Sage, 57–80.

Deneulin, S. and Bano, M. (2009) *Religion in Development: Rewriting the Secular Script*, London: Zed.

Deneulin, S. and Rakodi, C. (2011) Revisiting religion: Development studies thirty years on, *World Development*, 39, 1, 45–54.

Duschinsky, R. (2012) Fundamentalism and the changing religious field, *Social Compass*, 59, 1, 21–33.

Fountain, P. (2013) The myth of religious NGOs: Development studies and the return of religion, *International Development Policy/Revue Internationale de Politique de Développement*, 4, 9–30.

Fuller, C.J. (2004) *The Camphor Flame: Popular Hinduism and Society in India*, Princeton, NJ: Princeton University Press.

Gallup International (2012) *Global Index of Religiosity and Atheism*, www.Gallup-international.com

Goh, R.B.H. (2005) *Christianity in Southeast Asia*, Singapore: Institute of Southeast Asian Studies.

Held, D. and McGrew, A. (2000) The great globalization debate: An introduction, in Held, D. and McGrew, A. (eds) *The Global Transformations Reader*, London: Polity Press, 1–45.

Hsu, B., Reynolds, A., Hackett, C. and Gibbon, J. (2008) Estimating the religious composition of all nations: An empirical assessment of the World Christian database, *Journal for the Scientific Study of Religion*, 47, 4, 678–93.

Kaviraj, S. (2010) On thick and thin religion: Some critical reflections on secularisation theory, in Katznelson, J. and Jones, G.S. (eds) *Religion and the Political Imagination*, Cambridge: Cambridge University Press, 336–55.

Marranci, G. (2010) Sociology and anthropology of Islam: A critical debate, in Turner, B.S. (ed) *The New Blackwell Companion to the Sociology of Religion*, Oxford: Wiley-Blackwell, 364–87.

Marshall, K. and Keough, L. (eds) (2004) *Mind, Heart and Soul in the Fight Against Poverty*, Washington, DC: The World Bank.

Martin, D. (2001) *Pentecostalism: The World their Parish*, Oxford: Wiley-Blackwell.
Martin, D. (2005) *On Secularization: Towards a Revised General Theory*, Aldershot: Ashgate.
Martin, G. (2010) Resacralization, in Turner, B.S. (ed) *The New Blackwell Companion to the Study of Religion*, Oxford: Wiley-Blackwell, 160–78.
Munson, H. (2006) Fundamentalism, in Segal, R. (ed) *The Blackwell Companion to the Study of Religion*, Oxford: Wiley-Blackwell, 255–69.
Murray, W.E. and Overton, J. (2014) *Geographies of Globalization*, London: Routledge.
Norris, P. and Inglehart (2004) *Sacred and Secular: Religion and Politics Worldwide*, Cambridge: Cambridge University Press.
Pew Forum (2006) *Spirit and Power: A 10-Country Study of Pentecostals*, Washington, DC: Pew Forum, http://pewforum.org/uploadedfiles/Orphan_Migrated_Content/pentocostals-08.pdf
Pew Forum (2010) *Tolerance and Tension: Islam and Christianity in Sub-Saharan Africa*, Washington, DC: Pew Forum, http://pewforum.org/uploadedFiles/Topics/Belief_and_Practices/sub-saharan-africa-full-report.pdf
Pew Research Center (2011) *Global Christianity: A Report on the Size and Distribution of the World's Christian Population*, Washington, DC: Pew Research Center. Pew-Templeton Global Religious Futures Project, http://assets.pewresearch.org/wp-content/uploads/sites/11/2011/12/Christianity-fullreport-web.pdf
Pew Research Center (2012) *The World's Muslims: Unity and Diversity*, Washington, DC: Pew Research Center, http://assets.pewresearch.org/wp-content/uploads/sites/11/2012/08/the-worlds-muslims-full-report.pdf
Pew Research Center (2015) *The Future of World Religions: Population Growth Projections, 2010–2050*, Washington, DC: Pew Research Center, http://assets.pewresearch.org/wp-content/uploads/sites/11/2015/03/PF_15.04.02_ProjectionsFullReport.pdf
Pew Research Center (2016) *The Gender Gap in Religion Around the World*, Washington, DC: Pew Research Center, http://assets.pewresearch.org/wp-content/uploads/sites/11/2016/03/Religion-and-Gender-Full-Report.pdf
Rakodi, C. (ed) (2014) *Religion, Religious Organisations and Development: Scrutinising Religious Perspectives and Organisations*, Abingdon: Routledge.
Rakodi, C. (2015) Development, religion and modernity, in Tomalin, E. (ed) *The Routledge Handbook of Religions and Global Development*, London: Routledge, 17–35.
Ram-Prasad, C. (2003) Contemporary political Hinduism, in Flood, G. (ed) *The Blackwell Companion to Hinduism*, Oxford: Wiley-Blackwell, 526–50.
Riesebrodt, M. and Konieczny, M.E. (2005) Sociology of religion, in Hinnells, J. (ed) *Penguin Companion to the Study of Religion*, New York: Penguin.
Ruffle, K.G. (2013) Islam in South Asia: Practising Islam today, in Pechilis, K. and Raj, S.J. (eds) *South Asian Religions: Tradition and Today*, London and New York: Routledge, 192–221.
Schultz, D.E. (2011) *Muslims and the New Media in West Africa: Pathways to God*, Bloomington, IN: Indiana University Press.
Sheth, D.L. (2010) Political communalisation of religions and the crisis of secularism, in Mahajan, G. and Jodhka, S.S. (eds) *Religion, Community and Development: Changing Contours of Politics and Policy in India*, New Delhi: Routledge, 36–64.
Shupe, A. (2009) Religious fundamentalism, in Clarke, P.B. (ed) *The Oxford Handbook of the Sociology of Religion*, Oxford: Oxford University Press, online. Doi: 10.1093/oxfordhb/9780199588961.013.0027, www.oxfordhandbooks.com
Stonawski, M., Potančoková, M., Cantele, M. and Skirbekk, V. (2016) The changing religious composition of Nigeria: Causes and implications of demographic divergence, *Journal of Modern African Studies*, 54, 3, 361–87.

Ter Haar, G. (ed) (2011) *Religion and Development: Ways of Transforming the World*, London: Hurst.

Thomas, S.M. (2005) *The Global Resurgence of Religion and the Transformation of International Relations: The Struggle for the Soul of the Twenty-first Century*, Basingstoke: Palgrave Macmillan.

Tibi, B. (2005) *Islam between Culture and Politics*, Basingstoke: Palgrave Macmillan, 2nd ed.

Tibi, B. (2008) *Islam's Predicament with Modernity: Religious Reform and Cultural Change*, London and New York: Routledge.

Tomalin, E. (2013) *Religions and Development*, Abingdon: Routledge.

Tomalin, E. (ed) (2015) *The Routledge Handbook of Religions and Global Development*, Abingdon: Routledge.

Turner, B.S. (2010) Introduction: Mapping the sociology of religion, in Turner, B.S. (ed) *The New Blackwell Companion to the Sociology of Religion*, Oxford: Wiley-Blackwell, 1–27.

Turner, B.S. (2014) Religion and contemporary sociological theories, *Current Sociology*, 62, 6, 771–88.

Tyndale, W. (2006) *Visions of Development*, Aldershot: Ashgate.

Voas, D. (2015) Religion, sociology of, in Wright, J.D. (ed) *International Encyclopedia of the Social and Behavioral Sciences*, London: Routledge, 2nd ed, 359–65.

Waines, D. (2009) Islam, in Woodhead, L., Kawanami, H. and Partridge, C. (eds) *Religions in the Modern World: Traditions and Transformations*, London: Routledge, 2nd ed, 237–63.

WIN/Gallup International (2017) Global Tables Q1–Q3, 11 January.

Winchester, D. (2016) Religion as theoretical case, lens and resource for critique: Three ways social theory can learn from the study of religion, *Sociology of Religion*, 77, 3, 241–60.

Woodhead, L., Kawanami, H. and Partridge, C. (eds) (2009) *Religions in the Modern World: Traditions and Transformations*, Abingdon: Routledge, 2nd ed.

Zaidi, A.H. (2006) Muslim reconstructions of knowledge and the re-enchantment of modernity, *Theology, Culture and Society*, 23, 5, 69–91.

2 Towards a framework for analysing the links between religion and society

The analysis in this volume examines the links between religion, people's lives and social characteristics and dynamics in contemporary Sub-Saharan Africa and southern Asia, in order to tease out how religious teachings influence beliefs, religious practices, values and attitudes and social behaviour. It takes account of religious and social diversity and identifies the social forces that influence the nature and effects of the links identified, focusing partly on the ways in which social forces and dynamics influence religions but mainly on the effects of religion on lives and societies. The same theme will be picked up in the companion volume, which explores the implications of the links between religions, societies and states for social and political change.

The nineteenth-century founders of social theory (Durkheim, Marx and Weber) and their twentieth-century successors argued that particular themes (forms of social integration, relations of production and rationalisation respectively) have "a primary explanatory relevance over a vast range of social phenomena" (Davis, 1999, p. 30).[1] While grand theory may be mined as a source of concepts that suggest ways in which the world can be perceived and analysed, the belief that it is possible to develop a single metanarrative has been discredited (though it is not dead). Instead, social research can better be seen as starting from and generative of middle-level theory, based on and testable by empirical investigation. The analytical frame adopted here therefore seeks to rotate between theorising, generating research questions and analysing empirical data to increase understanding and develop explanations (Davis, 1999). This requires a framework which suggests ways of seeing the world, identifies the key components of the social phenomenon/a under study, provides conceptual tools to guide initial understanding and identify the relevant characteristics of these components, guides examination of the available evidence to build up a 'picture' (or 'image') of these dimensions and draws conclusions which enable the frame to be confirmed or amended. Such a flexible frame can be used to identify the key features or

1 The key social theorists will not be systematically presented and their ideas critiqued here, but see, for example, Beckford (2003), Coleman (2010), Furseth and Repstad (2006), Hamilton (1995), O'Toole (2001), Turner (2010).

aspects that differentiate between contexts and cases in particular social settings and to make sense of the phenomenon/a of interest (Ragin and Amoroso, 2011).

The various steps that need to be taken in order to develop such a frame will be considered in this chapter. First, the overall perspective adopted is identified. The building blocks of the analysis are the concepts commonly used to understand the links between religion, people's values and attitudes, and their everyday lives and basic social relationships, especially the central concept of religion but also those used to understand the contexts in which religion is manifest in everyday lives: culture and society. Each complex and contested, it is important to set out my conceptualisation at the outset. This will pave the way for identifying relevant dimensions of the social field, the levels at which social analysis is required, the key elements of religion in society and the characteristics of the anticipated links between those elements.

Lived religion

Some analyses assume that religion is the *main* factor shaping people's lives and identities and that it is all pervasive. Such a view fails to recognise not only that no religion is monolithic, even if it has a common history and belief system, but also that religion is only one of many influences on people and societies. The analysis will, therefore, seek to tease out the ways in which religion influences individuals and social groups, the factors that explain the nature and extent of its influence and the ways that it interacts with local, national and transnational social, cultural, economic and political organisations and processes. In this volume, the focus will be on individuals and small social and religious groups rather than wider movements for socio-political change, the roles of religious organisations in society and their interactions with states, or religious engagements with politics.

The central pillar of the analysis is the concept of everyday lived religion. This distinguishes the experience of religious persons from the prescribed religion of institutionally defined beliefs and practices (McGuire, 2008; Orsi, 2003). The texts, their official interpreters and the established structures of authority within religious traditions are important and need to be recognised and respected. However, lived religion is expressed in local interpretations of religious teachings and individuals' responses to them, and these interact with other influences on the ways people understand the world, relate to others and exercise agency.

For many, all the domains of life have a spiritual meaning. Even though their everyday lifeworlds are mostly concerned with practical matters, they have transcendent concerns (Verhelst and Tyndale, 2002). To the extent that their beliefs and values are based on (their interpretations of) religious teachings, they "take their religious sensibilities with them in ways that shape their everyday relationships and behaviour" (Ammerman, 2014, p. 196). However, religion is not just located in individual consciousness. It is also expressed in social processes and practices, "in the conversational spaces both in religious organisations and beyond – where sacred and mundane dimensions of life are produced and negotiated" (Ammerman, 2014, p. 196). Thus religion lives in the world through individuals who participate in various social groups and organisations, have multifaceted identities and are part of more than one community. It is revealed by how people carry out their religious duties, envision a cosmological order in which those duties are meaningful and interact with each other in religious and non-religious settings at a variety of scales, including the family, social group, locality and wider society (Pankhurst and Houseknecht, 2000).

In seeking to understand how religion influences people's lives and social relationships, it is crucial to focus primarily on the everyday, ordinary and enduring rather than uncommon phenomena, dramatic changes and extreme positions, even though there is often only limited research on the everyday and mundane. As noted above, researchers tend to be drawn to studying uncommon phenomena, whether small breakaway religious organisations or unusual beliefs, leading to the neglect of quotidian views and practices. Similarly, analytical attention is often caught by radically different rather than mainstream perspectives. Here, however, the analysis will, for the most part, focus on the mainstream in order to throw light on the religious lives of the majority. Researchers also tend to be drawn to studying rapid change rather than enduring attitudes and institutions. While continuity may be taken for granted or seem dull, persistence should be documented and understood. Nevertheless, there have been significant economic, social and religious changes in recent decades, and so it is also important to identify and discuss the drivers, manifestations and implications of some of these. In particular, as discussed in Chapter 1, there are a variety of movements for religious revival and reform. The scale and significance of some of these for the lives of ordinary people is such that they will be considered in this volume, although because of its focus on the everyday and ordinary, the analysis will concentrate on so-called 'piety' movements rather than political, 'extremist' or violent movements.

Conceptual building blocks

Concepts are used to develop understanding of complex social phenomena in different cultures and rival discourses. The most central to the concerns of this book are religion, society and culture, which I conceive of as dimensions of social life. All are constructed concepts which have, over time and in different places, been contested and reconstructed. None have precise or universally accepted meanings and definitions – not only do the theoretical interpretations of each vary between disciplinary traditions and individual scholars, they have gained ideological loadings and changed over time, while the boundaries between them are fuzzy. I hold that, while they should not be essentialised, they nevertheless have analytical value, and so it is important to clarify them at the outset so that the ways they are used in the subsequent analysis are consistent and understandable to readers with varying cultural, religious and disciplinary backgrounds. In addition, while definitions are often problematic, they are potentially useful, helping to identify the most characteristic elements of a phenomenon and creating sufficient consensus for a productive discussion.

Religion

The concept of 'religion' is "the historical product of ideologically charged, discursive processes" (Asad, 1993, p. 29). It emerged in the West out of a long historical process during which the churches and religious specialists sought to consolidate their power to define 'true religion' and claim for themselves roles as the sole legitimate arbiters of the Christian tradition. They sought to undermine popular religion, to differentiate the religious sphere from that of science and, following a long period of conflict between churches and states, to separate religion from political power (Asad, 1993; McGuire, 2008). Their conception of religion as a distinctive space of human belief and practice is closely associated with Christianity, was tied up with the formation of a modern European identity and has predominated in Europe and North America. Regarded by some as the only religion capable of becoming universal, Christianity was also linked with colonialism. In the process, other religious traditions were '(re-)discovered' by Europeans, and the label 'world religion', which views religions as originating in particular locations and then spreading to different parts of the globe, came to be applied to other religions. The idea of world religions had the merit of giving other traditions (except indigenous or folk religions) the same weight as Christianity, although it also led to a tendency for them to be regarded as conceptually the same, i.e. primarily as a set of beliefs and practices defined by religious authorities and distinct from the secular sphere (Bender et al., 2012).

In practice, not only is "the very idea of neatly classified 'world religions' (e.g. Hinduism, Buddhism, Christianity) that have a linear individual history and that can be sharply differentiated from other world religions by virtue of the nature of their distinct beliefs and practices . . . considered by many contemporary scholars of religion to be a relatively recent (and originally western) invention" (Tomalin, 2013, p. 63), it also puts artificial boundaries around traditions, assumes people can only belong to one (with a change to another requiring conversion) and ignores the syncretistic nature of much religion. For example, it is argued that prior to the eighteenth century there was no such thing as a religion called 'Hinduism'. Colonial scholars, missionaries and administrators, influenced by a conception of religion based on Christianity, gave a name (borrowed from the Arabic name for India, *al-hind*) to the diverse practices followed by the majority of the population in India (Narayanan, 2013; Pechilis, 2013; Tomalin, 2013), while indigenous thinkers and movements contributed to the construction of Hinduism and the firming up of boundaries between it and the traditions of Buddhism and Sikhism. In doing so, the beliefs and practices associated with the priestly Brahmin caste (the elite Sanskrit tradition) were focused on, rather than the village or popular religion in which most people participated (Pechilis, 2013). Subsequently, these constructions have been central to the emergence of distinct ethno-religious identities in India. In addition, people, ideas, rituals, values and organisational structures have in the past and continue today to travel internationally, leading to the emergence of particular religious configurations in different places, where they interact with existing religious configurations, both changing in the process (Levitt, 2012; Vásquez, 2012).

The European conception of religion is theist. In it, "religious faith includes or implies beliefs about a supreme metaphysical reality, and these beliefs offer the foundation of religion as a whole. The philosophical core of Christianity – and of the other monotheistic religions – consists of beliefs about the supreme being" (Koistenen, 2006, p. 169), so that religious language is held to refer to "objectively existing transcendent reality" – God is real. A monotheistic conception of God is often extended to 'gods', 'the supernatural' or the 'sacred' in an attempt to make this concept of religion more generally applicable. The 'sacred' may be used theologically to refer to manifestations of the supernatural or transcendent, operating outside the ability of rationality (i.e. science) to explain them (Evans and Evans, 2008). It can also have a social institutional interpretation that refers to temporal, relational and spatial areas of life which are marked off from the mundane and are valued (usually by specific communities) because of their religious significance or because they symbolise the values of the community (Fitzgerald, 1997). Thus sacred, meaning set apart, is generally defined in relation to the profane, meaning ordinary, the world that can be observed, operates rationally and is explainable by human reason (Saler, 2008). This interpretation may regard religion and science as competing ways of knowing and establishing truth or as complementary and compatible (Evans and Evans, 2008). Koshul (2006), distinguishing between practical, formal and value rationality, argues that Islam provides such a bridge. 'Spirituality', often used in parts of the world where institutional religion has become less prevalent, refers to "a state of being related to a divine, supernatural or transcendent order of reality or, alternatively, as a sense or awareness of a suprareality that goes beyond life as ordinarily experienced" (Wuthnow, 2001, p. 307).

The central meaning of the word 'faith' is similar to 'belief' or 'trust'. In the context of religion, it implies a relationship with transcendental reality, the existence and nature of which cannot be logically proven or objectively known – thus faith refers to belief(s) in the truths of religion and often 'a religion' may be conceived of as the group of people who express a belief in a divine power and regard a particular source of knowledge or teaching about that power as authoritative. While these ideas of faith and divinity might characterise Christianity, it is doubtful whether they are equally applicable to religions that are not theistic, for example Buddhism (see Chapter 3). In addition, the widespread use of the term 'faith' in contemporary America, where it is linked with the rise and political influence of the religious (Christian) Right, has given it new and controversial connotations in many parts of the world.

Thus the ideas and terminology associated with the concept of 'religion' give rise to conceptual and practical difficulties when comparing different traditions and different cultural contexts (Fitzgerald, 1997). For example, in the major

Indian languages, 'religion' might be translated as *dharm* and faith as *astha*. However, the latter is close to 'belief' in English: conviction of the truth of a proposition. In Arabic (and, by derivation, languages such as Urdu or Kiswahili), there is more than one word for 'belief': *aqeedah*, meaning any religious belief system or creed, and *eeman*, referring to personal faith or belief. This is regarded as distinct from 'religion', which refers to 'a way of life' – the sum total of a believer's faith and practices (*deen* in Arabic, *dini* in Kiswahili). While much of Christianity was preoccupied with distinguishing between the spheres of religion and the state, "Muslims contend that Islam is not 'merely' a religion but also a seamless whole encompassing politics, economics and morality" (Benthall, 2008, pp. 8–9), and this view of religion as a way of life is true of other religious traditions also. However, Saler (2008) comments that the lack of concepts or terms for various beliefs or phenomena in different languages and societies does not mean that they do not exist, rather that, instead of relying on etic perspectives and concepts (those advanced by outsiders, here for analytical purposes) every effort needs to be made to understand the emic conceptualisations held by adherents and revealed by their own discourses and practices (Beckford, 2003; Benthall, 2008; Lincoln, 2007).

There have been many attempts to conceptualise and define religion and a number of recent reviews of these efforts (see, for example, Deneulin and Bano, 2009; Droogers, 2009; Flood, 2006; Furseth and Repstad, 2006; Saler, 2008; Tomalin, 2013; Woodhead, 2011). A comprehensive review is unnecessary here. Commonly, conceptualisations of religion are divided into substantive and functionalist. The former concentrate on what religion *is*, often focusing on the cross-cultural attributes that distinguish it from other social phenomena, particularly belief in a transcendental reality or spiritual being(s) (the sacred), religiosity (signified by adherence to a set of core beliefs and engagement in religious or ritual practices) and affiliation with a religious organisation. Such definitions are widely used, although they have been criticised for being guilty of deciding in advance what a religion is, by making assumptions about the nature of the phenomenon being studied and excluding some forms of religiousness, privileging belief over practice and implying that the boundaries between traditions, sects, schools of thought or denominations are clear, whereas in practice they are blurred.

Functional conceptualisations, in contrast, are concerned with what religion *does* – the roles it plays in the construction of people's worldviews, identity and social relationships, and in wider socio-political organisation. For example, it is said that religion generally offers answers to questions about the nature of being, the universe, humanity and the divine and the meaning of life, suffering and death, as well as ways of sustaining the cosmological, social and moral order, both by specifying principles for human interaction and ethical behaviour and by religious organisations playing practical roles in society and politics. Critics of purely functional definitions think that they may lose sight of what distinguishes the sacred from the secular or mundane – not only may they "share an incredulity to religious truth claims" (Flood, 2006, p. 49), they are alleged to pay insufficient attention to religious claims about the nature of reality and humans' relationship to the divine and the ways in which these inform the everyday lives of adherents and provide legitimacy for the actions of many social groups (Deneulin and Bano, 2009).

Here, therefore, I will adopt a conceptualisation of religion that contains both substantive and functional elements, while seeking to avoid the trap of essentialist or family resemblance approaches that simply list features considered to be necessary or typical (Beyer, 2001). Of the various attempts to develop such a conceptualisation, Lincoln's provides a useful starting point. He suggests that religion has four inter-related domains, all of which should be considered to develop a full understanding of its nature and roles for individuals and in society:

- A *discourse* whose concerns "transcend the human, temporal and contingent, and that claims for itself a similarly transcendental status" (Lincoln, 2006, p. 5). Thus a discourse is religious not just because of its content but also because of the nature of its claim to truth and authority.
- A *set of practices* which embody the religious discourse and render it operational. These include ritual and ethical practices. They acquire a religious character by being connected to a discourse that constitutes them as such and which they move from the realm of speech and consciousness to action.
- A *community* whose members associate themselves with a religious discourse and its attendant practices. What they have in common is reference to the same texts (whether written or oral), adherence to the same precepts (taken from the texts and associated commentaries) and engagement in the same sorts of practices (which are grounded in the texts and precepts). This does not preclude disagreement over interpretations, values and practices, but a community is held together by a sense of affinity fostered by specific aspects of discourse and practices, which serve to delineate the religious and social boundaries between one group and another, and also by disciplinary mechanisms intended to enforce compliance.
- "An *institution* that regulates religious discourse, practices, and community, reproducing them over time and modifying them as necessary, while asserting their eternal validity and transcendent value" (Lincoln, 2006, p. 7). This involves widely varying formal or semiformal structures, housing the leaders, religious specialists and authority figures who assume responsibility for interpreting and disseminating a group's defining discourse, supervising its rituals, enforcing its ethics and defending and advancing the interests of the community.

To understand any religious group, Lincoln argues that all four domains must be analysed, although he recognises that the extent to which they are developed and their relative importance vary between and within groups and over time. In addition, they may be coherent and well-coordinated or in competition (Lincoln, 2006). The analysis, he suggests, must identify whether each of the four has features and a distinctive character and logic that justify it being conceived of as religious (Lincoln, 2007).

Elaborating on the role of discourse in replicating social groups and identities and constructing, deconstructing and reconstructing society, he notes that it can be used for ideological persuasion, to mystify social inequities and win the consent of those over whom power is exercised and to transform simple power into legitimate authority by obviating the need for direct coercive use of force. However, it can also enable subordinated groups to demystify, delegitimate and deconstruct the established norms and institutions that play a role in their subordination. Ritual, he suggests, can be understood as an authoritative mode of symbolic discourse and a powerful instrument for the evocation of those sentiments (affinity and estrangement) out of which society is constructed. Although he does not argue that all rituals are religious, this conception of ritual seems to fit the practices that are associated with religious participation. Ethical practices, in contrast, are more to do with the derivation of values from religious teachings and the ways in which they are translated into attitudes, morals and behaviour (Lincoln, 2014). Finally, a religious community is held together by boundary-defining discourses and practices, which are refined, reproduced and regulated by an organisational structure. Thus organised religion can be defined as "a way of tying together multiple experiences and memories of the sacred into a single system of belief and practice", with the revelations and pronouncements of intermediaries integrated into "a more or less comprehensive and authoritative view of the world" (Fenn, 2001, p. 6).

Culture

Complicating understanding of the nexus between religion and society is the struggle to distinguish between religion and culture, itself a contested concept. Beckford (2003), for example, acknowledges that culture may be regarded as a way of life, a social institution, a framework for action or a set of meanings "that are generated, transformed, challenged and transmitted in the course of social interaction over time and codified in symbolic forms" (p. 186), while Riesebrodt (2006, p. 604) defines it as "the cognitive, moral and emotional repertoire of a given society through which practices are regulated". The concept may be applied to social groups at different scales, including kinship networks, residents in a locality or country and groups linked by shared interests or a common identity rather than locality. However, care must be taken not to assume that social groups with a common cultural identity (such as shared 'ethnicity') are homogeneous in other respects, for example socio-economic status.

Verhelst and Tyndale (2002) suggest that culture has three dimensions that overlap: the symbolic (e.g. values, symbols, archetypes, spirituality, religion), the social (organisational patterns for family and community linkages and support, systems for decision making and conflict resolution etc.) and the technological (skills, expertise, technology, food preparation, architecture etc.). Myths and symbols are key tools through which culture is revealed and expressed, suggesting how people understand their place in the world and explain why they behave as they do, not least in their religious practices. For example, ideas about spirit possession are said to reveal how individuals and groups experience their personhood, identity and social world, offering insights into their perceptions of power and marginalisation, internal personal struggles and community conflicts (Bradley, 2007; see also Lincoln, 2014).

Cultural affinity may be expressed through participation in distinctive group activities (e.g. speaking a common language, with its associated repertoire of myths, stories and rituals) and shared nonverbal systems of signification (e.g. dance, music, architecture, etiquette). These serve to identify insiders and outsiders, although there is often an internal hierarchy, based on varying degrees of compliance with the values and practices that define the group and valorisation of certain works and genres by those who seek to portray themselves as custodians and arbiters of values for the group as a whole. The latter potentially has adverse implications for understanding if the key interlocutors in research are part of the dominant element. Moreover, the distinction between insiders and outsiders is unlikely to be clear. While legitimating discourses remain persuasive, Lincoln suggests, social integration persists, but if they are challenged and undermined, social competition and fluidity increase. Thus no cultures are stable and unchanging, even supposedly 'traditional' ones (Lincoln, 2006, 2014).

Lincoln emphasises the role of religion by suggesting that culture has three dimensions: aesthetics, ethics and religion, with the latter investing human preferences relating to the former two with transcendent status, insulating them against criticism and assisting their transmission from one generation to another (Lincoln, 2006). A context in which religion seeks to colonise culture and a cultural group seeks to buttress itself by appealing to religion he terms 'maximalist'. Although in Europe, the power of religion to maintain hegemony over culture has declined, the process of secularisation is, he emphasises, neither unidirectional nor universal.

In practice, social interaction is regulated by social and cultural institutions, and these in turn are influenced by religious beliefs and ideas, while culture influences the shape of religious spaces and the identities and practices associated with them. Overall, therefore, it is difficult to conceive of culture and religion as distinct spheres and probably impossible to analyse them separately. In institutionally undifferentiated societies in particular, they are often more or less coterminous – as has already been mentioned, in contemporary societies where folk religious traditions persist, they may be seen as 'culture' rather than 'religion'. In contrast, in heterogeneous, institutionally differentiated societies, religion may be regarded as only one element of culture (Beckford, 2003).

Society

Social theorists can be broadly divided into those who argue that the basic elements of society are individuals and that these should therefore be the unit of analysis and those who assert that individuals are the products of social and cultural forces and so society must be the unit of analysis. Psychology and certain theoretical and methodological orientations in economics and sociology are primarily associated with the former position (e.g. Boyer, 2001; Wulff, 1997, 2001). Psychology is concerned with "behaviour, acts or mental processes, as well as the mind, self or person who behaves or acts or has mental processes" (Corsini, 1999, p. 784). Key concepts used in attempts to understand people from this perspective include motivation, personality, cognition, need, identity and well-being. Economics has also been dominated by approaches based on the assumption that individual producers and consumers seek to further their own interests by rational choices so that economic behaviour at the group or societal level can, it is asserted, be understood by aggregating individual decisions. Neoliberal economic theory and its underlying assumption of individualism has influenced sociological approaches that start with individual characteristics and actions and seek to analyse patterns of individual behaviour (including religious behaviour) according to the same logic.

In contrast, anthropology and much of sociology view society as made up of the social units that make political, social and economic life possible. These groups are governed by social institutions, the norms and rules that regulate human interaction. Thus they are shaped by and shape the behaviour of their members in accordance with roles and rules which may be regulative, indicating which behaviour and relations are permitted or prohibited, or constitutive, making certain forms of activity possible and thus enabling behaviour (Fay, 1996). In a functionalist view, society may be seen as made up of interdependent parts (cultural beliefs, legal institutions, patterns of family organisation, political institutions and economic organisation) that are inter-related, each performing functions that support each other and maintain the whole. In the mid-twentieth century, this was set out in Talcott Parsons's theory of structural functionalism, which in its most extreme version holds that social structures *determine* how their members behave and relate to each other (Fay, 1996). In this perspective, religion plays functional roles in the social system, providing norms that influence actors' behaviour and the functioning of society; promoting moral values and rituals that give meaning to life and enable individuals to live with uncertainty; and either sustaining social institutions or facilitating socio-economic change (Lidz, 2010).

However, it is argued that individualism ignores the fact that people need others to be what they are, while structural functionalism downplays agency, in contrast to actor-oriented theories that explain social reality by reference to purposive social action (Furseth and Repstad, 2006). This critique was taken up by, amongst others, Giddens, who in the 1980s developed a concept of social structure as the enabling and constraining rules and roles that are produced and reproduced by agents in a process of structuration, which refers to the interplay between structure and action, producing, reproducing and changing the social system. In this volume, society is understood as the structure, organisation and functioning of the collectivity/ies in and through which people live. It is based on a view that individuals and society are "mutually constitutive in the sense that individuals construct social collectivities and cultural meanings, and, conversely, social and cultural configurations shape the properties of individuals" (Beckford, 2003, p. 199). Religion may play important roles in these mutually constitutive processes.

A society or social group is not necessarily homogeneous or cohesive – most are differentiated in a variety of respects, related to socio-cultural characteristics such as age, gender and ethnicity. Ethnicity refers to "a sense of group belonging, based on ideas of common origins, culture, experience and values" (Castles, 2004). Socially constructed and changeable, it joins the subjective dimension of 'peoplehood' with the more objective dimension of common economic and social interests. Initially associated with co-residence and similarities of culture, tradition and legal and economic practices, the salience and rigidity of ethnic grouping was increased by colonial attempts to make sense of and codify social divisions. Historically, the boundaries between ethnic groups were malleable, but the use by colonial and postcolonial states of ethnic identity to differentiate between groups with respect to their access to state resources, by particular ethnic groups to advance their own interests vis-à-vis their rivals or by politicians to manipulate ethnic identity to build their own political base has resulted in the instrumentalisation of these constructed identities and also in their reconstruction over time. Often people have quite vague and changeable ideas about their ethnic identity, but experience of discrimination, struggles for recognition and equality between ethnic groups and the mobilisation of ethnic identity for other purposes may harden members' sense of identity, even though dominant fractions may seek to privilege their own interests. In addition to being associated with the perceived shared characteristics of a cultural group, including language, culture and/or territory, ethnicity may be associated with religion. Identities based on both ethnicity and religion have a strong emotional load and potential for instrumentalisation (Castles, 2004; Handelman, 2005).

A similar marker of social identity and stratification is caste, which is largely confined to South Asia. It is based on *jatis* (the basic social units that govern marriage, social networks, rituals and food taboos) and may be seen as a social system based on principles of inclusion/exclusion, hierarchy and interdependence, in which castes follow a rule of endogamy, have a place in the hierarchy of castes and follow traditional occupations. The association between caste, class and Hinduism is much contested. The system is changing in some respects, associated with wider socio-economic change. However, not only is caste in India the basis for wider socio-cultural stratification, with Muslims, Buddhists, Christians and Sikhs also observing some aspects of caste-related behaviour, but also efforts to address the disadvantage associated with membership of certain caste groups has reinforced and politicised caste identity and hardened boundaries between castes (Mitra et al., 2006; Srinivas, 1996). Differences related to ethnicity and caste (and other axes of social difference such as gender or age) thus interact both with each other and with religious differences and also are associated with differential access to economic, social and political resources, resulting in social stratification and inequality.

Sociologists and anthropologists have tended to focus on individual societies. The former seek to explain how societies evolve and change, in particular during periods of widespread economic, political and social change, such as industrialisation or colonialism and their aftermath. Many are interested in contributing to the realisation of development objectives. In the past, anthropologists tended to be more concerned with understanding places and groups perceived as having characteristics typical of 'traditional' societies. Their studies tended to focus on a small group or area (often a village), a social group living in relative isolation and a location geographically and culturally distant from that of the researcher. For the most part, they were concerned with the disruptive and destructive effects of external actors and influences on the societies and social groups under study. Although there continue to be differences between the disciplines in terms of their perspective, level of analysis, methodological approach and interaction with policy processes, there is considerable overlap between their concerns and approaches.

While in the past, some social scientists treated societies as isolated or autonomous social groups, today the influence of external processes on societies and social groups is universally acknowledged. Long-term trends associated with trade and colonialism have involved movements of ideas, people, information, technology, goods and cultural influences between societies. They have led to the development of social relationships across national borders and been associated with the emergence of international forms of governance. These have been reinforced and accelerated by contemporary globalisation processes, so that social change cannot be fully understood without examining the interactions between internal and global trends and influences.

Religion and society: disentangling the links

The framework sketched here is intended to provide the conceptual basis for the analysis that follows, identify the domains that need to be considered in order to develop an understanding of contemporary religion and its interactions with other social phenomena, delineate the dimensions of the social field in which religion and society are intertwined, specify the levels at which analysis must be carried out, and make an initial identification of some of the links between them. These may be seen as crosscutting axes along each of which differentiation occurs, producing a matrix within which the analyses that follow can be situated and drawing attention to the key elements that need to be considered in the discussion.

Dimensions of the social field

Social science seeks to understand and explain human action in terms of personal qualities (the characteristics and practices of individuals), social relations (the interplay between individuals and social groups) and social systems. Explanations of individual and social behaviour may rest on the characteristics of individuals and social units (such as households, enterprises or communities), the patterns or regularities identified and the relationships detected between specific variables or characteristics. However, many believe that individual views and actions and patterns of social allegiance and behaviour cannot be understood without reference to deep-rooted social structures and forces, although views differ on the extent to which these *determine* human actions (Beckford, 2003; Furseth and Repstad, 2006; Hamilton, 1995). Often, analyses focus primarily on one of these dimensions, but in my view, each has shortcomings. Here, therefore, the framework I will employ incorporates all three:

- A personal dimension that conceives of religion as involving cognitive, experiential and bodily engagement on the part of individuals, for whom it has meaning, informing their views, motivations and actions
- A social dimension that sees religion and social groups and spheres as mutually constitutive, as a result of religion's interaction with social and political processes, through the institutions (norms and rules) with which it is associated and that underpin social relationships
- An organisational dimension that recognises that religions are organised in different ways and at different levels, with implications for the ways in which religious groups seek to sustain, reproduce and sometimes expand themselves, as well as the ways in which religious and other socio-political organisations interact with each other.

Level of analysis

Linked to the above, analyses often seek to identify and explain the characteristics and dynamics of social phenomena at one level, whether individual religiosity; small-scale or local organisational structures, such as the family, 'community' or congregation; national social, cultural and political systems; or transnational flows and trends. I will seek to understand not only the interactions between the personal, societal and organisational dimensions at each level but also the links between levels (Smith, 2008). For example, historical and contemporary interactions between religious traditions and groups shape the traditions themselves and give rise to hybrid assemblages of beliefs and practices. In addition, individual views and actions are shaped not only by small-scale social organisations, especially families, but also by religious organisations and institutions at the local, denominational, national and international levels. To understand these interactions and their outcomes, Wong and Levitt (2014) suggest, it is necessary to analyse the characteristics of the carriers of religion, the geographies of circulation of ideas, people and practices and the sites of encounter in which carriers and receivers interact.

Religion in society: domains and links

The starting point for the analytical framework adopted here is, as suggested above, Lincoln's conceptualisation of the role of religion in society as being understandable in terms of the four domains he identifies: discourse, practices, community and institutions, recognising that each of these is contended rather than coherent or homogeneous and dynamic rather than static. With respect to the first, he emphasises the importance of asking who is speaking (what person, group or institution), to which audience, in what context, with what system of mediation and with whose interests in view (Lincoln, 2006). In Chapter 3, the analysis of beliefs will seek to distinguish, in particular, between the official/authoritative interpretations of religious teachings and those of local religious leaders and ordinary adherents. Lincoln groups all the practices associated with religion in a single domain. Instead, I suggest that it is useful to distinguish between ritual and ethical practices. Thus in the analysis that follows, despite the inevitable questions such a distinction raises about overlaps, links and contradictions, religious practices will be considered in Part III and values, attitudes and behaviour in Part IV.

Lincoln uses the term 'institution' to refer to formal and informal authority structures and the individuals on whom they confer authority. In my view, this collapses two concepts into one: institutions (formal and informal rules, which shape but do not determine human behaviour and social relations), and organisations (formal or informal vehicles for the aggregation of interests and ideas, which have their own internal rules and systems of hierarchy and authority). Recent social scientific thinking finds it useful to distinguish the two, and I will do so here. As with other concepts, there are problems when the idea of 'an organised religion' travels. For example, in India 'faith communities' might be translated as *sampradaya*, but this means 'sects' in English – a narrower concept. Here, I will use the term 'religious tradition' to refer to a movement that to a greater or lesser extent shares a history, culture and common body of teachings but is comprised of more than one branch, sect or denomination, each with more or less distinctive teachings, practices and organisational forms.

Some of the possible links between the domains, dimensions and levels of analysis outlined above are discussed briefly below, drawing attention to issues that may arise during the course of analysing them. Social practices involve beliefs, values, attitudes and behaviour, but tracing the links between them is difficult. First, it cannot be assumed that beliefs can be either isolated or defined as a stable set of collective beliefs which are transferred to individuals who act on them to produce certain kinds of behaviour, because it is unlikely that religion will provide an entire set of norms for an individual or society (McGuire, 2008). Second, while religion may be regarded as something that can be understood at the individual level, in terms of personality traits or cognitive systems, most recognise that there is more to it than that – individuals' religious experience is more than simply cognitive, most do not try to achieve logical coherence between their beliefs, experiences, values, practices and actions, and they shape their way of being in the world not as self-contained individuals but together. While one pathway between religion and everyday life depends on individual religious consciousness, often developed and cultivated in explicitly religious places and carried from there to other social fields, religion also travels through social relationships in which the idiom and practice of religion is part of the discourse and interaction. These social fields and the relationships, ideas and values they reflect infuse religious spaces and vice versa (Ammerman, 2014), although Osella and

Osella (2014) warn that religion should not be essentialised and privileged over other aspects of the everyday, because this may reduce complex subjectivities and social relations to religious orientations and practices.

Both religion and the family are crucial to the organisation of primary-group relations. Religious bodies and families are seen to be the key social institutions that deal with persons as a whole, overcome individualism and connect individuals with the wider community by constructing buffer zones that help people cope with stress and change, promising stability and order as societies adapt to changing circumstances and possibly providing public goods. They are closely intertwined: "Religion provides the symbolic legitimation for family patterns . . . and the family is a prerequisite for a vigorous religious system because it produces members and instills them with religious values. In fact, numerous familial events (such as weddings and funerals) are marked in religious contexts, and many religious observances . . . take place within the familial setting. . . [while] religious factors have shaped family dynamics. . . [and] provide ritual support for family and kinship structures" (Pankhurst and Houseknecht, 2000, p. 7). In addition to schools and places of worship, families are potential institutional contexts for socialisation into a set of moral and religious values (Guest, 2009). The characteristics and relative importance of these social institutions for the inculcation and expression of beliefs and norms varies between contexts and over time. Religious and family values are often regarded as being inherently socially conservative, especially with respect to gender and sexuality, supportive of oppressive social relationships and resistant to changes in attitudes and practices. Whether this is the case and how it plays out in different circumstances will form an important focus of the analysis that follows.

Methodological implications

The task of developing an appreciation of everyday lived religion and its roles in society raises a number of methodological issues. An assumption that religion is "a universal supra-historical phenomenon" (Droogers, 2009, p. 8) leads analysts to identify and potentially quantify the characteristics of religious groups or adherents. Such indicators often relate to beliefs and organisational belonging, an approach that has dominated much sociology of religion in Europe and beyond.

Religion is said to provide an idiom that makes it possible for people to describe their reality, formulate beliefs and develop a repertoire of values on which they can draw to deal with experiences and crises. For many, the idiom or discourse is embodied in narratives or texts. For 'religions of the book' (especially Islam and Christianity), religion is indeed a set of core beliefs and teachings that among other things specify (or suggest) how to live in accordance with the principles of the faith tradition and how society should be ordered (Bender et al., 2012). In this case, the task is to identify the official formulations of those beliefs and examine their contemporary interpretations, even though this risks reifying a particular conception of religion and measuring only one type of religious commitment

(Beckford, 2003). It may be less appropriate for other religions (e.g. Hinduism, Buddhism), where the texts provide a worldview, a philosophy and guidelines for living. It may also be less appropriate in the contemporary world, in which people conceptualise, experience and engage with religion in different ways. These range from the individualist choices emphasised by rational choice theory, in which the central metaphor for religion is 'the market' and religion a matter of choice (see, for example, Furseth and Repstad, 2006; McKinnon, 2011), to the hybrid or syncretic melding of beliefs from more than one tradition. The written and oral discourses and texts contribute to religious understanding and practice through the stories and teachings they contain and their use of myths, symbols and metaphors (Droogers, 2009). For the most part, they reflect more than one voice from the past, are filtered through interpreters and so have been inherited by contemporary religious authorities and adherents through a chain of oral and/or written memory (Flood, 2006).

Each religious tradition has evolved institutions, authority structures and forms of organisation to ensure that the relevant texts are reproduced and their interpretation safeguarded and that they meet the needs of adherents and regulate their thinking and behaviour. Thus, while acknowledging that religion might be a universal human impulse, need or experience, constructivists believe that it is always socially and culturally constructed, so that religious teachings are constantly interpreted, contested and acted upon in the contexts in which people live. For constructivists, therefore, religion has variable meanings which are subject to negotiation and contextualised. It is, therefore, valuable to conceive of it as involving processes and actions rather than a set of given beliefs and established institutions and, in addition, to recognise that boundaries between religious traditions, sects and denominations are human products resulting from contested meanings and interests (Beyer, 2003; Maduro, 2004).

It is generally assumed that those who associate themselves with a particular religious tradition or organisation share a set of beliefs, values and norms, especially if that religious group regards its beliefs as 'true' and 'exclusive'. Association with a religious tradition or organisation may, however, take a variety of forms, ranging from loose affiliation to formal membership. The assumption that beliefs, norms and behaviour can be read off from such an association underlies a methodological approach based on, first, ascertaining religious affiliation and, second, gauging the strength of that affiliation by assessing religiosity (or religiousness) (a basic form of which has been used in Chapter 1). This may be taken further by adherents self-reporting the beliefs they hold, the frequency with which they observe various religious practices, their values and norms and their behaviour with respect to personal conduct and relations with others. One of the ways in which religious traditions and organisations seek to ensure that the relevant texts are known, interpreted 'correctly' and guide the lives of adherents is by providing guidance on correct practices – Lincoln's second domain. They do this by specifying religious practices, especially rituals, and social practices – those related to people's actions and interactions with others. The extent to which the texts and their official interpretations determine the belief and value systems and actions of individual adherents and religious bodies (and the perceived desirability of them doing so) varies between religious traditions, and also within traditions between different strands and locations over time, in ways that are matters for empirical analysis (Lidz, 2010).

The methodological approach adopted in this analysis is based on the view that the nature and content of the texts and how they are interpreted do influence adherents' beliefs and values and that self-reported affiliation or membership is often linked to shared beliefs, religious practices, values and social practices. However, to understand the interactions between religion, people's lives and their social relationships, the everyday practices and experiences of lived religion also must be examined (Ammerman, 2014; Orsi, 2003). In addition, it is assumed that, if we are to develop a cross-cultural understanding of the nexus between religion and society, attention must be paid to both the substantive and functional aspects of religion. These can be ascertained from the descriptions and interpretations offered by adherents and observers without assuming "that the units of study are anything more than matters of convenience or accommodation to the conscious self-identification of the subjects" (Beyer, 2001, p. 427). Thus it is acknowledged throughout that the forms and functions of both religion and society are constructed, influenced by history and the cultural context and vary over time and between contexts.

The analysis in this volume will use the results of surveys that have generated quantitative data where these are available and relevant and draw on ethnographic research within the interpretivist tradition while recognising that every methodological approach faces challenges in generating valid and reliable data and interpretations. First, the analysis would ideally be based on case studies and evidence from each major religious tradition under study (and its major subdivisions) and from a range of countries to represent different geographical and cultural contexts and enable systematic comparative analysis. In practice, there are relatively few sets of case studies driven by a defined theoretical starting point, the same research questions and a standard design.[2] Instead, while the individual case studies available often enable an appreciation of complexity, provide in-depth analysis and suggest propositions that can be refined and tested by additional cases, general conclusions must be drawn with great caution. In addition, not only are there relatively few recent ethnographies that seek to capture the understandings of ordinary urban or rural people and interpret religious beliefs and practices from their perspective, those that are available tend to focus on new, surprising or alarming religious groups. For example, there are

2 In the Religions and Development Research Programme, for various practical reasons, research on the topics relevant to this volume did not use a comparative case study design for all four focus countries – thus much of the good-quality material on what religion means for adherents emerged out of the research in India, while research on attitudes to corruption was only undertaken in Nigeria and India.

increasing numbers of studies of religious revivalist movements, especially Pentecostal churches and Muslim pietist movements, but few of the mainstream Christian denominations or mosque congregations.[3] Often, the research subjects in available studies are highly educated, primarily because it is easier for both local and foreign researchers to access these individuals and groups. However, there are dangers in assuming that groups such as university students are typical or representative. Also, studies may focus on religious specialists and leaders, who often present the official/orthodox perspective, which is not necessarily identical to that of their followers. Further, the quantitative data that are available tend to be based on national sample surveys in only some of the countries relevant to this volume. They provide useful national and international comparative data. However, in the selection and design of questions, they reflect the assumptions of their designers, who are influenced by their own religious and national backgrounds and theoretical perspectives. In addition, they have, for the most part, been used to generate fairly simple descriptive statistics and contain few explanations of the similarities and differences between the different contexts on which they report.[4] Rather than quantitative and qualitative approaches being regarded as different and incompatible, the value of using a combination of methods is increasingly recognised.

Second, because texts and narratives play such a central role in religion, people's lifeworlds and societies, the most commonly used analytical method is narrative or discourse analysis. Formations of ideas, images and practices, discourses contribute to an individual's sense of personhood and bind people together, legitimising and reinforcing some ideas and practices and discrediting others (Müller, 2006). They are 'languages in use', reflecting and shaping individuals' social relations and the social order (Jaworski and Coupland, 1999). They aim to recreate the world described by informants as narrators and interpreters of their own histories and experiences. However, they may contain deeper meanings, including some of which informants may not be aware, and the points of view they reveal often reflect what is considered 'normal' or 'appropriate' in particular religious, social and institutional settings. Such viewpoints may appear to those who hold them to be natural, concealing the ideology and power relationships that underlie their collective production (McGuire, 2008). Moreover, studies that rely on the analysis of discourse, observation and interpretation inevitably filter their selection of what is perceived to be relevant and their interpretations through the lenses of the researchers involved, who may be local or foreign, and insiders to

3 There are also some studies of extremist groups, which mostly try to understand the backgrounds and motivations of their members, especially their fundamentalist worldviews and violent tactics. Not only is entry to such groups for research very difficult, they are outside the scope of this volume.

4 The result may be inconsistencies between the results of surveys and ethnographic material or between different types of surveys. For critiques of such surveys, see, for example, Berzano and Riis (2012); Halman (2002); McKinnon (2011).

the relevant religious tradition or outsiders,[5] as well as having their own gender, class, ethnic and caste identities.[6]

Third, people's ability to articulate their beliefs, values and ethics varies enormously, and the vocabulary and concepts they use also vary between individuals, groups and societies. For example, the educated research subjects referred to above may be more able to articulate and explain their beliefs, values and attitudes than people who are not accustomed to doing so in their everyday lives.

Fourth, the more deeply embedded people are in a religious milieu, the more likely they are to acquire and employ a religious idiom to describe who they are and how they live their lives, whereas if informants are not deeply embedded in social sites where religious discourse is pervasive, it is less likely that their self-understanding will be informed by religious ideas. In a religious milieu, people have not necessarily *learned* a set of beliefs, doctrines and prescribed behaviours; rather they are more likely to 'speak religion' as a way of talking about their lives (Ammerman, 2014, p. 200). Thus people's religious experience and values may result from explicit religious teaching and be, at least in part, clearly articulated, but they are also integral to a person's mental and emotional being, so it may seem quite foreign to articulate them.

Fifth, and related to this, as noted above, it is notoriously difficult to trace the links between beliefs, values, attitudes and behaviour. In part this is because the self-reporting of behaviour and its attribution to religious or other ideas, teachings and influences is subjective and may not be very reliable. Analysis of dimensions of religion other than texts, especially performance (ritual etc.), can complement the attention paid to narrative. I have made a case for conceptualising religion as a relatively autonomous social field, i.e. that religious beliefs, practices and organisations can be studied in their own right, even though they can only be fully understood in relation to the social, historical, political and economic contexts in which they operate. However, this does not mean that religion is perceived by people as a separate sphere. In practice, for many, religious life does not operate separately from social life as a whole, so they may not conceive of a belief, attitude or practice as being 'religious' (Riesebrodt, 2006).

The analysis that follows seeks to provide a general overview rather than a comprehensive explication and comparison of the different faith traditions, let

5 In this field, where there is disciplinary competition between the insider disciplines that seek to interpret the teachings, values and histories of a particular religion for the benefit of adherents and religious groups within it and the intentionally objective standpoint of the social sciences, the issue of whether only 'insider' scholars can fully understand a religion is controversial (see, for example, Bell and Taylor, 2014; Droogers, 2008; Fay, 1996; Knott, 2005; Neitz, 2013; Riesebrodt, 2006; Tomalin, 2013). For both 'insider' and 'outsider' approaches, reflexivity is crucial.

6 In addition, researchers may undertake research in order to advocate social changes or particular policies, which influences the research questions, design and analysis. In this book, although it is impossible, even undesirable, to avoid judgements about the social impacts of religion (e.g. on gender equality), my emphasis is on improving understanding rather than arriving at either judgements about whether religion plays positive or negative roles in society or engaging in advocacy.

alone all the differences between and within them or the ways in which they are rooted in widely varying cultural and socio-political contexts. Its primary purpose is to review and synthesise the evidence available in Sub-Saharan African and southern Asian contexts. In some instances, it will be possible to compare religious traditions, countries and localities, but for the most part the studies will be used for illustrative purposes. The roles religion plays in people's lives and identities vary depending on their social position, and so variations by gender and other dimensions, such as class, caste and ethnicity, will be recognised where possible. However, generalisations will only be advanced with care.

In the next part of the book, religious beliefs will be examined, distinguishing as far as possible between the authoritative teachings of each major religious tradition and what their affiliates in Sub-Saharan Africa and southern Asia say they believe. Religious practices are important in daily lives and societies where they are widely practised by adherents and religious organisations. They are examined in Part III. Finally, the available evidence on the ways in which religious beliefs and practices affect a range of values and attitudes (and, where possible, behaviour) is reviewed.

References

Ammerman, N. (2014) Finding religion in everyday life, *Sociology of Religion*, 75, 2, 189–207.

Asad, T. (1993) *Genealogies of Religion*, Baltimore: Johns Hopkins University Press.

Beckford, J.A. (2003) *Social Theory and Religion*, Cambridge: Cambridge University Press.

Bell, E. and Taylor, S. (2014) Uncertainty in the study of belief: The risks and benefits of methodological agnosticism, *International Journal of Social Research Methodology*, 17, 5, 543–57.

Bender, C., Cadge, W., Levitt, P. and Smile, D. (2012) Introduction: Religion on the edge: De-centring and re-centering the sociology of religion, in Bender, C., Cadge, W., Levitt, P. and Smile, D. (eds) *Religion on the Edge: De-centring and Re-centering the Sociology of Religion*, Oxford: Oxford University Press (online version 2013).

Benthall, J. (2008) *Returning to Religion: Why a Secular Age Is Haunted by Faith*, London and New York: IB Tauris.

Berzano, L. and Riis, O. (2012) Preface: New methods in sociology of religion, in Berzano, L. and Riis, O. (eds) *Annual Review of the Sociology of Religion Vol 3 New Methods in the Sociology of Religion*, Leiden: Brill, vii–xvii.

Beyer, P. (2001) Contemporary social theory as it applies to the understanding of religion in cross-cultural perspective, in Fenn, R.K. (ed) *The Blackwell Companion to the Sociology of Religion*, Oxford: Wiley-Blackwell, 418–31.

Beyer, P. (2003) Social forms of religion and religions in contemporary global society, in Dillon, M. (ed) *Handbook of the Sociology of Religion*, Cambridge: Cambridge University Press, 45–60.

Boyer, P. (2001) *Religion Explained: The Human Instincts that Fashion Gods, Spirits and Ancestors*, London: Vintage House.

Bradley, T. (2007) *The Relationships between Religion and Development: Views from Anthropology*, Birmingham: University of Birmingham, Religions and Development WP 5.

Castles, S. (2004) *What Is Ethnic Identity and How Can It Be Studied?* Oxford: Centre for Research on Inequality, Human Security and Ethnicity, newsletter no 1.

Coleman, S. (2010) Recent developments in the anthropology of religion, in Turner, B.S. (ed) *The New Blackwell Companion to the Sociology of Religion*, Oxford: Wiley-Blackwell, 103–21.

Corsini, R.J. (1999) *The Dictionary of Psychology*, London: Taylor Francis.

Davis, C.A. (1999) *Reflexive Ethnography: A Guide to Researching Selves and Others*, London and New York: Routledge.

Deneulin, S. and Bano, M. (2009) *Religion in Development: Rewriting the Secular Script*, London: Zed.

Droogers, A. (2008) As close as a scholar can get: Exploring a one-field approach to the study of religion, in de Vries, H. (ed) *Religion: Beyond a Concept*, New York: Fordham Press, 448–63.

Droogers, A. (2009) Defining religion: A social science approach, in Clarke, P.B. (ed) *The Oxford Handbook of the Sociology of Religion*, Oxford: Oxford University Press, online ed.

Evans, J.H. and Evans, M.S. (2008) Religion and science: Beyond the epistemological conflict narrative, *Annual Review of Sociology*, 14, 87–105.

Fay, B. (1996) *Contemporary Philosophy of Social Science*, Oxford: Wiley-Blackwell.

Fenn, R.K. (2001) Editorial commentary: Religion and the secular: The sacred and the profane: The scope of the argument, in Fenn, R.K. (ed) *The Blackwell Companion to the Sociology of Religion*, Oxford: Wiley-Blackwell, 3–22.

Fitzgerald, T. (1997) A critique of 'religion' as a cross-cultural category, *Method and Theory in the Study of Religion*, 9, 2, 91–110.

Flood, G. (2006) Reflections on tradition and inquiry in the study of religions, *Journal of the American Academy of Religion*, 74, 1, 47–58.

Furseth, I. and Repstad, P. (2006) *An Introduction to the Sociology of Religion: Classical and Contemporary Perspectives*, Aldershot: Ashgate.

Guest, M. (2009) The reproduction and transmission of religion, in Clarke, P.B. (ed) *The Oxford Handbook of the Sociology of Religion*, Oxford: Wiley-Blackwell, online ed.

Halman, M. (2002) Theory and method in the comparative study of values: Critique and alternative to Inglehart, *European Sociological Review*, 18, 139–58.

Hamilton, M. (1995) *The Sociology of Religion: Theoretical and Comparative Perspectives*, London: Routledge, 2nd ed.

Handelman, H. (2005) Ethnicity and ethnic conflict, in Haynes, J. (ed) *Palgrave Advances in Development Studies*, Basingstoke: Palgrave Macmillan, 160–80.

Jaworski, A. and Coupland, N. (1999) Introduction: Perspectives on discourse analysis, in Jaworski, A. and Coupland, N. (eds) *The Discourse Reader*, London: Routledge, 1–44.

Knott, K. (2005) Insider/outsider perspectives, in Hinnells, J.R. (ed) *The Routledge Companion to the Study of Religion*, London: Routledge, 243–58.

Koistenen, T. (2006) Does the word 'God' have the same meaning in every context? in Boeva, L., Schrijvers, J., Stoker, W. and Vroom, H. (eds) *Faith in the Enlightenment? The Critique of the Enlightenment Revisited*, Amsterdam and New York: Rodopi, 169–81.

Koshul, Basit Bilal (2006) Scriptural reasoning and the philosophy of social science, *Modern Theology*, 22, 3, 483–501.

Levitt, P. (2012) Religion on the move: Mapping global cultural production and consumption, in Bender, C., Cadge, W., Levitt, P. and Smilde, D. (eds) *Religion on the Edge: De-centring and Re-centering the Sociology of Religion*, Oxford: Oxford University Press (Online version published 2013).

Lidz, V. (2010) The functional theory of religion, in Turner, B.S. (ed) *The New Blackwell Companion to the Sociology of Religion*, Oxford: Wiley-Blackwell, 76–102.

Lincoln, B. (2006) *Holy Terrors: Thinking about Religion after September 11*, Chicago: University of Chicago Press, 2nd ed.

Lincoln, B. (2007) Concessions, confessions, clarifications, ripostes: By way of response to Tim Fitzgerald, *Method and Theory in the Study of Religion*, 19, 163–8.

Lincoln, B. (2014) *Discourse and the Construction of Society: Comparative Studies of Myth, Ritual and Classification*, New York: Oxford University Press.

Maduro, O. (2004) 'Religion' under imperial duress? Post-colonial reflections and proposals, *Review of Religious Research*, 45, 31, 221–34.

McGuire, M. (2008) *Lived Religion: Faith and Practice in Everyday Life*, Oxford: Oxford University Press.

McKinnon, A.M. (2011) Ideology and the market metaphor in rational choice theory of religion: A rhetorical critique of 'religious economies', *Critical Sociology*, 39, 4, 529–43.

Mitra, S., Wolf, S.O. and Schöttli, J. (2006) *A Political and Economic Dictionary of South Asia*, London and New York: Routledge.

Müller, M. (2006) Discourses of postmodern epistemology: Radical impetus lost? *Progress in Development Studies*, 6, 1, 306–20.

Narayanan, V. (2013) Hinduism: Practicing tradition today, in Pechilis, K. and Raj, S.J. (eds) *South Asian Religions: Tradition and Today*, London and New York: Routledge, 39–72.

Neitz, M.J. (2013) Insiders, outsiders, advocates and apostates and the religions they study: Location and the study of religion, *Critical Research on Religion*, 1, 2, 129–40.

Orsi, R.A. (2003) Is the study of lived religion irrelevant to the world we live in? Special Presidential Plenary Address, Society for the Scientific Study of Religion, Salt Lake City, November 2, 2002, *Journal for the Scientific Study of Religion*, 42, 2, 169–74.

Osella, F. and Osella, C. (2014) Introduction, in Osella, F. and Osella, C. (eds) *Islamic Reform in South Asia*, Cambridge: Cambridge University Press, xi–xxviii.

O'Toole, R. (2001) Classics in the sociology of religion: An ambiguous legacy, in Fenn, R.K. (ed) *The Blackwell Companion to Sociology of Religion*, Oxford: Wiley-Blackwell, 133–60.

Pankhurst, J.G. and Houseknecht, S.K. (2000) Introduction: The religion-family linkage and social change – a neglected area of study, in Houseknecht, S.K. and Pankhurst, J.G. (eds) *Family, Religion and Social Change in Diverse Societies*, New York: Oxford University Press, 1–40.

Pechilis, K. (2013) Introduction. In a shared world: Selva J. Raj and the study of religion, in Pechilis, K. and Raj, S.J. (eds) *South Asian Religions: Tradition and Today*, London and New York: Routledge, 1–20.

Ragin, C.C. and Amoroso, L.M. (2011) *Constructing Social Research*, Los Angeles: Sage, 2nd ed.

Riesebrodt, M. (2006) Religion in global perspective, in Juergensmeyer, M. (ed) *The Oxford Handbook of Global Religions*, Oxford: Oxford University Press, 597–609.

Saler, B. (2008) Conceptualizing religion: Some recent reflections, *Religion*, 38, 219–25.

Smith, C. (2008) Future directions in the sociology of religion, *Social Forces*, 86, 4, 1561–89.

Srinivas, M.N. (ed) (1996) Introduction, in Srinivas, M.N. (ed) *Caste: Its Twentieth Century Avatar*, New Delhi: Penguin, ix–xxxviii.

Tomalin, E. (2013) *Religions and Development*, Abingdon: Routledge.

Turner, B.S. (2010) Introduction: Mapping the sociology of religion, in Turner, B.S. (ed) *The New Blackwell Companion to the Sociology of Religion*, Oxford: Wiley-Blackwell, 1–27.

Vásquez, M.A. (2012) Grappling with the legacy of modernity: Implications for the sociology of religion, in Bender, C., Cadge, W., Levitt, P. and Smile, D. (eds) *Religion on the Edge: De-centring and Re-centering the Sociology of Religion*, Oxford: Oxford University Press (online version 2013).

Verhelst, T. and Tyndale, W. (2002) Cultures, spirituality and development, in Eade, D. (ed) *Development and Culture*, Development in Practice Reader, Oxford: Oxfam.

Wong, D. and Levitt, P. (2014) Travelling faiths and migrant religions: The case of circulating models of *da'wa* among the Tablighi Jamaat and Foguangshan in Malaysia, *Global Networks*, 14, 3, 348–62.

Woodhead, L. (2011) Five concepts of religion, *International Review of Sociology*, 21, 1, 121–43.

Wulff, D.M. (1997) *Psychology of Religion: Classic and Contemporary*, New York: Wiley-Blackwell.

Wulff, D.M. (2001) Psychology of religion: An overview, in Jonte-Pace, D. and Parsons, W.B. (eds) *Religion and Psychology: Mapping the Terrain*, London: Routledge, 15–29.

Wuthnow, R. (2001) Spirituality and spiritual practice, in Fenn, R.K. (ed) *The Blackwell Companion to Sociology of Religion*, Oxford: Wiley-Blackwell, 306–20.

Part II

Religious beliefs and their implications for people's lives

The aim of this section of the book is to examine what religion means to (ordinary) people in contemporary Sub-Saharan Africa and southern Asia. The analysis focuses on what adherents say they believe and how they interpret the texts and teachings, although it is acknowledged that official interpretations of the key religious texts, sources and myths are important. Both are concerned with the first of Lincoln's four domains of religion. Some of the beliefs of the faith traditions concerned are outlined – those most likely to influence the meanings people attribute to religion, their perspectives on their own lives and their attitudes towards other people, religions and societies.

Classically, a belief is an assertion, claim or expectation about reality that is assumed to be true or false, perhaps on the basis of experience, although often its truth or falsity cannot be practically determined. It is commonly assumed that religious beliefs are personal (and they may indeed imply individual convincement/commitment), but people do not live in isolation from either the social dimension in which they interact with others in society or the organisational dimension that brings them into contact with the mechanisms and processes used by religious traditions and organisations to socialise, teach and ensure compliance by adherents. In the two chapters that follow, an introduction to the main beliefs associated with each of the traditions of concern is provided, their prevalence among adherents assessed where possible and the latter's interpretations explored. In this introduction, the notion of 'belief' will be explored, to help frame the analysis that follows.

Official religious belief systems typically involve "identification of what is considered sacred and meaningful, a set of theories about how and why the world was created, and an explanation for suffering and death" (Kurtz, 1995, p. 51). Cosmogenies (stories about how and why the world was created) usually include a theory about the creation of humans and how they should think of themselves, which is often linked to the cultural context within which the religion was born – they describe the nature of the world and explain people's place in it, providing information about the meaning of the world, the self and life. The creation stories of the faith traditions will not be analysed here, although some of their implications for people's perceptions of the world and their place in it will be considered. Theodocies seek to explain the presence and nature of evil, suffering and death, including ways in which it is believed that they can be alleviated and transcended (Kurtz, 1995, pp. 59–64). These will be analysed in Chapter 4. Collections of orthodoxies imply that belief systems are stable and coherent and also that people absorb and adopt them in total. However, in practice, they are contested within the religious traditions, overlap between them and also evolve over time as religious specialists and ordinary adherents respond to changing intellectual influences, interpret events and try to make sense of their experiences.

Both the nature of religious beliefs and the roles they play in people's worldviews and social relationships vary. Such variability is likely to be related to the prevalence of affiliation with particular religious traditions and groups, how important religion is to individuals and in societies (see Chapter 1) and how distinct the traditions and groups are in particular cultural contexts. As discussed in Chapter 2, a focus on 'beliefs' can be dangerous, because it is less relevant to some religious traditions than others, the interpretation of beliefs by both religious specialists and adherents is far from straightforward, and there is considerable variation within the traditions.

First, a focus on beliefs tends to direct attention to those who hold relevant beliefs and are able to articulate, systematise and justify them, generally with respect to the (written or oral) narratives and texts. It thus leads to a focus on 'religious specialists', for example theologians, clerics or *ulama* (Muslim scholars, the interpreters and transmitters of Islamic doctrine and law). These accounts and interpretations are important, but they need to be balanced with an understanding of how adherents articulate their belief systems, which depends on their exposure to religious teaching, their ability to read and interpret religious texts for themselves, and the wider social and cultural context in which they live. Beliefs are not propositions to which people give their intellectual consent – what they refer to and mean is shaped by the language and practices of particular religious and cultural traditions, including the prevailing social and cultural conceptions of what counts as religion (Beckford, 2003).

Second, as noted above, people are not individual and autonomous agents but part of narratives, social groups and histories, through which beliefs are passed on between the generations, not only by religious organisations but also through processes of socialisation within the family and society. As noted above, a belief implies a conviction that a proposition is true, although it may be taken for granted rather than being consciously adopted/accepted at a particular point in time. The extent to which beliefs are articulated and reinforced by religious specialists and adherents varies between and within religious traditions, depending on whether or not the latter are creedal, use religious teaching to explicate the meanings of beliefs and their implications for values and practices and/or have explicit rituals that require assent to and regular reiteration of beliefs.

Third, both metaphorical and literal language may be used in the texts and by informants. However, the dividing line between imagination and 'reality' is always blurred – metaphorical language may be used to describe actual events and feelings, while literal language may be used to describe events which have not occurred 'in reality' but are imagined (Coleman, 2010). Often, religions originate in myths (from the Greek *mythos*, narration or storytelling) – "tales which narrate human and supernatural events, and which . . . are ostensibly explanatory or historical in nature, generally with reference to events in the distant past" (Gindro, 2003). Such myths often refer to a group's origins and are used to develop a feeling of belonging, a common identity and a sense of continuity. The events may or may not be objectively or probably true, but they provide stories and symbols that are important for a group or, if used pejoratively, to denigrate another religious story as untrue or unreal. They offer an understanding of the world in which a sacred or mystical reality pervades the natural one, providing an alternative to scientific explanations or standing alongside them (Segal, 2005; Lincoln, 2014).

Fourth, while disciplines such as theology, philosophy, Islamic studies and religious studies bring a range of long-established and specialised methodological approaches to the study of texts and commentaries, social scientists generally rely on responses to simple survey questions or a limited number of ethnographies to analyse the interactions between religions and societies. Here, it is argued that social scientists need to develop a sound understanding of the nature and content of religious beliefs in order to improve social scientific understanding of the complex roles played by religion in cognition, identity formation, action motivation and so on and to enable more and better comparative research to be carried out (Smith, 2008).

No attempt is made here to provide a comprehensive account of the beliefs that are important to each of the main religious traditions in Sub-Saharan Africa and southern Asia. Instead, in Chapters 3 and 4, eight dimensions of belief that research and experience show to be particularly salient to adherents' understanding of their religion and how it is expressed in religious practices and values, attitudes and behaviour are considered. In each case, mainstream teachings of the relevant religious traditions are outlined, the available evidence on local understandings and interpretations in various contexts is reviewed and some illustrations are provided. Inevitably, the outlines do not capture all the variations within faith traditions, whether these are denominational/sectarian or relate to the spectrum of specialists'/adherents' views, nor is there space to discuss the beliefs in depth. Rather, the main aim of these chapters is to contribute to basic literacy in each of the religious traditions under consideration, in order to seek out and review the available evidence on the prevalence and interpretation of these beliefs in Sub-Saharan Africa and southern Asia.

References

Beckford, J.A. (2003) *Social Theory and Religion*, Cambridge: Cambridge University Press.

Coleman, S. (2010) Recent developments in the anthropology of religion, in Turner, B.S. (ed) *The New Blackwell Companion to the Sociology of Religion*, Oxford: Wiley-Blackwell, 103–21.

Gindro, S. (2003) Myth, in Bolaffi, G., Bracalenti, R., Brahman, P. and Gindro, S. (eds) *Dictionary of Race, Ethnicity and Culture*, London: Sage, 190–2.

Kurtz, L. (1995) *Gods in the Global Village: The World's Religions in Sociological Perspective*, Thousand Oaks, CA: Pine Forge Press.

Lincoln, B. (2014) *Discourse and the Construction of Society: Comparative Studies of Myth, Ritual and Classification*, New York: Oxford University Press.

Segal, R.A. (2005) Theories of religion, in Hinnell, J.R. (ed) *The Routledge Companion to the Study of Religion*, London: Routledge, 49–60.

Smith, C. (2008) Future directions in the sociology of religion, *Social Forces*, 86, 4, 1561–89.

3 Religious beliefs

Sources, interpretations

In this chapter, some of the key beliefs of the main faith traditions in Sub-Saharan Africa and southern Asia are introduced. Many of these are drawn from texts – written (and sometimes oral) sources that are identified and used by religious specialists to define the central tenets of a religion or denomination and provide authoritative interpretations and drawn on first- or second-hand by adherents to specify or guide their beliefs and practices. The first part of the chapter will, therefore, examine the roles and status attributed to the religious texts, which influence their perceived authenticity and use, as well as influencing adherents' attitudes to their own and others' religious beliefs. The texts are a primary source of beliefs in and conceptions of transcendental reality. These are outlined in the second half of the chapter, alongside research that reveals adherents' beliefs with respect to God or the divine, their views on the existence and roles of supernatural forces or beings other than God, and their reports of religious experiences. In the following chapter, their conceptions of heaven, hell and the purpose of religion, the extent to which they believe in fate as against the ability of individuals to determine their own destinies and finally, their views on the rewards of good religious practice, will be examined.

The status and use of texts

Important to understanding religious discourse and the sources of religious teachings are the status attributed to key texts, how they are used and how they are interpreted. I will make no attempt here to comprehensively list or analyse the key texts in each faith tradition, instead concentrating on identifying their perceived status and its implications, with an emphasis on those which are regarded as 'sacred', i.e. they have special status arising from the belief that they come from or are inspired by a transcendental source. This status may or may not imply that they have a more direct influence on the beliefs and practices of adherents than those considered to be mythical, although they are still subject to interpretation and internalisation (Flood, 2006).

The Qur'an and ahadith

The religious basis of Islam is a series of what are believed to be the final revelations from God, received and memorised by Muhammad over a period of 25 years and later dictated by him and written down. These were collected into the Qur'an, which is considered by Muslims to be the literal word of God. Because of this belief, special significance is attached to the fact that the Qur'an (recitation) is written in Arabic, with many Muslims believing that it would be wrong to translate it into any other language and that it is incumbent on them to recite, memorise and read it in Arabic. The obligation to do this has shaped Islamic scholarship, education and teaching in important ways and is a significant driver of contemporary Islamic piety movements in some parts of Sub-Saharan Africa and southern Asia. The initiators and leaders of these movements have often been those who have been able to study in and travel to madrasas and universities in Arabic-speaking countries, for example leaders of the Islamic revivalist movement in Nigeria (Obadare, 2016), returning male and female Malaysian students who studied overseas on government scholarships in the 1960s and 1970s and who constituted the core membership of the Angkatan Belia Islam Malaysia (ABIM, Islamic Youth Movement of Malaysia) (Stivens, 2006) and female religious leaders in Burkina Faso (Gomez-Perez, 2016) and Mali (Schultz, 2008).

The Qur'an provides instructions on how to worship God, perform the ritual obligations of Islam and treat one another (Ruffle, 2013). It contains the laws of God, the rules of the religion and moral guidelines. Many Muslims believe that it is indeed God's final book of law and revelation (see also Waines, 2009), completing and correcting the books of law that preceded it, especially the Torah of Moses, the psalms of David and the gospel of Jesus. "Because of the non-linear ordering of the Qur'an, its variety of messages, its complexity of language, and the multiple contexts in which it was revealed, the study and interpretation of this script is a central part of the Islamic religious sciences (*'ulum al-din*). *Tafsir* is the act of interpreting and explaining the meaning of particular verses in the Qur'an" (Ruffle, 2013, p. 203). Other scholarship is concerned with allegorical or mystical interpretation, the hidden meanings of a particular revelation.

As well as interpretation of the Qur'an, the formation of a politico-religious community in Medina gave rise to a need for further guidance and a system of law (*shari'a*). This is provided in the *sunnah* (trodden path), the example provided by Muhammad's life and the *ahadith* – stories and reports about what Muhammad and his close companions said or did, later vetted to ensure only valid ones are used. Gradually, six of the many collections of *ahadith* were recognised and accepted. *Shari'a* is based on the *sunnah* and *ahadith*, although it became clear that other modes of legal thought were also needed. *Fiqh* (study, mainly of the Qur'an and *sunnah*, jurisprudence) can be used for the purpose of *ijtihad*, personal intellectual grappling with individual texts, involving *qiyas* or analogical reasoning (meaning that some matters for which there was no precise rule are to be treated in accordance with the rule for an analogous matter), on the basis of *maslaha* (the public interest), *darura* (necessity) or *ada* (prevailing local custom), in order to arrive at *ijma* or a consensus of jurists and eventually the community. Four legal schools emerged, which recognise each other. For example, in South Asia most Muslims follow the less rigid Hanafi school of jurisprudence, although Islamist groups such as the Jamaat-e-Islami are influenced by the Hanbali school (Jamal, 2010).

Shari'a (Islamic law) regulates the external relations between human beings, God and other people (not inner consciousness, which is held to be a matter of faith and based on one's own relationship with God). It

> seeks to protect life (*'uqubat*) through corporal and capital punishments; property (*mu'amalat*) by regulating contracts and commercial transactions; honour (*adab*) through the control of interpersonal relationships, sexuality, morality, and etiquette; belief (*i'tiqadat*) in God; and worship (*'ibadat*) by establishing rules for fasting, prayer, paying alms, and pilgrimage. The degree of permissibility of an action exists within a spectrum that ranges from obligatory (*wajib* or *fard*) to absolutely forbidden (*haram*).
>
> (Ruffle, 2013, p. 205)

The most severe prescribed punishments are for crimes such as fornication, murder, theft and slander, although the sects and schools of law vary in their interpretation of *shari'a*, the permissibility of various human activities and punishments for transgressions. *Shari'a* is used to order society, preserve religious tradition and provide guidance to enable Muslims to stay on the 'straight path' to God and receive the rewards of paradise on the day of judgement (Ruffle, 2013). The rituals and values prescribed in the texts and some of the ways in which contemporary Muslims interpret and implement them will be explored in subsequent chapters.

The Bible

Rather than a single book, the Bible is a collection of histories, prophecies, poems, letters and other writings concerning God and his dealings with people compiled over more than a thousand years. It contains the Old Testament (the Hebrew Bible and the Jewish scriptures) and the New Testament, material produced in the first century CE and gathered together in the second century. In addition, some inter-testamental writings (also known as the Apocrypha) may be included.

Like the Qur'an and *ahadith*, which are rooted in Middle Eastern pastoral societies with cultural and economic contexts that resonate strongly with many Muslims, even today, the Bible is thought to appeal to Sub-Saharan African and southern Asian Christians because the world it describes, which is characterised by famine and plague, poverty and exile, clientelism and corruption, is familiar to them. A largely poor readership of rural dwellers and relatively recent rural–urban migrants can, it is suggested, readily identify with both the Old and New Testament social models. Societies in the Old Testament are characterised by nomadism, polygamy, the importance of genealogy for situating a person, the association between a person's kin group of origin and claims to land, the prioritisation of community over individual identity and responsibilities over rights and possession by spirits. The New Testament portrays a "society of peasants and small craftsmen dominated by powerful landlords and imperial forces, by networks of debt and credit" (Jenkins, 2006, p. 68; see also Vásquez, 2009).

The religion portrayed in the Bible also contains features recognisable to many of its readers in the global South, including elements familiar from folk religious traditions, for example ideas about sacrifice, sacred places, visions, dreams and revelation, prophecy and respect for proverbial wisdom (Jenkins, 2006). In addition, Jenkins argues, it provides a sense of the unpredictability and transience of life in the face of disaster, conflict and experiences of displacement and migration, teaches ways to cope in hostile environments and holds out hope of salvation and prosperity. The latter encourages a turn to the Christian faith for assistance and protection and provides backing for appeals for justice, forgiveness of debt and greater equality. In addition, the Bible supplies many stories and parables which are, he suggests, appealing in oral cultures that rely on storytelling and proverbs. The main missionary movements coincided with and encouraged the spread of literacy. The Bible was widely disseminated by means of translation and printing and today is available in translation, usually from the original Hebrew and Greek, in more than four hundred languages (Kim, 2007). For many Africans and Asians, the Bible was the first book they read (often the only one available), leading to considerable veneration for the written text.

However, the most commonly used translations are replete with language that puts men and maleness at the centre of reality, and men are foregrounded in the narratives. The way Christians interpret the Bible and the relative weight they give to different parts of it and to other literature varies. Some see the stories and teachings as valid for all times and across cultures, while others see them as the products of particular contexts (Jenkins, 2006; McGuire, 2008). The Bible is regarded as authoritative by all Christians, but most recognise that it represents more than one voice (for example, there are four gospels – records of the life of Jesus) and contains both historical accounts and later reflections on events. Reliance on the Bible does not preclude other sources of authority. Nor does it imply a belief that God's revelation ended with the New Testament (Jenkins, 2006). Although the Bible testifies to God and Jesus, most Christians believe that there are other ways God may be revealed or encountered. In particular, knowledge of God, experience of the resurrected and ascended Jesus and proper interpretation of scripture are considered to be mediated through the Holy Spirit. While the prayer most commonly used by Christians is taken from the New Testament accounts of the teachings of Jesus, the creeds formulated in the first four centuries after Jesus's death, which most Christians profess as containing their central beliefs, draw on the Bible but are not part of its contents.

Hindu sources

As noted above, the label 'Hinduism' is essentially a 'family resemblance' category applied to a number of folk traditions collected under the umbrella of Brahmanical Hinduism, especially by the scholars and administrators of the colonial period. Unlike the other 'world religions' Hinduism has no founder, nor can it be said to have originated at a particular time. In the view of orientalist scholars, Christian missionaries and colonial administrators, popular religion in the eighteenth and nineteenth centuries was a corruption of an authentic religious tradition that could be traced back to the religion of the Aryan authors of the Vedas, the oldest Sanskrit texts. What is typically recognised as Hinduism today is an elite version that focuses on the Vedas and other Sanskrit texts (e.g. the Upanishads, the Puranas, the epics) and the beliefs and practices of the priestly Brahmin caste (Tomalin, 2009; see also Narayanan, 2013; Oliver, 2010; Smith, 2009; Wolfe and Gudorf, 1999). Possibly the most influential of the indigenous scholars who contributed to this construction of Hinduism as a religion was Ram Mohan Roy, who formed the Brahmo Samaj in 1828 and defined Hinduism by referring to the 'golden age' of the elite Sanskrit traditions, casting the Upanishads as foundational texts (Jaffrelot, 1993).

The original oral texts of the Aryans were later written down as the Vedas, a set of four overlapping collections of hymns and verses (about 20,500 in total), which started to be assembled between about 1500 and 1000 BCE. Hindus refer above all to the Rig Veda, a collection of over 1,000 hymns and verses believed to date from about 4000 BCE.[1] The Vedas are known as *sruti* (scripture), while later literature (*smriti*) includes the epics, mystical texts and law books. The epics include the Ramayana (the story of Rama) and Mahabharata, which includes eighteen chapters that are often published separately as the Bhagavad Gita ('song of the lord'). Among other things, the texts contain instructions for conducting religious ceremonies and rituals and the Upanishads or Vedanta ('end of Veda'), composed between 1500 and 500 BCE. The latter deal with mystical awareness of the divine and the ideal of an individual who seeks to unite the human soul with that of the all-pervading spiritual force of the universe (Brahman) through living a reclusive, ascetic, meditative life.[2] Among the law books, the Manusriti, dating from the second and third centuries BCE, is a composite produced by many authors, who drew on earlier documents. It is concerned with *artha* (statecraft and legal codes) and *dharma* (moral codes with respect to duties, rights, virtues, conduct etc.). Although influential in India and beyond prior to colonialism, scholars doubt that it was ever codified or implemented as law. However, one version was one of the first Sanskrit texts to be translated during British rule (in 1794). Rather than being treated as a discourse or commentary on law and morals, the translation was used selectively by the colonial government to formulate 'Hindu law' to parallel *shari'a*, and so it has had lasting influence.

The Vedas and the Upanishads are regarded by some as the foundational canon, on which are built many other mystical, philosophical and ethical texts, although they are considered to be aids to knowing truth rather than the truth per se. Within Upanishad thinking, there are six distinctive 'orthodox' systems of philosophy and ten principal schools of *Vedanta*. These tackle various questions related to the nature of the absolute and the individual, the relationship between them and how to escape from *samsara* (future rebirths). These and

1 The dates are not precise and are disputed.
2 There are said to have been perhaps 108 Upanishads, but most have been lost.

subsequent schools and movements have continued to add to the body of texts, myths, commentaries and guidance available to Hindus (Oliver, 2010; Smith, 2009; Wolfe and Gudorf, 1999). Amongst these, the Bhagavad Gita has been particularly influential because it contains "the essence of Upanishadic philosophy and Vedic/Upanishadic ethics in a simple and lucid style" (Nadkarni, 2008, p. 139). It thus made the new paths towards realising God and achieving spiritual happiness accessible to ordinary people by stressing their direct access to God, providing practical, ethical and ritual guidance, emphasising selfless work for the welfare of the world and preaching a universal rather than sectarian religion. Because of the impracticality of deriving guidance from the vast Hindu literature, the Gita (together with many commentaries on it) has come to be accepted as the main canon by many Hindus (Nadkarni, 2008). As well as this great variety of sources, Hindus believe that truth can be accessed through persons (gurus, saints, prophets) and can also be realised intuitively.

Buddhist sources

Buddhism (along with Jainism) emerged as a reform movement within early Hinduism (Cantwell and Kawanami, 2009).[3] It started with the life and teachings of Siddhartha Gautama (who died c. 400 BCE), the Buddha (enlightened one), who emerged from a renouncer tradition that emphasised the abandonment of worldly life in pursuit of spiritual enlightenment. The Buddha rejected the caste system that placed power and authority in the hands of the Brahmins, the highly ritualised sacrificial practices of the day and fatalism arising from a Brahmanic interpretation of *dharma* (duty or role) (Wolfe and Gudorf, 1999). Originating in northeast India, Buddhism became the religion of the Mauryan Empire of Ashoka that united most of India in the third century BCE, from where it spread to much of Sri Lanka and southeast Asia. By the first century AD, it had begun to split into different strands, especially Theravada Buddhism (the oldest strand, today including over a third of all Buddhists and prevalent mainly in Sri Lanka and southeast Asia), Mahayana Buddhism (the largest strand, embracing over half of all Buddhists, in central and east Asia and Vietnam) and Vajrayana Buddhism (a much smaller branch of the latter, which adds esoteric practices and is now dominant in Tibet, Nepal and parts of northern India) (Cantwell and Kawanami, 2009; Loy, 2010; Oliver, 2010; Tomalin, 2007; Wolfe and Gudorf, 1999). By the tenth century CE, Buddhism had largely died out in India, although Tibetan exiles established a government in exile in Dharamsala in the northern State of Himachal Pradesh in 1959 following the Chinese invasion of Tibet, and Ambedkar, a prominent Marathi lawyer and political leader, led a large-scale

3 In the fifteenth century, Sikhism also emerged, based on an amalgam of ideas associated with Hinduism (especially the bhakti movement of northern India), tantric Buddhism and Islam (Shackle, 2009; Tatla, 2008). Today there are about 15 million Sikhs, mostly in India, compared to 376 million Buddhists, 965 million Hindus, more than a billion Muslims and about two billion Christians worldwide.

conversion of Maharashtran Dalits to Buddhism in the late 1950s and 1960s (for a comparison of Buddhism in Dharamsala and Pune, Maharashtra, see Bhatewara and Bradley, 2012).

Pali, a classical Indic language, was the original language of Indian Theravada Buddhism, which bases its teachings on the Pali texts (including the *suttas*, Sanskrit *sutras*), the oldest scriptures, while Sanskrit is the language of Indian Mahayana Buddhism. Buddhists believe that the Buddha's words are recorded in the *Tipitaka* or 'three baskets' of doctrines: the basic doctrines in the *Sutta pitaka*, the rules of conduct for monks and nuns in the *Vinaya pitaka* and analyses of doctrines in the first *pitaka* in the *Abdihamma pitaka*. In addition, the *Jataka* stories that describe the Buddha's previous births are considered to be accounts of actual events. They exemplify the Buddha's ethical conduct that guided him towards enlightenment in his final birth and provide guidelines for conduct in the present day. The Buddhist scriptures were, like many other ancient texts, transmitted orally by schools of reciters until they were written down around the beginning of the Common Era, "in the separate but parallel collections of the early Buddhist orders, including sections on monastic discipline (*Vinaya*), discourses (*Sutta*) and, in the case of most but not all schools, Higher Teachings (*Abhidhamma*). The Mahāyāna . . . movement added a further set of sutras, not recognised by all Buddhists" (Cantwell and Kawanami, 2009, p. 70). Because the Buddhist tradition, like all other religious traditions, arose in a particular socio-cultural context, it drew selectively on many previous traditions, especially the Vedic tradition, the mythical grand narratives of the Mahabharata, Ramayana and Puranas, and the Upanishads, resulting in a syncretic tradition involving cycles of time at various scales, from the cosmic to the human, and a pantheon of both deities and demons (see below) (Goonasekera, 2013). Nevertheless, the central teachings are contained in the Buddha's first and second sermons, the content of which was disseminated largely by the *sangha* of wandering monks and nuns (Oliver, 2010).[4]

Textual sources: their perceived authenticity and interpretation

Christian fundamentalism implies belief in the divine inspiration and inerrancy of the Bible, and today, at least half of all Christians in African countries surveyed by the Pew Forum believe that the Bible is the word of God and should be taken literally (ranging from 87% in Nigeria and Liberia to 52% in Guinea Bissau, median 76%). Most Muslims have a similar view of the Qur'an (ranging from 90%+ in Cameroon, Nigeria and Ghana to 54% in the DRC, median 80%) (Pew Forum, 2010, pp. 25–6).

Access to the Bible initially widened as a result of translation, printing and increased literacy and recently has expanded further with the emergence of new

4 From the outset, the Buddha envisaged the need for a four-fold *sangha* (community) comprised of fully ordained men and women (*bhikkhus* and *bhikkunis*) as well as lay men and women (*upasaka* and *upasika*). The former renounce the world to join the monastic *sangha* in order to practise meditation and live by the eightfold path in order to achieve enlightenment (Tomalin, 2007).

communications technologies. In southern Christianity today, conservative religious stances are more prevalent than liberal ones, implying

> a much greater respect for the authority of scripture, especially in matters of morality; a willingness to accept the Bible as an inspired text and a tendency to literalism; a special interest in supernatural elements of scripture, such as miracles, visions, and healings; a belief in the continuing power of prophecy; and a veneration for the Old Testament, which is considered as authoritative as the New.
>
> (Jenkins, 2006, p. 4)

Evidence from Kenya, Nigeria, South Africa and India confirms that the proportion of Christians who believe the Bible is the literal word of God is higher amongst Pentecostals and charismatics than other Christians (the Philippines is an exception) (Pew Forum, 2006, pp. 6, 23). Vásquez (2009) suggests that Christians generally treat the texts as sacred artefacts – not just sources of doctrines and ethics but tools believed to have the power to purify and transform the self and society, although the former typifies mainstream Christians and the latter Pentecostal and charismatic Christians. Today, acceptance of Biblical truth prevails, not because Christians in SSA and southern Asia are ignorant of textual scholarship or critical approaches to the Bible but because they reject them (Jenkins, 2006).

In addition, many Christians live in contexts where nearby Muslim communities regard the Qur'an as God's word, to be learned and recited, a source of guidance for social order and everyday life, the yardstick against which people will be judged. The evidence available from surveys in southern Asia confirms that contemporary Muslims believe the Qur'an and *shari'a* are the revealed word of God rather than a body of law based on the word of God but developed by men (81% and 8% respectively in Pakistan, 65% and 25% in Bangladesh, 52% and 20% in southern Thailand and 41% and 35% in Malaysia) (Pew Research Center, 2013, p. 42). Further, they tend to believe that there is only one true understanding of *shari'a*. More religiously committed Muslims (those who pray several times daily) are more likely to believe *shari'a* is the revealed word of God, with only one valid interpretation, and to say that they follow the *ahadith* and *sunnah*. Overall, three-quarters or more of Muslim respondents in these countries claimed that their lives reflect the Qur'an to some (often a considerable) extent. Nevertheless, many believe that *shari'a* is a changing body of law, based on the Qur'an and *sunnah* but interpreted in rulings and through legal reasoning. Thus a substantial proportion of Muslims in some countries think that the written texts can be reinterpreted in contemporary contexts and indeed that they can be open to multiple interpretations (see Table 3.1) (Pew Research Center, 2013).

Responses from a recent survey in Sri Lanka illustrate the extent to which religious authority is grounded in the texts. Overall 63% of respondents in this multi-religious country (two-thirds plus Buddhist, with Hindu, Muslim and Christian

Table 3.1 Muslim views on interpretation of the Qur'an

	There is only one interpretation of the Qur'an	*There can be multiple interpretations of the Qur'an*	*Don't know*
Thailand	51	29	20
Indonesia	56	44	-
Malaysia	43	35	20
Pakistan	61	17	20
Bangladesh	38	57	5

Source: Pew Research Center, 2013, p. 44.

minorities) identified local religious leaders as their main source of guidance on religious matters, 16% reported that they refer to 'holy books', 7% to a religious scholar, 5% to other religious literature and 1% to the internet (Asia Foundation, 2012, p. 7). The proportions seeking these sources of religious authority are similar for all the religious groups, except for Muslims, half of whom identified 'holy books' as their first choice and only a third (36%) religious leaders. This reflects the central role of the texts in Islam, the wide availability of a Tamil translation of the Qur'an and high literacy rates.

Various Islamic reform movements have emerged in the modern period. These attack local custom and 'superstition' in favour of a scriptural faith in order to produce 'proper' Muslims who adhere to 'true' Islam.[5] A prominent early example is the Deobandi movement, which emerged in mid-nineteenth-century India (Robinson, 2008). Like others, this sought to rid Islam of the many Hindu accretions that were believed to have weakened Muslims and facilitated the British colonial takeover. In 1867 a madrasa was established at Deoband in north India to promote reform by establishing the *ulama* as the guardians of 'correct' Islam and providing religious education to the educated classes, both men and women, who could then, it was thought, provide examples of pious and moral behaviour to the poor and uneducated. Members of this and similar reform movements advocate a more scripturalist approach to Islam, using *ahadith* to regulate many aspects of Muslim belief, religious practice, everyday life and bodily comportment. This madrasa has spawned both thousands of other madrasas and large-scale religious movements in India and beyond. These rest on the increasing availability of knowledge of how to be a Muslim, based on education, literacy and the print (and more recently electronic) media (see Box 3.1). In the absence of political power (although this path is favoured by some reform movements), their emphasis on the individual pursuit of 'true Islam' is perceived to be key to fashioning an Islamic society.

5 As Becker (2008) notes, large-scale conversion to Islam in southeast Tanzania during the colonial period was encouraged by the codified nature of Islamic law, which appealed to local followers of folk religion because, rather than replacing their previous religious practices, it provided 'proper' ways of conducting burials, making sacrifices, praying etc.

Box 3.1 Learning about Islam

Contemporary Islamic revival is marked by a proliferation of opportunities for ordinary Muslims to learn more about their religion, including increased numbers of madrasas and similar educational institutions, the spread of study circles or Qur'anic reading groups, and the increased availability of modes of learning that use different media, teachers and sources, including fiction, telenovelas and radio debates (Schultz, 2011).

In Pakistan, for example, opportunities for girls and women to access religious education have increased dramatically in recent years. Madrasa education for girls has become increasingly available and popular (Bano, 2010). A prominent example of a religious education provider for women is Al Huda in Islamabad, which seeks to make a particular version of scriptural knowledge available to women. There have now been thousands of graduates from its courses (Ahmad, 2008; Bano, 2011).[6]

The growth and spread of study circles, particularly those organised by and for women, is even more remarkable. Reflecting the belief that individuals are responsible for developing their own relationship with God, they recognise the need for ordinary Muslims to understand the written sources, as well as to cultivate individual religious dispositions and practices on a daily basis (Schultz, 2011). Institutionalised to different degrees in different places, they were little studied in SSA and southern Asia until the 1990s. Women's study circles typically meet in members' houses or a local mosque to read the texts, especially the Qur'an, in Arabic and/or the local language and to be instructed in their significance by male or female religious teachers. How groups operate depends on participants' level of education and the teacher's interpretation of his or her role – to instruct, provide commentary, admonish and sometimes to encourage debate. Teachers are often self- or madrasa educated and may be given authority by a mosque to run such classes, the purpose of which is primarily to enable participants to move beyond formulaic interpretations of Islam to understand how Qur'anic teaching provides a guide to everyday life (Ladbury and Khan, 2008).

Elite women participating in a Qur'anic reading class in Dhaka, for example, seek to return to a literal interpretation of the Qur'an and *ahadith*

6 Indonesia is in a rather different position, since its long-established mass Muslim organisations, Nahdlatul Ulama and Muhammadiyah, have offered extensive educational opportunities for many years. The Wahhabi-influenced NU runs a network of nearly 7,000 Islamic boarding schools (*pesantren*) for both girls and boys, as well as 44 universities. The more modernist Muhammadiyah also prioritises education, both religious and secular, operating nearly 6,000 schools and nearly 130 universities.

to provide themselves with a guide to the norms of Islamic conduct. They reported that as a result, they had become more meticulous in observing the Five Pillars and were more able to lead 'rightly guided' lives (Huq and Rashid, 2008). Originally participants in such study groups in Bangladesh were middle class and literate urban women, but they have spread to the rural areas (Riaz, 2013). Since the 1990s in Colombo neighbourhoods and elsewhere in Sri Lanka, middle-class women have participated in similar groups, in which *bayans* (lay sermons) are delivered. Participants state their motive as being to improve their knowledge of Islamic teaching to enable them to follow correct religious practices and encourage their engagement in *da'wah* (Haniffa, 2008). Similar groups and practices have proliferated in West Africa. For example, in Mali, Arabic and religious education were previously restricted to a few women from the merchant and scholarly elite. More recently, the core texts have become available to non-elite women. Schultz's in-depth study shows how middle-status women in southern Malian towns meet to learn and pray together, acquiring literacy in Arabic, learning ritual and moral prescriptions, providing each other with moral and financial support and engaging in *da'wah*. Like women elsewhere, they regard reading and learning as the foundation of a process of self-making, which is then expressed in proper ritual practice and moral behaviour (Schultz, 2008, 2011).

Mostly, the study circles focus on learning 'correct' interpretations of the texts, but in some instances participants do seek to reinterpret them, either to purge local interpretations of elements considered inauthentic or to ensure that interpretations are relevant to contemporary life, perhaps drawing on international Islamic (or occasionally secular feminist) discourses to do so.

As discussed above, the nature and status of the texts in Hinduism and Buddhism is very different to the monotheistic 'religions of the book'. Although (unlike folk religions) their oral traditions were eventually recorded in writing, these take a variety of forms. Thus, while in Buddhism the central teachings are generally accepted as those contained in the Buddha's sermons (the Four Noble Truths, including the Eightfold Path), the closest Hinduism comes to a similar collection is the Bhagavad Gita (see above). Perhaps for this reason, attempts to ascertain adherents' views on the status of the 'key texts' appear to have been confined to Islam and Christianity.

The only/true religion?

Some religions hold that they are the 'true' religion or that particular beliefs and interpretations are 'true'. However, there are many understandings of what constitutes 'truth'. Most but not all religions have a body of 'truths' that are considered to have been spiritually revealed and, in many cases, subsequently elaborated, although some hold that truths are constructed socially (agreed by the consensus of a group, say a denomination/sect or religion) and so are historically and socially specific, relative rather than absolute. Variations on the relevant question have been included in a number of international surveys. For example, in 1999, 31% of respondents in Gallup opinion polls worldwide (62% in West Africa, 31% in Southeast Asia) considered their own religion 'the only true faith' (Gallup International, 2000), while of respondents in World Values Surveys, the proportion who agreed or strongly agreed that their religion was 'the only acceptable religion' varied from two-thirds or more in Pakistan (91%), the Philippines (66%) and Malaysia (67%) to about 20% in Singapore, Rwanda and Zimbabwe (www.worldvaluessurvey.org/WVSOnline.jsp accessed 10/8/2017). Amongst other factors, including the religious composition of the context in which adherents live and whether there is a socio-political culture of tolerance, the role and status of key texts is important in determining whether a religious tradition, sect or denomination considers itself the only valid or true religion. This is particularly true of the monotheistic religions, especially the strands within them that interpret the central texts literally.

Often the phrasing of the question links it to ideas about salvation and heaven/hell (see below). For example, within Christianity, the vast majority of Pentecostals believe that faith in Jesus Christ is the *only* path to salvation from damnation (Pew Forum, 2006, p. 7).

The vast majority of Muslims in most countries share this belief that their own religion is the only one that leads to salvation (eternal life in heaven), ranging from 90% or more in Malaysia, Pakistan and Niger to about half in Guinea Bissau, Chad and Mozambique. The remaining Muslims in these samples believe that adherents to other religions can also achieve salvation, although this is less likely to be believed by pious Muslims (those who pray several times daily) (Pew Research Center, 2012, pp. 110, 112). Indeed, in all the 15 African countries with a substantial Muslim population included in the Pew Forum's surveys, a fifth or more of respondents (varying from 44% in DRC to 19% in Cameroon) favour the death penalty for Muslims who leave Islam (Pew Forum, 2010).

While more pious Muslims are, in general, more likely to believe that there is only one true interpretation of their religion, their specification of which sects/groups they consider to adhere to this interpretation varies. For example, they often exclude the Ahmadiyya sect, a reformist movement founded by Mirza Ghulam Ahmad in India in the late nineteenth century, most adherents of which live in southern Asia. In Bangladesh, 40% of Muslims accept them as Muslims, but only 12% in Malaysia and 7% in Pakistan, while in southern Thailand and Indonesia, most Muslims had never heard of them or had no view (Pew Research Center, 2012, p. 93).

Belief that one's own religion is the only true one may, along with other factors, have implications for the value placed on religious freedom, levels of tolerance and engagement in proselytising (see Chapter 7).

A transcendent reality?

Religious pluralists regard the world religions as different responses to a transcendent reality that is believed to be universally present, although beyond the scope of our human conceptual systems. The different traditions were 'founded' in some sense by spiritual figures, who were perceived as being exceptionally open to the transcendent (Hick, 1999). Central to their revelations were conceptions of a God or Absolute. In the monotheistic religions (Islam and Christianity), belief in God is central, although apart from a general conception of God as transcendent, creator and personal, views about the nature of God and God's relationship to humankind differ.

The *shahadah* is the ***Islamic*** declaration of belief in the oneness of God (*tawhid*) and acceptance of Muhammad as the final Prophet or Messenger. The first of the Five Pillars of Islam, it is translated from Arabic into English as 'There is no god but Allah and Muhammad is the Messenger of Allah', statement of which denotes conversion to Islam and is often repeated during *salat* (daily prayers or worship). Shi'a Muslims may add 'Ali is the *wali* [viceregent] of God', reflecting their belief that Muhammad designated his cousin and son-in-law as his successor (Rahman, 1987; Ruffle, 2013). "Muslim doctrine centres on God's uniqueness and incomparability with any other divine figure (*tawhid*), which is affirmed by the absolute prohibition of associating any being or attribute to God (*shirk*) . . . [and] in God's role as Creator (*al-Khaliq*) and Judge (*al-Hakam*) of humanity" (Ruffle, 2013, p. 194).

The opening chapter of the Qur'an succinctly outlines the fundamentals of Islam: God is unique and all-powerful, the Creator of the universe and humanity and the lord of the day of judgement but also merciful and a source of guidance for living a good life centred on God-consciousness (*taqwa*) (see also Cornell, 2005; Rahman, 1987). These are then elaborated on throughout (and in the *sunnah*). The Qur'an is believed to provide revelations from Allah and about Allah, including an argument for the existence of the one God, regarded as above number and any likeness, omniscient and both transcendent and immanent. As an aid to understanding the complexity of *tawhid*, the Qur'an employs (at least) 99 terms or names that convey different (metaphorical) aspects or attributes of the divine nature, each contributing to a partial description of a divine reality that is believed to be ultimately indescribable. The most commonly used are probably the 'most compassionate' (*Al-Rahman*) and the 'most merciful' (*Al-Rahim*). Sunni (and Wahhabi) Islam relies on a literal reading of the Qur'an and avoids speculation on the nature of God and his attributes, while in contrast, Sufism holds that believers 'find' God in the world (Cornell, 2005; see also Anawati, 2005). Statements of belief in Islam exist but do not have the status of official creeds, although Sunnis (by far the majority) have come to a large measure of agreement about the articles of belief, with each legal/theological school (especially the Hanafi and Hanbali schools) developing creedal statements.

Islam is essentially a religion of orthopraxy – for Muslims faith in and a consciousness of God are considered to be sufficient, and there is relatively little discussion of the nature of transcendental reality/God, other than repetition of some of the 'beautiful names'. Speculation about Allah is discouraged, with most Muslims appearing to be content to repeat the *shahadah*. Perhaps because of this, there seem to have been few attempts to understand how individual Muslims interpret the basic statement of belief. Nevertheless, as part of the recent fascination with piety or renewal movements, it is referred to in a couple of recent studies.

For example, in urban Mali, as described in Box 3.1, more and more women are organising themselves into groups to learn to read and write Arabic and thereby to develop a better understanding of the rules of Islam (Schultz, 2011). These women seek to renew themselves and their society in accordance with the authentic teachings of Islam: "first and foremost, [they share] the objective to achieve a 'greater closeness to God' and to experience his presence in their daily lives; second, they jointly embark on the path to attain this goal, namely, through an unconditional 'submission' (the literal meaning of Islam) to His will" (Schultz, 2011, p. 4). The pathway to God (*alasira*), they believe, is through correct ritual performance and appropriate behaviour. They seek to return to what they regard as the original and authentic teachings of Islam, entailing individual responsibility for developing a personal relationship with God. The notion of a pathway is closely associated with Sufism, where it refers to spiritual and meditative exercises through which the spiritual leader (*shaykh*) of a Sufi order and his disciples seek to experience the presence of Allah in the world and achieve communion with God (Schultz, 2011).

Similar interpretations were advanced in a very different setting by a group of men and women in their twenties studied by Blom in Pakistan. These young people are seeking to overcome the perceived deficiencies of both their parents' tradition and modernity by 'returning to Islam' as the central principle of their lives. They seek to become 'true believers', surrendering to God's will through 'going on the path of Islam', self-reform and a desire to revitalise society through establishing a new moral order (Blom, 2017). In Bangladesh, participants in Qur'anic study sessions for female students reported that their aim was to progress from being 'a basic believer' (*mumin*) to greater intimacy with God, in order to become a "God-loving person whose entire being is a constant longing for the presence of God (*muhsin*)" (Huq, 2008, p. 475), expressed in greater devotion, improved behaviour and ongoing self-surveillance.

Although ***Christianity*** is also a monotheistic religion, it has developed the concept of the Trinity to denote that God, while of one substance, is thought to exist in three forms: the Father, the Son and the Holy Spirit. The attributes of God the Father include his transcendence and timelessness – he is considered to be powerful and creator of the universe and all things in it but also immanent in that he protects, cares for and sustains all living things. Christians believe that God sent his Son, Jesus Christ, to be born on earth as a human being, to live as a person and eventually to be condemned to death, followed by his resurrection and taking into heaven to be with God the Father. Through his life, death and resurrection, they believe, human beings' sins can be forgiven, and they can be reconciled with God. Thirdly, in the form of the Holy Spirit, they believe that God is at all times within human beings who believe in him and are faithful to Jesus and his teachings, helping them to lead Christian lives. The idea of salvation from sin is therefore central to Christianity. While in theory God is regarded as neither male nor female, the doctrine of the Trinity, the symbolism used and the focus on Jesus as God made flesh in the male sex have given rise to a predominantly masculine conception of God (Furseth and Repstad, 2006; Kwok, 2004). The complexity of these ideas means that attempts to develop theological understandings of them play a huge role in Christianity, with alternative interpretations often being highly contested. Various creeds have been developed to provide statements of essential/central Christian beliefs, although there is disagreement over their detailed content, and they are not subscribed to by all Christians (Oliver, 2010). Although practice is important, theology plays a major role in the Christian tradition, which is seen as devoting more time and energy to interpreting beliefs than other religions. This also seems to open up more scope for individuals to develop what they consider to be personal relationships with God and to reinterpret the texts in order to develop understandings which can depart from official teaching. For example, a group of young Christians in Indonesia is using this possibility to resist the dominant discourse that positions them as moral if they abstain from sexual activities until marriage and immoral if they do not by reinterpreting religious morality on the basis of their own understanding of a loving God with whom they can have a personal relationship (Mulya, 2018).

Recent international surveys confirm the centrality of a belief in a single God amongst Christians and Muslims in Sub-Saharan African and southern Asian countries. For example, the 2010–14 World Values Surveys include 11 countries in the regions of interest to this analysis. Between 97% and 100% of respondents in nine of these countries claim that they believe in God, including not only predominantly Christian and Muslim countries (Ghana, Nigeria, Zimbabwe, South Africa, Rwanda, Pakistan, the Philippines, Malaysia) but also predominantly Hindu India. Only in two is such a belief less common (Singapore with 83% of respondents and predominantly Buddhist Thailand with 33%).[7] A Gallup International poll in 1999 specifically asked respondents whether they think of God as a person – 45% responded in the affirmative (64% in West Africa and 27% in Southeast Asia), while only 8% (11% in Southeast Asia and fewer than 1% in West Africa) did not think there is any sort of spirit, God or life force.[8] The limitations of simple survey questions are clear from these figures, which indicate the widespread prevalence of belief in God or a divine reality but tell us little about how believers conceive of that/those deity/ies or ultimate reality.

7 www.worldvaluessurvey.org/WVSOnline.jsp. Accessed 10/8/2017.

8 Seventeen percent responded that they do not know or did not respond at all. Note that general users can access the WVS data online, but for non-subscribers only brief reports on the findings of Gallup polls are accessible.

One might assume that in religions in which the attributes of God are spelt out in some detail in the texts and teachings, adherents believe that the deity has those attributes, but evidence about the views of ordinary adherents is scarce.

While there may well be gaps between official/textual statements of belief and what adherents actually believe in Islam and Christianity, this is likely to be even more characteristic of religious traditions in which there is no single authoritative written source, for example, ***Hinduism***. As described above, the label 'Hinduism' was applied to the beliefs and practices of a number of folk traditions in India collected together during the colonial period. In addition to portraying these beliefs and practices as a single tradition, the conception of Hinduism that emerged sought historicity for particular incarnations of the deities and even encouraged the idea of a central sacred book, claiming that these would allow it to transcend local interpretations and caste identities (Thapar, 1989). The Vedas reveal a ritualistic culture that emphasised worshipping and placating a variety of deities, many of whom were linked to natural phenomena (Agni/fire, Vayu/wind, Surya/sun, Indra/thunderbolt etc.), requiring regular sacrifices to maintain *rita* (cosmic order). In the later texts, especially the Upanishads, the emphasis shifted from the deities and the performance of rituals to contemplating 'Brahman' (the supreme god-head or impersonal absolute) and its relationship to individuals. Many schools of thought emerged, all concerned with the nature of the absolute and the individual self or *atman*, the relationship between them, and how to escape from *samsara* (the cycle of death and reincarnation) to a state of spiritual freedom or liberation (*moksha*) (Tomalin, 2009).[9]

Contemporary Hinduism embraces orthodox philosophical Brahmanical religion and devotion to various divine figures, some of whom are recognised across Hindu India and beyond (e.g. in Sri Lanka and Malaysia, see Smith, 2009, p. 39) and some of whom are local (Tomalin, 2009). The former includes contemplation of Brahman, who/which can be conceptualised in any way adherents find helpful – impersonal or personal, formless or with the form of a deity, gendered or beyond gender, with or without attributes, transcendent and/or immanent in the universe. The great male deities (Vishnu, Krishna, Shiva), manifestations of the goddess (Kali, Durga, Lakshmi etc.) and the lesser deities (often portrayed as avatars or incarnations of the great deities) are all seen as ways of accessing the ultimate reality. Most Hindus claim to be monotheistic, but their conceptions of the ultimate reality vary. Most believe that the Brahman is ineffable (beyond expression), but that he/it becomes manifest periodically to protect the good and destroy evil (Narayanan, 2013). Commonly, the absolute is seen to be manifest as Vishnu or Krishna, although some hold Shiva to be the manifestation of the supreme being, while Tamil-speaking people in India, Sri Lanka and Malaysia

9 An *atman* (individual self or soul) is believed to be independent of the bodies it inhabits, with each progressing through various bodies and only becoming aware of itself when it comes into a human body. The *atman* may have many rebirths before finally attaining *moksha*, spiritual freedom (Wolfe and Gudorf, 1999, p. 13). The concept of 'soul' may be too closely associated with Christian theology to be appropriate.

view Murugan, the son of Shiva and Parvati, in this way. Devi, the great goddess, is regarded as the mother of the universe, the all-knowing source of all existence, although the Shakti, divine cosmic energy, also female, may also be seen as the absolute (Furseth and Repstad, 2006). While some of the goddesses are cast as the wives or consorts of male deities, reflecting and reinforcing the model of women as wives and homemakers, others are associated with independence and sexuality (Tomalin, 2009). In Nadkarni's view, the deities reflect people's need or preference for a personal God, who may be seen as creator, merciful and full of compassion and grace (*shivam*), and with whom personal communication is possible. Worship forges bonds of mutual love between deity and devotee (Nadkarni, 2008). In particular, popular worship of Vishnu, Krishna, Shiva and the Shakti (personified in a range of goddesses with different characteristics) has come to dominate. The mythologies associated with these deities are found in the Sanskrit texts, for example the Puranas, the epics and vernacular devotional poetry (Narayanan, 2013). Some see divinity not only in the deities but also in other shapes and forms (human, animal, places, natural features or combinations) (Nadkarni, 2008).

To understand contemporary Hinduism, therefore, it is necessary to place it in its historical context, refer to the Sanskrit texts, the epics and the myths associated with both the main and local deities and examine how one or more of the aspects of belief outlined above characterise local expressions. Today, the Hindu Right is promoting Hindutva, an ideology that seeks to define and gain acceptance of a central set of religious and social practices, thereby seeking to impose uniformity on considerable diversity in order to promote Hinduism as the basis of Indian identity and nationhood (see Chapter 1).

Buddhist teachings do not begin with the divine: rather than seeking to transcend the world to experience another reality, they focus on the need to awaken to the true nature of this world and oneself (Loy, 2010). The core teachings are set out in the Buddha's first sermon, often labelled the Four Noble Truths. The teaching begins with the fact of human suffering (Pali *dukkha*, Sanskrit *duhkha*). This results from physical pain as a result of illness, infirmity or old age, fear of death, dissatisfaction with material things and the impermanence of relations between humans – it is less to do with the nature of the world than with the reaction of people to it. Suffering is believed to be caused by insatiable craving and desire (*tanha/trsna*) for things external to the self. The Buddha taught that attachment to things in the world is a source of *kamma/karma*, which refers in the Indic religions to action and the results of one's actions. It is what ties individuals to future rebirths and influences the nature of each rebirth. For better *kamma/karma*, the *dukkha/duhkha* which is caused by greed, ill will and delusion must be transformed into generosity, loving kindness and replacement of the notion of a separate self with wisdom (*panna/prajna*), marked by freedom from greed and hatred and realisation of our interdependence with others (Harvey, 2000; Loy, 2010; Tomalin and Starkey, 2013). Ultimately, for suffering to cease (*nirodha*) and liberation from the cycle of rebirths (*samsara*) to be achieved (*nibbana/nirvana*), craving must cease, and the discrete self, to which we are attached, has to be recognised as a delusion. *Nibbana*, release from *samsara*, remains the ultimate goal in Theravada Buddhism, whereas in the Mahayana (and Vajrayana) tradition, the highest goal is Buddhahood, implying that the Buddha (awakened one) does not abide in *nirvana* but continues to take rebirths in the world to help liberate other beings by teaching the Buddhist path.

However, the focus in Buddhism is not on metaphysical speculation about the nature of the universe, which is regarded as a diversion from the demanding work of seeking enlightenment. Instead, it is on steps that can be taken in this life to achieve release from the pervasive anxiety and insecurity of ordinary life, for which the Buddha provides spiritual and ethical guidance. This is set out in the fourth of the Noble Truths, the Eightfold Path of right understanding, speech, action, livelihood, effort, mindfulness and concentration/contemplation, with the aim of becoming an *arahant/arhat*, one who has gained insight into the true nature of existence and dwells in perfect serenity because s/he has put an end to all craving (Hick, 1999; Tomalin, 2007; Wolfe and Gudorf, 1999). In practice, although Buddhism is not based on a concept of the divine, Buddhists may combine local animist beliefs in the supernatural with Buddhist practice, and in popular Buddhism, the Buddha and various *bodhisattvas* fulfil the role of transcendent deities. Within Mahayana and Vajrayana Buddhism, *bodhisattvas* are individuals who "are selflessly devoted to helping all beings awaken to their true nature" (Loy, 2010, p. 56), based on an understanding that they cannot become fully enlightened and attain *nirvana* themselves until all have done so. Theravada Buddhism, in contrast, emphasises the individual's own enlightenment, without such a focus on helping others to become enlightened (Tomalin, 2007, p. 34; Loy, 2010).

Folk religions persist worldwide and are still widespread and influential amongst not only tribes and ethnic groups living in relatively isolated areas but also people living in villages, towns and the largest cities. Even more than 'world religions', folk religions and local cultures are so inextricably linked that they may be perceived by both outside observers and those who adhere to their beliefs and practices alongside Christianity or another religion as 'culture' rather than 'religion' (Kujur, 2013).[10] Despite their pervasiveness and many common elements, they are essentially local. Even if they have travelled, usually with migrants, they are associated with particular socio-linguistic groups.

Folk religions conceive of the material and spiritual spheres as interdependent "such that events in the material world are predicated on relations to the spiritual world, but open to individual, and sometimes contested, interpretation" (Gore, 2009; see also Alhassan Alolo, 2007; Alhassan Alolo and Connell, 2013). They are generally animistic:

10 In Tanzania, for example, Islam and Christianity are known as *dini* (religion) and indigenous religious practices as *jadi* (tradition). "'Tradition' lacked the core of explicit, systematic, codified and literate knowledge whose existence, even if far away, was seen as the foundation of 'religion', therefore it was practical, ritual and implicit" (Becker, 2008, p. 156).

> A form of religion centred on relating to spiritual powers or beings who [are believed to] permeate the world. The spiritual power may be conceived of as an impersonal force running through everything and capable of being used for good or evil. The unseen power may also be understood as numerous spirits, some of whom are friendly, some 'tricksters', and some dangerous. Spirits and humans are [regarded as] interdependent parts of a single cosmos.
>
> (Tamney, 1998, pp. 22–3)

Whether folk religions have a concept of a supreme spiritual being or are polytheistic is contended. Kujur (2013), for example, asserts that most of the tribal communities in India believe in one supreme being, who has different names in different traditions. Groups may or may not attribute common characteristics to the supreme being. In ATR, the supreme being is believed to be omnipotent, transcendent and immanent (Mbiti, 1970 in Alhassan Alolo, 2007; Wiredu, 2011). It may, however, be conceived of as gendered or beyond gender (Parrinder, 1974), personalised or the force responsible for cosmic order (Wiredu, 2011). Often all life is believed to flow from and return to the supreme being, although s/he is not thought to intervene in everyday life, which is believed to be governed by other invisible forces which have access to spiritual power, as do some human beings, who are gifted with, can acquire or can tap into it (Alhassan Alolo and Connell, 2013). The relationships between the powerful supreme being (the benevolent creator interested in the welfare of his/her creatures, giver of knowledge and wisdom and guardian of morality), the natural world and human beings are believed to be harmonious but constantly threatened by the forces of evil and the actions of human beings (Kujur, 2013). The supreme being may be venerated, worshipped directly (through ceremonies, sacrifices or offerings) or worshipped through intermediaries such as lesser deities, since s/he/it may be considered too distant and too feared to be approached directly. Even if a group's myth of origin posits a supreme being, s/he may not figure as prominently in contemporary beliefs and rituals as other spirits or deities.

For example, a study of Ndwumizili, a village in southern Ghana, found that "the central question for residents . . . is not whether spiritual beings exist, but what they are like, how much power they have and how far one can and should interact with them" (Graveling, 2010, pp. 36–7). Although fewer than a third of adults in the village are members of churches and only 20% regularly participate in church services, many more identify themselves as being Christian. Nevertheless, "to varying degrees. . . [they] engage in customs such as pouring libation to ancestors, seek . . . out practitioners of 'juju' and uphold . . . rules and taboos ordained by 'small gods'" (Graveling, 2010, p. 34). Christianity has become so embedded in Ghanaian society that it is hard to distinguish between local and Christian concepts of God: Nyame (and variants) is used for both the Christian God and the Akan God, who is considered to be male, supreme above all other spiritual beings and the rest of creation, omnipresent and omniscient.

Other supernatural forces

Even among adherents of the monotheistic religions, various other supernatural forces are thought to be at work in the world. However, spirits are central to the belief systems of folk religions, and so the discussion here will focus on these beliefs to start off with while also exploring the extent to which beliefs that there are supernatural forces other than God are also held amongst Christians and Muslims.

Deities and spirits

In folk religions, other supernatural powers may include lesser deities and spirits, who are generally thought to have been created by and to be dependent on and subordinate to God. They can be broadly divided into nature and human spirits. In Africa, both are concerned with ensuring life and fertility. The former are personifications of natural objects and forces, subdivided into spirits of the sky and spirits of the earth. They are believed to reside in natural objects and forces (e.g. the sun, mountains, trees, metals, animals, water, thunder, wind), or these are themselves regarded as spirits or sources of spiritual power, if not deities. Human spirits include the long dead (ghosts), whose stories are told in myths and legends, and the recently deceased (ancestors), who live on in the memories of the living. Commonly, the spirits of an individual or group's ancestors must be recognised and are sometimes venerated, although their role with respect to the living varies. Ancestral spirits may be either male or female. They may have "'parental' relations of absolute power over the succeeding generations" (Gore, 2009, p. 274) and are usually regarded as good and just, although they can inflict harm as a corrective or punitive measure for transgressions (Ellis and Ter Haar, 2004). Spirits are regarded as watchful beings who possess powers that can harm or help living humans, thus making it necessary to seek their blessing or avert their anger (Alhassan Alolo, 2007; Mahdi, 2006; Mbiti, 1991; Parrinder, 1974; Wiredu, 2011). Similarly, amongst tribespeople in India, benevolent, malevolent and neutral spirits are recognised. The former include ancestor, village, clan or household spirits (the last have to be kept tamed), while malevolent spirits are believed to be those who have died an unnatural death and have to be propitiated (Kujur, 2013).

That such beliefs are central to the worldviews of many contemporary Africans is illustrated by the study in southern Ghana referred to above. In this region, while *Nyame* is considered to be the supreme God, a range of 'small gods'/fetishes (*abosom* in Fante) are also thought to exist.

> These vary in importance and characteristics: some are related to specific ethnic groups, clans, lineages or smaller social divisions; some are imported from other groups owing to their perceived power and still others are associated with particular locations, such as [an] island and lagoons. Although usually invisible, [it is believed that] they can appear in human form if they

> so wish. . . [They] are traditionally considered neutral rather than malicious, and a moral reason is usually attributed to any harm they cause.
>
> (Graveling, 2010, p. 38)

If their moral norms or taboos are violated, they are believed by villagers to punish the offender, sometimes with death; otherwise they are considered unlikely to cause harm, and the places they inhabit are not feared or avoided in everyday life. Relationships between human beings and these deities tend to be utilitarian: if the former understand how the deities function and can establish satisfactory relations with them, they can exploit their powers to their advantage, but if deities' power to grant requests or punish is thought to have waned, they are likely to be disregarded or abandoned (Graveling, 2010).

Much folk religion centres on the well-being of a community or social grouping, with the beliefs and practices associated with religion providing a means of understanding, tackling and taking control over the bewildering forces by which people believe they are threatened. The nature and activities of the spirits and deities, and sometimes their human intermediaries, are believed to be the causes and explanations of good and ill fortune, ranging from large-scale disasters to everyday problems such as ill health or crop failure. They offer ways of ensuring the former and staving off or combating the latter. The cosmological understanding they represent and the associated religious beliefs and practices are not necessarily coherent, unchanging or shared by all members of a community, despite the efforts of missionaries, colonial agents, anthropologists and others to record and systematise the belief and ritual systems of particular groups (Gore, 2009).[11]

While some spiritual powers may be exercised by lineage heads, more generally intermediaries are religious specialists (diviners, healers, priests, priestesses), those who adopt such roles on occasion (e.g. performers in masquerades, those in charge of initiation rites) or those believed to be possessed by particular deities (spirit mediums). These intermediaries are thought to be more able than others to communicate with the spirits, seeking to please them, win their favour and avert their wrath through prayer, divination, dreams or possession (Ellis and Ter Haar, 2004). Spirit possession is thus considered to be a way in which communication with a world beyond ordinary human experience is possible. It may be considered either good (mediumship) or harmful (requiring exorcism) (Alhassan Alolo and Connell, 2013). For example, amongst the Hausa, such an individual's personality is believed to be possessed by one or more spirits in the pantheon, so that s/he takes on the characteristics of that spirit. In the past, women were particularly involved as mediums, despite Muslim efforts to eliminate the practice,

11 Alhassan Alolo (2007) outlines and critiques a variety of different historical and disciplinary approaches to studying ATRs, pointing out that scholars' historical/cultural positions, religious backgrounds and cultural-political projects influence their analyses, in ways that they frequently appear to be unaware of or do not acknowledge. In particular, understandings of ATRs may be refracted through the religious beliefs and purposes of scholars who are themselves Christian or Muslim.

the importance and acceptability of which declined as Islam spread, although during the twentieth century some women attempted to protect it, because it provided a medical system for treating female diseases, which were neglected by Muslim medicine (Mahdi, 2006).

In the past in Africa, according to Ellis and Ter Haar (2004), spirits were considered to be morally neutral rather than intrinsically good or evil. Today, they suggest, many have come to see evil as pervasive and spirits or other mystical agencies as evil and harmful, not least under the influence of Pentecostal Christianity. As well as various categories of evil spirits, these include human agencies of evil (witches and sorcerers), while evil may also be attributed to the evil eye, broken taboos, or perjured oaths.

Witches and sorcerers

Witchcraft is perceived to involve spiritual power employed by humans in an evil manner or for malign purposes (Gore, 2009; Iliffe, 1995). Witches may be women or men (varying between ethnic groups). Their attacks are seen as unprovoked, unjust or motivated by malice or envy. They may be distinguished from sorcerers, who use medicine and charms for evil purposes and are often men. When an explanation of misfortune or economic success is sought, witchcraft is often suspected. Accusations are frequently made against people close to the supposed victim(s) and may be used to obtain revenge. Witchcraft, it is thought, may also be used to achieve self-advancement. For example, in politics, witchcraft allegations "provide a potent mode of discourse about the acquisition of power, legitimate and illegitimate, and a means to critique the way it is exerted within a moral economy" (Gore, 2009, p. 284). Witchcraft accusations may also be used to enforce discipline, for example, to pressure for the redistribution of wealth within a family or community.

For example, in southern Ghana, witches are those seen as having spiritual wealth that they use to the detriment of society (Graveling, 2010). Usually thought of as female, their power is seen as being intrinsic to their person, acquired through the mediation of someone who is already a witch, inherited at birth, transferred from a witch at her death or given without the recipient being aware of it. The common idiom is that witches 'feed' on their victims, physically, financially or in other ways. Any kind of misfortune may be suspected of being caused by witchcraft, especially if it is serious, recurrent or ongoing, and alleged witches are feared, ostracised and even lynched. Mostly they are thought to attack their own family members, so the latter are prime suspects. The identity of alleged witches is only revealed through accusations and confessions, so it is difficult to either accuse anyone with certainty or to prove one's innocence.

A study in Dormaa-Ahenkro District on the western border of Ghana in the early 1990s also found that witches are female. They are regarded as both perpetrators and victims: believed to have been infected by an evil spirit, through her own misconduct or the malice of the spirit, a supposed witch is believed to attack others, bringing misfortune to them (Parish, 1999). Those accused of witchcraft are often young women accused of rejecting their friends or relatives or believed by older women or their fellow Christians of betraying their faith or church community. Once the supposed witch recognises that her conduct has made her susceptible to witchcraft, she may be ritually cleansed, or if her Bible is believed to have been contaminated by Satan, it may be replaced. Alternatively, alleged witches may be young men who appear to be wealthy because of their consumption of foreign goods but who have few friends and ignore their kin. Parish (1999) attributes the first type of witchcraft to the pressure older African Initiated Churches feel from growing Pentecostalism and the second to moral uncertainties about individualism, consumption, sociability and kinship obligations.

The current prevalence of such beliefs in SSA despite almost universal adherence to Islam or Christianity is demonstrated by Gallup polls in which respondents were asked whether they personally believe in witchcraft. In many, at least two-thirds responded that they do, although in some, fewer than a half replied in the affirmative (e.g. Nigeria 45%, Kenya 26%, Uganda 15%).[12] Overall, older, less educated and poorer respondents are more likely to believe in witchcraft (Tortora, 2010).

Periodic upsurges in witchcraft accusations seem to be associated with periods of stress and change, when interpersonal tensions are high and ways of resolving them weak. Most recently, there seems to have been an upsurge in accusations since the 1970s, which is often attributed to a period of disillusion following the realisation that many diseases cannot be cured by modern medicine, fear of childlessness, loss of employment etc. – i.e. a search for explanations of various kinds of misfortune (Tamney, 1998). In Niger, for example, witchcraft accusations seem to be associated with increasing inequality. The educated elite turned to Islam to unite the country after independence, so recent accusations appear to be responses to both Islamisation and the emergence of increasingly powerful and wealthy individuals. These have led to jealousy of Muslim administrators and businesspeople. While Muslims credit their wealth and good fortune to hard work and the blessings of God, non-Muslims seem to regard piety as a façade for reliance on spirit forces. Prosperous Muslims are therefore susceptible to allegations of witchcraft. They are thought to harbour blood-sucking spirits (*doguwa*) in their homes. A study in the town of Aresa in a Hausa-speaking area of south Niger found that victims of witchcraft are believed to have been attacked by a blood-sucking spirit, resulting in ill health or death if nothing is done. A suspected witch has to undergo a water ordeal in a distant village, a ritual which establishes his/her guilt, divests him/her of the *doguwa*, makes amends to his/her victim and also frees the latter. Supposed witches are thought to believe that the only way to appease the spirits is to keep supplying them with blood or to pass them on to a relative. As both the witch and his/her relative may then be under

12 The polls were based on face-to-face or telephone interviews of adults in 18 countries in 2009, mostly with samples of 1,000+.

suspicion, witchcraft accusations can be very destructive of kinship relationships. It is believed that victims can also be relieved of a blood-sucking spirit by exorcism (Masquelier, 2008).

While seeking to understand why a victim of misfortune might resort to a belief in witchcraft for an explanation, it must not be forgotten that accusations of witchcraft or spirit possession are often psychologically and socially devastating for the accused, who are disproportionately women and children and often, although not always, those who are already socially marginalised and vulnerable.

Spiritual beings in other religions

The processes by which indigenous worldviews have interacted with Islam, Christianity, Hinduism or Buddhism differ depending on the historical, geographical and political circumstances. What emerges from their interaction may be syncretism (the adoption or merging of elements of one religious tradition into another) or bricolage, whereby seemingly disparate bits and pieces of meaning and practice are pasted together by individuals and/or groups in ways that make sense to those concerned (McGuire, 2008). These are sometimes resisted (for example, by labelling beliefs or practices as heretical). Beliefs in supernatural or spiritual beings other than God or ultimate reality are not confined to followers of folk religions – they form an official part of some other traditions and also are held unofficially by many adherents. Available studies refer to Islam and Christianity.

Muslim *ahadith* include belief in the angels (*mala'ika*) (Kroessin, 2008, p. 18; Mbacke, 2016). The Pew Research Center's surveys in Sub-Saharan Africa and Asia found that a belief that angels exist is nearly universal in Asia and held by more than seven in ten Muslims in most of SSA, especially amongst those most committed to the faith (those who pray daily). The Qur'an and *ahadith* also mention supernatural beings (Arabic *jinn*), who are created by God along with angels and humans, and can be good, evil or morally neutral (Pew Research Center, 2012, p. 67). No question about *jinn* was included in the earlier surveys of SSA countries, but in Bangladesh 84%, Malaysia 77%, Indonesia 53% and southern Thailand 47% of Muslims said that they believe in *jinn*, even though most agree that Islam forbids appealing to *jinn* and hardly any respondents reported that they had personally seen any (Pew Research Center, 2012, p. 69) (see Box 3.2).

Both the Qur'an and *ahadith* refer to witchcraft and the evil eye, based on a belief that certain people can cast curses or spells that cause bad things to happen. The proportion of Muslim respondents in the Pew Research Center surveys who said that witchcraft/sorcery is real and capable of causing harm varies from two thirds or more in 5 of 19 countries for which figures are available, but 25% or less in 6 (See Figure 3.1). These responses demonstrate both the widespread nature of these beliefs and differences in their prevalence. Individually and collectively, therefore, Muslims have been influenced by both pre-Islamic and local indigenous beliefs, which, in addition to those mentioned above, include ancestral spirits and faith healing (Kroessin, 2008).

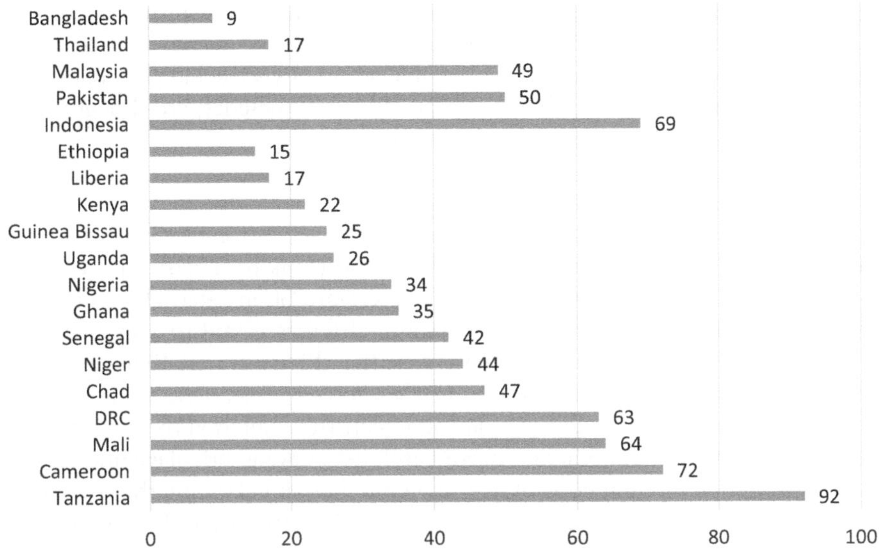

Figure 3.1 Proportion of Muslims who say witchcraft/sorcery is real

Source: Pew Research Center, 2012, p. 71, based on 2008/9 surveys of adults 18+ in 19 SSA countries and 2011/12 in Asian countries, median 37%.

Box 3.2 Spirit beings in Islam

In largely Muslim Zanzibar, off the coast of Tanzania, spirit beings, more often called *masheitani* than *majinni*, are widely recognised. A study by Larsen (2014) provides insights into local beliefs about the nature of spirits and their relationship to humans, although because the findings are based on only two cases, it is impossible to tell how widespread the beliefs are amongst Zanzibaris. Both humans and spirits are said to be subject to the will of God. The latter are considered to be beings of the air, created from fire, with bodies of their own that are invisible to the human world. It is believed that they can materialise by inhabiting a human body. They are thought to have lives and characteristics similar to humans, except that they are believed to live for up to 700 years and to understand all languages. The spirits decide who they wish to attach themselves to. The presence of a spirit in a person may be indicated by suffering and pain, although illness may also be regarded as indicating that one needs a relationship with a spirit. However, spirits may also protect their hosts, assist them in difficult relationships and help them acquire specialised knowledge. It is said to be possible for a host to realise that s/he is hosting a spirit because s/he acquires knowledge, for example of rituals or herbal medicines, perhaps while asleep. Such knowledge can enable an otherwise secluded woman to move around outside her house. Spirits, Larsen concludes, are thought to provide ways of managing difficult situations.

Many Christians in the global South today have a strong belief in evil and are willing to accept the reality of the diabolical (Jenkins, 2006). In the Christian gospels, angels, spirits and demonic forces are often referred to, although the attribution of evil to powerful supernatural forces is said to also reflect the continued influence of animist beliefs, whether this is in post-Reformation or contemporary Christianity. McGuire (2008), for example, gives an account of the Long Reformation in Europe, during which the church authorities tried to specify what beliefs and practices were valid, denouncing bricolage and syncretism, strengthening their control over the churches and claiming that they were the sole arbiters of the authentic Christian tradition. In SSA and southern Asia, people have come to Christianity relatively recently, generally from folk religions. In SSA, about a third of the population became Christian during the twentieth century (see Figure 1.1). Most Africans believe that "demonic forces exist, [and that] they exercise pervasive power over human beings, until and unless people opt for the Christian resources that alone can deliver them from evil" (Jenkins, 2006, p. 102). These resources include healing and exorcism, as apparently practised by Jesus himself. Individuals report that being 'born again' and delivered from the tyranny of evil spirits offers them not just personal strength but also entry into a church community that provides them with a sense of belonging and mutual support (Vásquez, 2009). The mainstream churches and missionaries condemned certain beliefs and practices as pagan. However, in the eyes of many these churches failed (and continue to fail) to offer plausible spiritual resources to overcome evil spirits, witches and sorcerers. In reaction to these and other perceived failings of the mainstream churches, first African Independent Churches and later Pentecostal churches and charismatic elements within the mainstream churches emerged. These all take spiritual menaces seriously and, drawing on both the gospels and indigenous religion, offer the means to combat them, including healing, exorcism, miracles, spells and charms. According to the Pew Forum surveys, two thirds of Pentecostals in five African churches believe that angels and demons are active in the world today (as do half or more charismatic Christians and between a third and a half of non-renewalist/other Christians) (Pew Forum, 2006, p. 26). A contemporary example of the persistence of beliefs in deities and spirits among Christians, based on a study of Ghanaian small-scale miners, is given in Box 3.3.

In popular Hinduism, the concept of evil spirits and forces (*rakshasas*) is also present, although sin is regarded as a result of ignorance rather than temptation by an evil force (Nadkarni, 2008).

Box 3.3 Artisanal miners in Ghana coping with uncertainty and misfortune

Based on narratives and observations of small-scale artisanal (and mostly unlicensed) gold miners in Tarkwa-Msuea, Ghana, Sarpong (2017) explores miners' conceptions of the nature of gold, the likelihood of success and the hazards encountered during mining operations. Although many of the miners are Christians, some see gold as a spiritual thing which can move about and disappear and so revere it. In addition, they may believe that some of their fellow miners have the power to draw gold into their own concessions from adjoining ones. These beliefs, alongside the unpredictability of success, have led to the evolution of rituals and taboos associated with prospecting and mining. It is common for miners to invoke deities/spirits to gain a profitable concession, counter interference by ancestral spirits or acts of sorcery by fellow miners and prevent accidents. They engage the services of spiritualists, Christian pastors or traditional priests to perform the rituals, which typically involve gifts and animal sacrifice. It appears that the frequency of these practices has increased in recent years, along with rumours of human sacrifices. The miners' beliefs and practices are derived from a widespread belief in the invisible world of spirits, who are regarded as real and powerful (although subject to God). Despite their understanding of geology, the difficult and unpredictable conditions and the dangers they face lead them to seek explanations for and means of tackling the apparently arbitrary distribution of success, failure and accidents. They believe that success can be achieved with the help of benign and powerful spiritual forces but also that others are able to harness malign forces to bring failure and accidents.

Religious experiences

Amongst other things, the concept of lived religion implies giving attention to experiences that are interpreted in religious terms, for example a sense of the sacred or divine, visions or voices. Some believe that such experiences are universal, because in their view human beings are "in tune with the ultimate reality that underlies, interpenetrates and transcends the physical universe" (Hick, 1999, p. 2). Both the founders of religions and many adherents report emotionally intense religious or mystical experiences, which may involve altered states of consciousness, change their perception of things and trigger revelations, although sceptics may suggest that the experiences are delusions. Experiences labelled as religious are generally shared through participation or personal accounts rather than being transmitted through formalised doctrine (Vásquez, 2009). The evolution of new media has made it possible to share such accounts widely. In a world of constantly recirculating images, Vásquez suggests, what stands out and is remembered is the spectacular. Such experiences include exorcism, divine healing, receiving a revelation from God and giving/receiving prophecy. Intense or mystical experiences are not confined to any one religion. Thus of respondents in recent surveys in nine African countries with substantial Christian and Muslim populations, on average 56% of Christians and 34% of Muslims reported that they had witnessed or experienced divine healing of an illness or injury and 40% or more that they had seen the devil or evil spirits being driven out of a person (two-thirds in Ethiopia) (Pew Forum, 2010, p. 13).

For individuals, deliverance from sin, the devil or evil spirits can produce a person who regards her/himself as being in control of her/his destiny. Such a person is neither purely individualistic (like economic man) nor trapped in kinship ties but considers her/himself to be a member of a family and congregation which offer meaning, belonging and support (Vásquez, 2009). Within Christianity, Pentecostalism emphasises the gifts of the Holy Spirit (including receiving direct revelations from God, giving/receiving prophecy, glossolalia, exorcising evil and healing through prayer). Currently, about a quarter of Christians in Ethiopia, Ghana, Liberia and Nigeria belong to Pentecostal denominations and at least one in ten in eight other countries (Pew Forum, 2010). Pew Forum surveys of Christians in a number of countries, including three in SSA and two in southern Asia, found that intense religious experiences are more commonly reported by renewalist than non-renewalist Christians. A majority of Pentecostals in the three SSA countries (Kenya, Nigeria and South Africa) reported that the church services they attend "frequently include signs of the spirit", as did a significant minority (40%+) in India and the Philippines. Witnessing a divine healing and/or exorcism is the most common experience in every country, especially among African Christians, even among non-renewalists (Pew Forum, 2006). That these findings are not exceptional is borne out by the results of more recent sample surveys in a larger number of countries: asked whether they had ever seen the devil or evil spirits being driven out of someone, up to a fifth of respondents in Asian countries responded in the affirmative, compared to a quarter in SSA, varying from fewer than 20% in Nigeria, Cameroon, Niger and Ghana up to 48% in Ethiopia (Pew Research Center, 2012, p. 75). An even larger proportion believe that "miracles are common today" (60%+ Pentecostals, 34%+ Charismatics, 30–46% other Christians, except in Nigeria, where 68% of other Christians believe this) (Pew Forum, 2006, p. 25). In addition, most renewalists reported that they had at least once received a definite response to a specific prayer request, a larger proportion than non-renewalists (Pew Forum, 2006, p. 16).

Early in Islam's history, the desire to become closer to God, to cultivate the inner life and to attain a deeper personal understanding of Islam (*tassawuf*) led to an increased emphasis on ascetic practices and the emergence of rites such as the invocation of saints or participation in dancing (Kroessin, 2008, p. 25). Hardy (1987, p. 396) describes the mystical strand in the religion, in which mystics seek to become one with God. Early Sufism, in particular, had strong mystical dimensions and a devotional element through which Sufis sought to progressively attain a deeper communion with God, often identified as detachment from the world, patience, gratitude for whatever God gives and pleasure with respect to whatever God desires (Rahman, 1987, p. 314).

Up to the twelfth century, Sufism consisted of a limited number of elite schools with different spiritual ideologies and techniques. Subsequently, large-scale orders developed, mostly within Sunni Islam, establishing *khangahs/dargahs* in which *shaykhs* lived and taught (Hardy, 1987; Rahman, 1987). In West Africa, for example, Sufi brotherhoods seek to demonstrate ways (*tariqa*) of reaching a personal religious experience through intermediation by a charismatic founder and his descendants. The founder and subsequent leaders (*marabouts*) have a knowledge of Islamic teaching and are also believed to have received *baraqa* (divine grace or blessing), which is transmittable to his (usually) or her successors and their followers. Those who are believed to have received *baraqa* promise salvation for their followers, guidance towards heaven and protection from misfortune. This system also has worldly dimensions – such leaders can attain wealth, social status and political power through tithing, the work their disciples perform for them and their social position. In turn, their disciples believe that the actions they perform for their *marabout* will earn them religious merit, in proportion to the *baraqa* attributed to the *marabout* (Bop, 2005; Sieveking, 2007). In South Asia, *shaykhs* or *pirs* (saints or spiritual guides) founded *khangahs*, mostly in towns and rural areas, while wandering mendicants resembled Hindu ascetics more than orthodox Muslims. The scale and importance of the Sufi orders in some parts of the Muslim world declined in the eighteenth and nineteenth centuries, but reformed orders have played (and continue to play) an important role in West Africa, especially in Senegal, where they have had a strong influence on Islamic practice, notwithstanding the emergence of fundamentalist reformist movements (Levtzion, 1987).

In Islam, therefore, the relative importance of *shari'a* and mysticism have varied over time and in different contexts. Mystical and other Sufi practices have consistently worried mainstream Muslims. They fear that the mystics' search for unity with God may blur the Qur'anic distinction between creator and created. In addition, the belief that the *pirs* are endowed with supernatural powers, which is linked to the practice of saint worship and intercession with the spirits of dead saints at shrines, is much disapproved of in 'orthodox' Islam.

Concluding comments

This chapter has started to explore some key elements of religious discourse, the first of Lincoln's four domains of religion, by providing a basic introduction to the sources and contents of the official teachings of the main religious traditions in SSA and southern Asia and reviewing evidence from surveys and qualitative studies to examine adherents' attitudes to the central religious texts, the extent to which official beliefs are accepted and how these beliefs are interpreted. To identify relevant evidence and assess its contribution to understanding, the analysis has had to consider all three dimensions of the social field (personal, social and organisational). People interact with the more or less formally organised bodies associated with their own religion and also with a variety of other religious and social groups, both locally, during their daily lives, and at the denominational level. Many adherents simply accept the sources and official teachings and beliefs of the religious tradition or denomination of their birth, through processes of socialisation in their families and social circles as much as formal religious education. However, they do not necessarily accept them wholesale – people do exercise agency, choosing whether to accept certain official beliefs and sometimes switching their allegiance if those of another denomination seem more convincing. Acceptance of the beliefs of the religious tradition with which they are affiliated does not necessarily mean that they abandon other beliefs. For many, the spiritual dimension or ultimate reality is not just conceived of in terms of the imagery of a particular religious tradition but is also influenced by other religio-cultural beliefs (e.g. spirits and their interactions with people) and their personal and social experiences (e.g. the Christian concept of God as father). Some evidence on how their beliefs influence people's perceptions of the world has emerged in this chapter, but this is probed further in Chapter 4, in which beliefs related to death, destiny and well-being are explored.

References

Ahmad, S. (2008) Identity matters, culture wars: An account of Al-Huda (re)defining identity and reconfiguring culture in Pakistan, *Culture and Religion*, 9, 63–80.

Alhassan Alolo, N. (2007) *African Traditional Religion and Concepts of Development: A Background Paper*, Birmingham: University of Birmingham, Religions and Development WP 17.

Alhassan Alolo, N. and Connell, J.A. (2013) Indigenous religions and development: African traditional religion, in Clarke, M. (ed) *Handbook of Research on Development and Religion*, Cheltenham: Edward Elgar, 138–63.

Anawati, G.C. (2005) Attributes of God: Islamic concepts, in Jones, L. (ed) *Encyclopedia of Religion*, Detroit: Macmillan, Vol. 2, 2nd ed, 616–22.

Asia Foundation (2012) *National Values Survey, Sri Lanka 2011*, Colombo: The Asia Foundation.

Bano, M. (2010) *Madrasas in Pakistan: A Response to Modernity*, Birmingham: University of Birmingham, Religions and Development WP 45.

Bano, M. (2011) Transforming faith: The story of Al-Huda and Islamic revivalism among urban Pakistani women, *Journal of Islamic Studies*, 22, 3, 461–3.

Becker, F. (2008) *Becoming Muslim in Mainland Tanzania, 1890–2000*, Oxford: Oxford University Press.

Bhatewara, Z. and Bradley, T. (2012) Buddhist engagements with social justice: A comparison between Tibetan exiled Buddhists in Dharamsala and Dalit Buddhists of Pune, *Economic and Political Weekly*, XLVII, 1, 69–76.

Blom, A. (2017) Emotions and the microfoundations of religious activism: The bittersweet experiences of 'born again' Muslims in Pakistan, *The Indian Economic and Social History Review*, 54, 1, 123–45.

Bop, C. (2005) Roles and the position of women in Sufi brotherhoods in Senegal, *Journal of the American Academy of Religion*, 73, 4, 1099–19.

Cantwell, C. and Kawanami, H. (2009) Buddhism, in Woodhead, L., Kawanami, H. and Partridge, C. (eds) *Religions in the Modern World: Traditions and Transformations*, London: Routledge, 67–102.

Cornell, V.J. (2005) God: God in Islam, in Jones, L. (ed) *Encyclopedia of Religion*, Detroit: Macmillan, Vol. 5, 2nd ed, 3560–7.

Ellis, S. and Ter Haar, G. (2004) *Religion and Development in Africa.*

Flood, G. (2006) Reflections on tradition and inquiry in the study of religions, *Journal of the American Academy of Religion*, 74, 1, 47–58.

Furseth, I. and Repstad, P. (2006) *An Introduction to the Sociology of Religion: Classical and Contemporary Perspectives*, Aldershot: Ashgate.

Gallup International (2000) *Religion in the World at the End of the Millennium*, www.gallup-international.com/ContentFiles/millenium15.asp

Gomez-Perez, M. (2016) Women's Islamic activism in Burkina Faso: Toward renegotiated social norms? *Canadian Journal of African Studies/Revue canadienne des études africaines*, 50, 1, 45–63.

Goonasekera, S. (2013) Buddhism in South Asia: Practicing tradition today, in Pechilis, K. and Raj, S.J. (eds) *South Asian Religions: Tradition and Today*, London and New York: Routledge, 101–42.

Gore, C. (2009) Religion in Africa, in Woodhead, L., Kawanami, H. and Partridge, C. (eds) *Religions in the Modern World: Traditions and Transformations*, Abingdon: Routledge, 2nd ed, 265–96.

Graveling, E. (2010) 'That is not religion, that is the gods': Ways of conceiving religious practices in rural Ghana, *Culture and Religion*, 11, 1, 31–50.

Haniffa, F. (2008) Piety as politics amongst Muslim women in contemporary Sri Lanka, *Modern Asian Studies*, 42, 2–3, 347–75.

Hardy, P. (1987) Islam in South Asia, in Eliade, M. and Adams, C.J. (eds) *The Encyclopedia of Religion*, New York: Macmillan, Vol. 7, 390–404.

Harvey, P. (2000) *An Introduction to Buddhist Ethics*, Cambridge: Cambridge University Press.

Hick, J. (1999) *The Fifth Dimension: An Exploration of the Spiritual Realm*, Oxford: One World.

Huq, M. (2008) Reading the Qur'an in Bangladesh: The politics of 'belief' among Islamist women *Modern Asian Studies*, 42, 2/3, 457–488.

Huq, S. and Rashid, S.F. (2008) Refashioning Islam: Elite women and piety in Bangladesh, *Contemporary Islam*, 2, 7–22.

Iliffe, J. (1995) *Africans: The History of a Continent*, Cambridge: Cambridge University Press.

Jaffrelot, C. (1993) Hindu nationalism: Strategic syncretism in ideology building, *Economic and Political Weekly*, 28, 12/13, 517–24.

Jamal, A. (2010) 'To enjoin virtues and restrain vice': Modernising discourses and engendered traditions in Pakistan's Jama'at-e-Islami, *Totalitarian Movements and Political Religions*, 11, 3–4, 327–40.

Jenkins, P. (2006) *The New Faces of Christianity: Believing the Bible in the Global South*, New York: Oxford University Press.

Kim, K. (2007) *Concepts of Development in the Christian Traditions: A Religions and Development Background Paper*, Birmingham: University of Birmingham, Religions and Development WP 16.

Kroessin, M.R. (2008) *Concepts of Development in 'Islam': A Review of Contemporary Literature and Practice*, Birmingham: Religions and Development WP 20.

Kujur, J.M. (2013) Tribal religions in India: Practicing tradition today, in Pechilis, K. and Raj, S.J. (eds) *South Asian Religions: Tradition and Today*, London and New York: Routledge, 101–42.

Kwok, P.-L. (2004) Feminist theology, Southern, in Scott, P. and Cavanaugh, W.T. (eds) *The Blackwell Companion to Political Theology*, Oxford: Wiley-Blackwell, 194–209.

Ladbury, S. and Khan, S. (2008) *Increased Religiosity among Women in Muslim Majority Countries*, Birmingham: Governance and Social Development Resource Centre, issues paper.

Larsen, K. (2014) Bodily selves: Identity and shared realities among humans and spirits in Zanzibar, *Journal of Religion in Africa*, 44, 5–27.

Levtzion, N. (1987) Islam in Sub-Saharan Africa, in Eliade, M. and Adams, C.J. (eds) *The Encyclopedia of Religion*, New York: Macmillan, Vol. 7, 344–57.

Loy, D.R. (2010) The karma of poverty: A Buddhist perspective, in Galston, W.A. and Hoffenberg, P.H. (eds) *Poverty and Morality: Religious and Secular Perspectives*, Cambridge: Cambridge University Press, 44–61.

Mahdi, H. (2006) *Gender and Citizenship: Hausa Women's Political Identity from the Caliphate to the Protectorate*, Göteborg: University of Göteborg, PhD dissertation.

Masquelier, A. (2008) Witchcraft, blood-sucking spirits and the demonization of Islam in Dogondoutchi, Niger, *Cahiers d'Etudes Africaines*, 48, 189/90, 131–60.

Mbacke, S. (2016) *The Mouride Order*, Washington, DC: Georgetown University, Berkeley Centre for Religion, Peace and World Affairs and World Faiths Development Dialogue.

Mbiti, J. (1991) *Introduction to African Religion*, London: Heinemann.

McGuire, M. (2008) *Lived Religion: Faith and Practice in Everyday Life*, Oxford: Oxford University Press.

Mulya, T.W. (2018) From divine instruction to human invention: The constitution of Indonesian Christian young people's sexual subjectivities through the dominant discourse of sexual morality, *Asian Studies Review*, 42, 1, 53–68.

Nadkarni, M.V. (2008) *Hinduism: A Gandhian Perspective*, New Delhi: Ane Books, 2nd ed.

Narayanan, V. (2013) Hinduism: Practicing tradition today, in Pechilis, K. and Raj, S.J. (eds) *South Asian Religions: Tradition and Today*, London and New York: Routledge, 39–72.

Obadare, E. (2016) The Muslim response to the Pentecostal surge in Nigeria: Prayer and the rise of charismatic Islam, *Journal of Religious and Political Practice*, 2, 1, 75–91.

Oliver, P. (2010) *World Faiths: An Introduction*, London: Hodder Education, 4th ed.

Parish, J. (1999) The dynamics of witchcraft and indigenous shrines among the Akan, *Africa*, 69, 3, 426–47.

Parrinder, G. (1974) *African Traditional Religion*, London: Sheldon Press.

Pew Forum (2006) *Spirit and Power: A 10-Country Study of Pentecostals*, Washington, DC: Pew Forum, http://pewforum.org/uploadedfiles/Orphan_Migrated_Content/pentocostals-08.pdf

Pew Forum (2010) *Tolerance and Tension: Islam and Christianity in Sub-Saharan Africa*, Washington, DC: Pew Forum, http://pewforum.org/uploadedFiles/Topics/Belief_and_Practices/sub-saharan-africa-full-report.pdf

Pew Research Center (2012) *The World's Muslims: Unity and Diversity*, Washington, DC: Pew Research Center, http://assets.pewresearch.org/wp-content/uploads/sites/11/2012/08/the-worlds-muslims-full-report.pdf

Pew Research Center (2013) *The World's Muslims: Religion, Politics and Society*, Washington, DC: Pew Research Center, http://assets.pewresearch.org/wp-content/uploads/sites/11/2013/04/worlds-muslims-religion-politics-society-full-report.pdf

Rahman, F. (1987) Islam: An overview, in Eliade, M. and Adams, C.J. (eds) *The Encyclopedia of Religion*, New York: Macmillan, Vol. 7, 303–22.

Riaz, A. (2013) The new Islamist public sphere in Bangladesh, *Global Change, Peace and Security*, 25, 3, 299–312.

Robinson, C. (2008) Islamic reform and modernities in South Asia, *Modern Asian Studies*, 42, 2–3, 259–81.

Ruffle, K.G. (2013) Islam in South Asia: Practising Islam today, in Pechilis, K. and Raj, S.J. (eds) *South Asian Religions: Tradition and Today*, London and New York: Routledge, 192–221.

Sarpong, S. (2017) Sweat and blood: Deific interventions in small-scale mining in Ghana, *Journal of Africa and Asia*, 52, 3, 346–62.

Schultz, D.E. (2008) (Re)turning to proper Muslim practice: Islamic moral renewal and women's conflicting assertions of Sunni identity in urban Mali, *Africa Today*, 54, 4, 21–43.

Schultz, D.E. (2011) *Muslims and the New Media in West Africa: Pathways to God*, Bloomington, IN: Indiana University Press.

Shackle, C. (2009) Sikhism, in Woodhead, L., Kawanami, H. and Partridge, C. (eds) (2009) *Religions in the Modern World: Traditions and Transformations*, Abingdon: Routledge, 2nd ed, 103–20.

Sieveking, N. (2007) 'We don't want equality, we want to be given our rights': Muslim women negotiating global development concepts in Senegal, *Africa Spectrum*, 42, 1, 29–48.

Smith, D. (2009) Hinduism, in Woodhead, L., Kawanami, H. and Partridge, C. (eds) (2009) *Religions in the Modern World: Traditions and Transformations*, Abingdon: Routledge, 2nd ed, 37–66.

Stivens, M. (2006) 'Family values' and Islamic revival: Gender, rights and state moral projects in Malaysia, *Women's Studies International Forum*, 29, 354–67.

Tamney, J.B. (1998) Animism, in Swatos, W.H. Jnr (ed) *Encylopedia of Religion and Society*, London and Altamira: Sage, 22–3.

Tatla, D.S. (2008) *Sikhism and Development: A Review*, Birmingham: University of Birmingham, Religions and Development WP 21.

Thapar, R. (1989) Imagined religious communities? Ancient history and the modern search for a Hindu identity, *Modern Asian Studies*, 23, 2, 209–31.

Tomalin, E. (2007) *Buddhism and Development: A Background Paper*, Birmingham: University of Birmingham, Religions and Development WP 18.

Tomalin, E. (2009) *Hinduism and International Development: Religions and Development Background Paper*, Birmingham: Religions and Development WP 19.

Tomalin, E. and Starkey, C. (2013) Buddhism and development, in Clarke, M. (ed) *Handbook of Research on Religion and Development*, Cheltenham: Edward Elgar, 31–50.

Tortora, B. (2010) *Witchcraft Believers in Sug-Saharan Africa Rate Lives Worse*, www.gallu.com/poll/142640/Witchcraft-Believers-Sub-Saharan-Africa-Rate-Lives-Worse.aspx?g_source=&g_medium=&g_campaign=tiles

Vásquez, M.A. (2009) The global portability of pneumatic Christianity: Comparing African and Latin American Pentecostalisms, *African Studies*, 68, 2, 273–86.

Waines, D. (2009) Islam, in Woodhead, L., Kawanami, H. and Partridge, C. (eds) *Religions in the Modern World: Traditions and Transformations*, London: Routledge, 2nd ed, 237–63.

Wiredu, K. (2011) An African religions perspective, in Meister, C. (ed) *The Oxford Handbook of Religious Diversity*, Oxford: Oxford University Press, online ed.

Wolfe, R.W. and Gudorf, C.E. (1999) Introduction, in Wolfe, R.W. and Gudorf, C.E. (eds) *Ethics and World Religions: Cross-Cultural Studies*, Maryknoll, NY: Orbis Books, 1–20.

4 Religious beliefs and people's lives

Death, destiny and well-being

Constantly faced with hardship and misfortune, people have often resorted to religion in a search for explanations and a means for overcoming their difficulties, personal and societal. To do so, they draw on both the official teachings of their own religious tradition and other religio-cultural beliefs to understand and give meaning to suffering, attribute evil and provide ways of combating its effects on their lives. Some of these were discussed in Chapter 3, especially the ideas of a transcendental reality, personified as an omniscient and compassionate but also judgemental deity (or deities) or conceived of as impersonal ultimate reality and other spiritual forces or entities which interact with human beings with neutral, benign or malign intentions. The ways in which religious beliefs provide adherents with understandings of evil, suffering and death are examined in more depth in this chapter, again drawing on the discourses offered by both religious texts and their interpreters and ordinary adherents of the main religious traditions in Sub-Saharan Africa and southern Asia. First, teachings and beliefs relating to heaven and hell will be discussed. Second, the question of whether people perceive their lives and experiences as being determined by religious factors or under their own control will be explored by examining teachings on destiny and fate and how people interpret those teachings in attempting to understand their own experiences. Finally, while the lives of many, especially in poor countries, seem to be predominantly characterised by hardship and misfortune, the quest for and experience of well-being and success is also important, and so in the last section of the chapter, alternative explanations for good fortune will be considered.

Dealing with evil, suffering and death: belief in heaven and hell

All religions offer 'salvation' as a way of dealing with evil, suffering and death if it is defined very broadly to include deliverance from harm or ill health, the attainment of harmony with universal laws, ultimate unity with some conception of the infinite and release from a cycle of suffering and rebirth. The belief that some form of human identity continues after death is fairly universal. Indeed, some argue that religion is itself a response to death. Whether that is the case or not, this belief has been the major means available to people to turn death into a positive event. Many religions offer a saviour figure, often said to be divine, who is thought to stand beyond death and offer a way through it. Funeral rituals effect the symbolic death of a material life and symbolic rebirth into an eternal life (Davies, 2006). The contradiction between the decay of the body and the hoped-for continuation of the individual is overcome by belief in an eternal soul, a resurrected body or a combination. Thus the body is often assumed to be powered by a soul, spirit or life force that departs at death and continues beyond it, perhaps as an ancestor in touch with the living, perhaps reincarnated on earth in a different body, or maybe living in a distant heaven.

There are concepts of heaven and hell in most religions, but they are most influential in the monotheisms, because if God is conceived of as all-powerful and all-good, then the question of how to understand and deal with evil becomes imperative. Even if God is considered to be all-powerful and all-knowing, actions that deliberately inflict suffering are deemed evil by most societies, and it is generally accepted that humans are a mixture of good and evil, implying that they have a degree of responsibility for their choices. No non-English words carry the exact meanings of heaven and hell. The ideas first emerged in the eighth century, initially meaning 'the sky' and 'the shadowy land of the dead', Christianised to 'a place of joy, where God and good spirits dwell, good humans go and where they are reunited with absolute reality and their loved ones; while hell is where evil spirits torment the wicked'. Both imply a life after death, usually permanent, and determined by the god(s), fate or the *karma* of one's previous lives (Russell, 2006). Heaven is conceived of as being in enduring joy, in God's presence, although views on which dimensions of a person's self (body, soul or both) this applies to differ. Hell, in contrast, is regarded as enduring misery and torment, where God is missing. A doctrine of hell is a way of reminding people of their fate if they stray too far from salvation on the terms offered by the religion – it is a place of punishment for the unworthy dead, those who violated the specified norms while alive.

Christianity centres on a belief in resurrection, related less to ethical judgement on individuals and more to faith in Jesus Christ, which is thought to free believers from sin and guarantee them an afterlife of joy in heaven. The Christian afterlife is thought to involve a new identity, as the old self is freed from sin and transformed by its new relationship with God, although what precisely lives on (the soul?) cannot be known. The counterposing of heaven and hell does, however, imply that the former is the destination for the obedient and the latter for the disobedient, leading to different views on the relative importance of God's grace and his judgement (Davies, 2006). Moreover, the idea of salvation as being solely concerned with heaven is complicated by the Pentecostal promise of earthly salvation (prosperity, healing and dominion over the occult) (Ukah, 2007).

Similar to Christianity, Islam emphasises obedience to God, and death is associated with judgement and either rewards or punishments in the afterlife (Davies, 2006). Individuals are believed to be judged according to their deeds in this world and their loyalty to the teachings of the Prophet, especially compassion and generosity, although Sufism may depart from the interpretation of the relationship between the individual and Allah through the idea of *baraqa* (divine grace) and the saint as a channel or mediator (see Chapter 3). While Islam focuses on the formation of a just society on earth, the Qur'an also asserts the resurrection of the body on the day of judgement, when the faithful are expected to enter heaven/paradise, coming either metaphorically or physically into the presence of Allah.

In Hinduism and Buddhism, as discussed in Chapter 3, one's position in life derives from one's *karma*, a kind of merit based on doing one's duty in previous lives. The potential to increase or decrease one's *karma* in order to have a better or worse rebirth implies judgement by, but also love for and from, the ultimate reality. For example, it is believed in Hinduism that the deity Krishna can enable devotees to escape *samsara* and enter into the realm of divine bliss. Love between the deity and the devotee, during life and after death, it is believed, goes beyond the merit resulting from obedience to religious rules. To inhabit an eternal realm, and because death can challenge an individual's identity and attachment, it is thought by some that a transformed body can be reunited with a person's 'self', provided appropriate rituals are performed (Davies, 2006). However, in these religious traditions, views about what exactly is reborn and seeks union with the divine differ: the human personality, the reborn person's consciousness, the divine image within each person or the universal *atman* in the depths of each person's being.

Surveys reveal that beliefs about death and the afterlife are widely held in SSA and southern Asia, amongst people with many different religious backgrounds, although such surveys do not provide any commentary on or explanations of the between-country (or between-religion) differences or explore what the beliefs might mean for people and their lives. For example, when asked whether the basic meaning of religion is "to make sense of life after death or to make sense of life in this world", respondents in the 11 relevant countries included in the 2010–14 wave of World Values Surveys had varied views – in four countries (Ghana, Nigeria, Malaysia, Pakistan), the majority of respondents considered that its meaning is to make sense of life after death; in five (India, the Philippines, Thailand, Rwanda, South Africa), the majority considered its meaning to be to make sense of life in this world; while in Zimbabwe respondents were evenly split, in Singapore the majority responded 'neither', and in India a quarter thought 'both'.[1]

In keeping with the teachings of Islam and Christianity, respondents in Pew Forum surveys expressed a belief that there will be a day of judgement when God will determine whether individuals are consigned to heaven or hell, with most believing that heaven/paradise awaits those who have lived 'righteous' lives

1 www.worldvaluessurvey.org/WVSOnline.jsp accessed 10/8/2017. The data are not available disaggregated according to the religion of the respondents.

and hell the fate of those who do not live righteously and do not repent (Pew Research Center, 2012). More than eight out of ten respondents in nearly every SSA country surveyed said that they believe in heaven (nearly all in Mali, Senegal, Nigeria, Ethiopia, Ghana and Kenya). Belief in hell is less common, although still 60%+ of respondents in every country (Pew Forum, 2010, p. 25). A majority of all Christians and an even higher proportion of renewalists hold this belief (Pew Forum, 2006). Figures from later surveys confirm that belief in heaven is almost universal. In Asia, this is also the case with respect to belief in hell. While fewer African respondents reported a belief in hell, the overall median was still 85% (Pew Research Center, 2012, p. 14).

Destiny and fate

What counts, in what happens both after death and during this life, is also subject to different interpretations and understandings. In the monotheistic religions and beyond, the understanding of God as all-knowing and all-powerful has given rise to ongoing debates about the apparent contradiction between destiny/fate and/or God's will as determining the lives of humans in this world and the next, influencing the scope for and obligation on individuals to exercise agency and thus their ability and responsibility to make choices. In practice, people's perceptions of the role of fate, God's will and their own ability to control their lives are complex, even muddled (Marranci, 2010). As well as the formal teachings of their religion, these perceptions depend on the context in which the issues are being considered and individuals' own position in life.

In Islam, "Religious language signifies the belief that man has little control over human destiny. The most popular formula is '*insh'a Allah (Inshallah)*/if Allah is willing'" (Tibi, 2008, p. 57), which denies the will of the self and reflects a belief that the destiny of individuals is in the hands of God (Pew Research Center, 2012). In addition, the articles of faith (*iman*), explained in a *hadith*, include belief in destiny (*qadar*), which includes a person's predestined position (*rizq*). Nobody, it is believed, has the ability to ascertain their destiny, not least the outcome of the Day of Judgement (*qiyama*) and the prospect of life after death (Kroessin, 2008, p. 18). Asked as a straightforward question, survey responses seem to indicate that the idea of predestination or fate is widely accepted by Muslims in southern Asia (Pakistan 91%, Indonesia 95%, Malaysia 89%, southern Thailand 85%, Bangladesh 74%) (Pew Research Center, 2012, p. 61).

However, both Muslim and Christian teaching is that people have free will and should take responsibility for their own lives and actions. Thus the "Christian emphasis on prayer and Bible reading [as well as deliverance and spiritual healing] defuses the fatalism inherent in a traditional system based on notions such as witchcraft, curses and the power of ancestors. Instead, Christians are taught to rely on faith, and on the role of the individual, who is no longer a slave to destiny or fate" (Jenkins, 2006, p. 100). The responses of Christians surveyed by the Pew Forum in selected countries demonstrate this sense that faith in God, backed up by other factors (such as one's parents' economic situation and supportive government policies) and accompanied by practical strategies, including education and hard work, are capable of outweighing fate. This view was expressed by the majority of respondents in most countries for which the data are available (81% Kenya, 65% South Africa, 66% the Philippines, 70% India, 45% Nigeria) (Pew Forum, 2006, p. 54).

Nevertheless, for many in the global South, life is so uncertain that it is impossible to plan: "One's fate and one's prosperity are in the hands of God" (Jenkins, 2006, p. 92). In the low-income settlement of Kibera, Nairobi, preachers in the Pentecostal churches Deacon studied often tell churchgoers that "there is no '*plan God has for you* [that] *cannot be changed*'" (Deacon, 2012, p. 671). In American Pentecostal churches, this implies that if a person receives the Holy Spirit and accepts that God is in control, s/he is able to fulfil his/her destiny, which is to be a wealthy and successful individual. However, in Kibera, it "seemed to allow people to view their circumstances as part of a plan for them rather than their own 'fault' . . . or sinful behaviour" (Deacon, 2012, p. 671), assuaging the damaging impacts of poverty, uncertainty and inequality. Freeman (2015) distinguishes between the poor, who, even if transformed into valued and agentic individuals, lack opportunities, and the middle class, who are able to invest in education and business activities to become prosperous.

In Hinduism, the idea of religion as *dharma* is arguably as central to understanding the tradition as any consideration of deities. *Dharma* brings together notions of individual paths in life and the sense of an underlying cosmo-moral order that structures them and enables them to be comprehended (White et al., 2011). *Samsara* (the endless cycle of rebirths) is a given for every person until s/he recognises the nature of the relationship between Brahman, the spiritual force of the universe, and *atman*, the individual self. Closely related to this belief and the concept of *dharma* is the notion of *karma*, which can include the work humans carry out, the actions they perform and the way they respond to other humans. "The central idea of karma is that if human beings behave unethically, they cannot avoid the consequences of their actions" (Oliver, 2010, p. 131; see also Smith, 2009). However, if they behave morally in a particular life, this will reduce the *karma* associated with a human self, eventually eliminating all *karma* and enabling the self to achieve salvation (*moksha*, or release from *samsara*).[2] *Karma* is then not fate or destiny (nor is it judgement by God). Rather, it is determined by a person's own actions – it urges human beings to use their relative freedom to move towards their own moral or material good through right conduct, fulfilling a proper role in life or living in accordance with *dharma*. Moreover, retribution for past actions can be lessened by the grace of God, selfless service to society, *bhakti* (devotion), adhering to moral norms, repentance and corrective action. *Karma* is inexorable, but it is regarded as reformative rather than retributive. Since desire induces attachment and attachment leads to the operation of the law of *karma*, acting without expectation of reward or fear of punishment and

2 The texts vary, and so do their interpretations, both over time and between sects and individuals.

the use of methods that can help break the link between desire and action (especially yoga) can provide liberation from bondage to the law of *karma* (Narayanan, 2013; Nadkarni, 2008).

Research in India and Bangladesh reveals both the centrality and the complexity of the concepts of *dharma* and *karma*. A study in rural and urban locations in Punjab and Orissa in India in 2008–9, for example, revealed that people regard 'religion', in the sense of a particular tradition or set of practices, as subordinate to a larger sense of the moral order. Respondents explained what they do, their duties and roles (*dharma*), in terms of this larger sense of what is right and appropriate for people like themselves. They believe that they must judge what the right *karma* is for themselves and act accordingly. Thus they are required to make an effort, although the outcomes of their actions are ultimately determined by God (White et al., 2011):

> Reflective choice, reviews of the state of one's life and intercessions to seek improvement all highlight the fact that, as much as our respondents may believe that everything is ultimately in the hands of a more powerful reality, the ability to choose and remain on the 'correct path' ultimately requires moral effort and human deliberation. God's grace may be paramount but conscious decision making and efforts are no less important . . . Destiny therefore is not a predetermined force but the result of people's actions, and not a predetermining force but an arena in which the prospects of a better future life can be established.
>
> (Devine and Deneulin, 2011, pp. 72–3)

This theory of moral causality is the basis for the idea of *karma*.

Even though 90% of Bangladesh's population are Muslim, ethnographic research in villages in two parts of the country found that the concept of *dharma* is central to the ways in which not only Hindus but also Muslims view the world and live their lives. Here again, *dharma* refers to 'the proper order of things' and is used to denote both 'religion' (religious affiliation and *dharma palon kora*, religious doctrines and practices) and the wider social and moral order of how things are and how they should be. The moral order is not confined to a separate 'religious' sphere but informs and structures everyday life. As well as prayers and rituals, it is manifest in conduct and behaviour, knowledge of what is right and wrong and social relationships within families and communities. Although not a Muslim term, it became so deeply rooted in Bengali Muslim culture between the thirteenth and eighteenth centuries that it provides the foundation of Bangladeshi Muslims' view of the world today (Devine and White, 2009). Within this moral order, outcomes are regarded as ultimately within the scope of God's gift, but there is a general acceptance of the need to make an effort oneself, with the poor being more likely to see what happens as due to fate than the middle or upper classes, because the latter have more resources and more control over their circumstances. Whatever actions are taken, the ultimate outcome, whether positive or negative, is put down to fate or God, the latter especially if the outcome is positive (White, 2012).

In Buddhism, the approximate equivalent of fate is that advanced in the First Noble Truth – that the human condition is characterised by suffering and misery, which is caused by desire for things external to the individual self, as well as anger, hatred, delusion and ignorance, as discussed in Chapter 3 (see also Cantwell and Kawanami, 2009; Goonasekera, 2013). However, the Buddha's teachings offer a diagnosis of this, explanations of its persistence and guidance to enable humans to gradually reduce their tendency to suffer (the Eightfold Path of right understanding, thought, speech, action, livelihood, effort, mindfulness and concentration), which can ultimately enable aspirants to achieve peace and tranquillity – *nibbana/nirvana* or enlightenment. Like Hinduism, Buddhist beliefs accommodate the theory of *kamma/karma* as a result of action, affecting certain aspects of life, including one's current rebirth, social class at birth, general character, the way an individual tends to see the world and certain good or bad things that happen to a person. However, this is not fatalism, as indicated in Bradley and Ramsay's study in two slums in the city of Pune, India (see Box 4.1). Buddhists seek to accumulate good and avoid accumulating bad *kamma/karma*. The Buddha taught that craving for things in the world is a source of *karma*, which 'binds' human beings to future rebirth and means that they continue to be reborn each time they die. *Punna/punya* (merit), obtained for example through giving alms, can improve one's chances of a better rebirth, but to escape *samsara*, human beings must abandon the belief that they are discrete individual 'selves'. In theory, Buddhists reject deity worship, ritual practices and extreme asceticism as means of achieving "soul purification and reaching *moksa*, their permanent resting place" (Goonasekera, 2013). In practice, they have accommodated many deity cults from other traditions and seek to propitiate these gods in return for benefits in mundane affairs, diluting the otherwise impossible stringencies of the Noble Eightfold Path for lay Buddhists.

Box 4.1 Hindus and Buddhists in the slums of Pune, Maharashtra

A short-term ethnographic study sought to explore how Hindu and Buddhist residents in the slums of Pune understand, relate to and challenge the world in which they live and the role played by religious beliefs, practices and values in their daily lives. Informants consistently made use of religious references in describing how they understand the world and their own circumstances, intertwined with their attitudes and responses to caste (Bhatewara and Bradley, 2013; Bradley and Ramsay, 2010).

Since the 1950s, conversion from Hinduism to Buddhism has been common among Dalits in Maharashtra, mainly because of the leadership provided by Dr B.R. Ambedkar, who in the late nineteenth century was born into the untouchable Mahar caste and spent his life challenging the dominant social processes sanctioning untouchability. A lawyer and influential political leader at the national as well as local levels, he continues to be revered by Dalits throughout India. His aim was to foster a sense of self-esteem and autonomy amongst so-called untouchables, who were subjected to economic, cultural, physical and psychological oppression, in particular by encouraging them, in his writings from the 1930s, to pursue education. Becoming convinced that caste discrimination was an integral part of Hinduism and, unlike Mahatma Gandhi, that reform was impossible, he turned to Buddhism, concluding that it was the most egalitarian of all the religions. In *The Buddha and His Dhamma* (1984[1957]), he presented his own interpretation of Buddhist teachings (Navayana Buddhism), which focuses on personal spirituality rather than dogma, ritual, religious organisation and hierarchy and draws out those aspects that emphasise the interconnectedness of all living beings, articulate an anti-Brahmin message and stress social equality, to provide a guide to his followers and enable them to pursue social action.

While *karma* is a core religious concept for both Hindus and Buddhists, therefore, "the Hindu understanding of *karma* guides Hindus toward acceptance of their lot in this life, whereas the Ambedkarite interpretation of the concept encourages a view of *karma* as enacted in a person's current lifetime, through removing the uncontrollable fatalistic version, [although] Both interpretations encourage the active pursuit of a better rebirth" (Bradley and Ramsay, 2010, p. 29). Rather than inner peace gained from meditative practices, this version of Buddhism provides a socially respectable language and identity that gives adherents hope and enables them to argue for social justice. Burra (1996) also notes the persistence of Hindu interpretations of *karma* and rebirth amongst Buddhist converts in the rural areas she studied. For Ambedkar and both urban and rural Mahar converts to Buddhism, therefore, liberation is as much to do with freeing Dalits from oppression as with spiritual liberation.

Most religions offer narratives for understanding and interpreting the natural world. These shape believers' attitudes, perceptions and practices in ways that are brought into sharp relief by disasters. 'Religion' is an important element in how people understand and explain natural phenomena, how they understand vulnerability to hazards and disasters and potentially how they respond to and recover from disasters and take steps to reduce future risks. Much of the relevant writing focuses on theodicy (attempts to reconcile the seemingly contradictory nature of God as loving, just and omniscient with the existence of evil and suffering)

(Joakim and White, 2015). Although some question the applicability of this to non-monotheistic religions with no creator God, such as Buddhism (Merli, 2010; see also Falk, 2010), around the world there is a long history of seeing disasters (whatever their physical explanation) as retribution from God, often for the general decline of a once morally well-ordered society. They are events that shake certainties and induce feelings of helplessness, guilt and fear of future unpredictable events. In practise, "general ways of living and thinking in relation to ideas about human life, divine justice, the agency of people and things, fate and purpose" (Simpson, 2011, p. 422) tend to be more important than the beliefs and prescriptions of formal organised religion.

Perceptions of and reactions to disasters have been researched in a number of SSA and southern Asian countries. The findings show that, although those affected often have at least a partial scientific understanding of a disaster event, this often co-exists alongside mythological, spiritual and religious interpretations and attributions of causality (Joakim and White, 2015). Some studies explore the implications of their findings for the extent and nature of resilience in the aftermath and preparation for future disasters. In addition, several examine the roles religious organisations play in response, recovery and risk reduction (for example, analyses of the aftermath of the tsunami in Aceh, Indonesia). A number of recent disasters in southern Asian countries have been extensively studied, but this discussion draws only on the most relevant accounts – those that consider religion – to illustrate what disasters and their aftermath reveal about people's beliefs.

Box 4.2 'Natural' disasters?

Relevant ethnographic accounts deal with earthquakes in Gujarat, India, in 2001 (Simpson, 2011) and Java, Indonesia, in 2006 (Joakim and White, 2015; Schlehe, 2010), as well as the tsunami that affected several countries around the Indian Ocean in 2004 (Falk, 2010; Merli, 2010). All the studies focused on selected case study areas or villages, combining semi-structured interviews with local people, community and religious leaders and representatives from various organisations involved, including faith-based organisations, NGOs and national and international agencies. During the years of recovery and reconstruction, Simpson (2011) reviewed newspaper reports from the immediate aftermath of the Gujarat earthquake and talked to many people he had known in Kutch before it happened. Joakim and White's 2013 research was in three villages in Bantul, the regency which suffered the heaviest impact of the 2006 earthquake in Java, which caused extensive damage in Yogyakarta and Central Java, especially the districts of Banten and Klaten. The figures vary, but it is estimated that the earthquake caused over 5,700 deaths, 40,000 to 60,000 injuries, the destruction of at least 150,000 buildings and damage to many more, including 3,000+ places of worship, leaving an estimated 1,345 communities without their place of worship. Schlehe (2010) researched the same earthquake in Yogyakarta and the surrounding area. In the countries affected by the tsunami, studies relevant to this discussion focus on southern Thailand. Merli (2010) carried out research in the four coastal districts of Satun, one of the four Muslim-majority provinces, largely among Muslims, while Falk (2010) worked among Buddhists in the coastal villages of Buddhist-majority Phang Nga, the worst-hit province, with a focus on recovery.

The 2001 earthquake in Gujarat, which killed about 14,000 people, was attributed by Simpson's informants to the wrongdoing of Man and God's displeasure, although they also acknowledged the geological explanation (Simpson, 2011). In particular, some respondents referred to the post-independence moral, ritual and political decline of the once-prominent Indian kingdom of Kutch and some Hindu respondents to the presence of Muslims, who they see as out of place, at odds with the design and direction of the Hindu gods. Moreover, both reformist and traditionalist Sunni Muslims interpreted the earthquake as a warning to other Sunni Muslims, especially those most adversely affected: reformist Muslims saw it as admonishing Muslims who do not adhere to their universalist form of Islam, while traditionalist Muslims regarded it as a signal to reformist Muslims to not neglect the religion's local roots, especially the saints, who traditionalists regard as "the conduits of God's love and intermediaries in divine intercession" (Simpson, 2011, p. 434). Although there were differences between Hindus and Muslims in the attribution of blame for the earthquake, Simpson noted that they shared the view that human beings must account to God for their actions and that "some group (usually one's own) was doing something against the prescribed order of things [for which] the earthquake served as a mechanism of corrective justice" (2011, p. 435).

There is no state religion in Indonesia, but every citizen must embrace one of the officially recognised religions (Islam, Protestantism, Catholicism, Buddhism, Hinduism and, more recently, Confucianism). However, nearly 90% of the Javanese population are Muslims. All the respondents in Joakim and White's research attributed the earthquake to the will of God, leading "some villagers to express the conviction that there was nothing that could or should be done to prepare for disasters" (p. 198), although the majority identified actions that could be taken to reduce the number of victims. In addition, many noted the importance of prayer and religious activities in preparing people emotionally and spiritually for future disasters. However, even if people prepare in this way, a disaster itself, they believed, will happen anyway if it is the will of God. The earthquake was interpreted as a warning to people to return to the glory and power of God, a trial to test people's faith and character and also, less commonly, a punishment for wrong deeds and actions.

Surprisingly, Joakim and White (2015) did not refer to indigenous religious explanations of the earthquake, despite the significance attached to these in Schlehe's analysis (2010). Many of the latter's respondents connected the event to the local myth of Ratu Kidul, the queen or goddess of the South Sea, ruler of all the spirits of Java, who is reputed to have promised the founder of a local Central Javan kingdom (later divided into two Sultanates) that she would protect him, his dynasty and realm against eruption of the nearby volcano, Merapi. Although the current Sultans do not support this myth, their courts regularly perform rituals connected with it. Following an eruption in 1994, increased volcanic activity preceded the 2006 earthquake, which this time affected the city of Yogyakarta, prompting many local people to interpret it as a reminder to the Sultan of his traditional role and responsibility to uphold harmony between the sea, the volcano and his court, and further, as a sign of the need to revive Javanese culture. While some criticised the Sultan himself for being too Muslim and too modern, others blamed modernity in general (globalisation, consumerism, moral decay, individualism, corruption and environmental exploitation), which is seen as a threat to Javanese values and cultural heritage. In this study, some identified Ratu Kidul as having sent the earthquake, others Allah, to remind people to live a moral life according to Islamic rules and to respect nature. Joakim and White and Schlehe's accounts of the earthquake show that there is variety within Islam in terms of whether Muslims proffer purely Islamic or syncretic explanations.

Differences in Muslim interpretations were also identified in Merli's (2010) study of the tsunami. Some of the latter's clerical informants attributed the disaster to God's retribution for human actions, referring to personal sin and moral pollution, at both the individual and social levels. Distinguishing between pious Muslims (those constantly aware of God in their everyday lives) and those who are Muslim only in name, the images used by these informants identified mosques and graveyards left standing near houses and undesirable buildings that had been destroyed. They interpreted moral pollution at the social level as *fitna*, a state of trouble, conflict and chaos. The earthquake that generated the tsunami was connected to oil drilling, seen as exploiting natural resources for greed and demonstrating moral corruption among the companies concerned. Just as Muslim clerics and politicians had interpreted the 1883 Krakatoa eruption and tsunami as a divine judgement on colonial rule, moral corruption associated with tourism in Aceh and elsewhere was advanced to explain why tourist areas were particularly badly affected by the 2004 tsunami. The Muslim requirement for personal cleansing with water before prayer rituals led some of the locals in the Satun coastal areas to interpret the vast amounts of water poured over 'dirty [unobservant, tourist] places' as cleansing in both a material and moral sense (Merli, 2010).

Religious explanations of the tsunami were also advanced by Buddhists. Most saw it as ultimately being caused by foolish human encroachment on the interconnectedness and natural equilibrium of nature (Falk, 2010). However, other explanations Merli recorded referred to local myths (e.g. the existence of a mythical creature living in the ocean depths, which causes colossal waves when it moves) and the idea that a natural cataclysm would announce the birth of a person with abnormally great amounts of religious merit, the next Buddha (Merli, 2010). People turned to monks (and sometimes nuns)[3] for explanations and to assist in recovery.

For Buddhists, funeral rites are the most important of the life cycle rituals because to be prepared for the transition from one existence to another is considered crucial for the next life. Death caused by accident or violence is particularly problematic. After the tsunami, many turned to temples for rituals and counselling and to listen to Buddhist teaching "that life is unpredictable and inevitably involves suffering of various kinds" (Falk, 2010, p. 99). Local monks, while focusing on *dhamma (dharma)* and *kamma (karma)*, said that they had encouraged people to seek practical help in the short term but had also tried to give people hope. Between death and cremation of a body, perceived as a dangerous liminal period before the dead person is believed to have transcended into a new existence, relatives and friends make merits for the deceased by presenting offerings to monks and temples. During this period, the spirit is regarded as still adhering to the body, but with the potential to be a malevolent ghost and wander among the living, so those dying unexpectedly, especially if young, are thought to need special rites. In southern Thailand during the aftermath of the tsunami, when many believed the areas affected to be infested with such ghosts, the number of *thambun* ceremonies increased markedly. In addition, many people were missing, so many families could not hold normal funeral rites. In response, previously rare or unknown 'counterfeit funerals' were conducted by some monks to enable families to move on with their lives. On the day of a cremation, it is common for sons and grandsons of the deceased to receive time-limited ordination as monks and novices, enabling them to grieve and to play an ongoing role in transferring merits to their dead relative. These ordination retreats were an important part of

3 There is no *bhikkuni* (fully ordained female order) in Thailand, only provision for women to enter the monastic order as white-robed *mae chi*, who live in men's monasteries, cannot be fully ordained and are mainly occupied in caring for the day-to-day needs of the monks, with little opportunity to study (Tomalin, 2006, 2007).

the recovery process for many, especially those with ambiguous loss i.e. a missing person but no body. Monks and nuns therefore played significant roles in explaining life and death, consoling people, performing ceremonies and restoring continuity and order (Falk, 2010).

In all the examples found, there was an element of fatalism in people's explanations of the disaster. While recognising the geological explanations of earthquakes and tsunamis, it was also common amongst Muslims, Hindus and Buddhists alike to attribute the events to human error in relation to divine intention, at both the collective and individual levels. What emerges from the studies, however, is that while the ways in which most people see the world are religious, in the sense that ideas about relationships between humans and the natural and divine worlds are central to the explanations advanced by adherents of all the religions, these are intertwined with explanations and responses that acknowledge social and political elements and dimensions, such as social inequalities and divisions and struggles for political power. Muslims, for example, often advanced both scientific and religious explanations for disasters, interpreting them as warnings and punishments for sin (not being 'good Muslims', individually and collectively). Such interpretations, Merli (2010) suggests, may help to enforce religious discipline with respect to observance of ritual and behavioural rules. However, explaining a catastrophe in terms of fate or divine retribution did not stop informants from attributing the lack of practical preparation in regions prone to earthquakes and tropical storms to socio-political factors.

Religion also played positive roles. For example, respondents in Joakim and White's study, like people in many other post-disaster settings, referred to their religious faith as a source of strength and comfort in dealing with the effects of the event, second only to family support, not least because it enabled people to see the deaths as part of God's plan, with those killed assured of a life in the hereafter. Local religious leaders, as well as encouraging their followers to increase their religious activity (often reflected in the performance/reinvention of traditional rituals and sometimes ritual innovation), provided emotional and psychological support for recovery from the trauma and were often more trusted sources of information and guidance than local government leaders. In addition, the analyses reveal how a disaster can enable different groups to speak about and attempt to advance their political and social interests.

Well-being and success: the fruit of faith or hard work?

Religious interpretations of God's will or fate are also relevant to beliefs about well-being, good fortune and success. Conceptions of and attitudes towards well-being will be explored further in Chapter 8, so the discussion here will focus on beliefs about how it may be secured.

Hindus and others in Punjab and Orissa talk about well-being in terms of satisfying material needs but also stress that good family relationships and community harmony are essential. They may or may not explicitly refer to religion – if they do, it is in terms of life being a blessing, a gift from God. Personal piety, doing one's duty and fulfilling one's responsibilities are seen as central to achieving well-being, although ultimately this is regarded as being in the hands of God (White et al., 2011). Thus good outcomes are attributed to divine intervention even when respondents accept that the primary explanation lies elsewhere (for example, successful medical treatment) – the explanations are supplementary and compatible, rather than competing (Devine and Deneulin, 2011).

While participants in Muslim revivalist movements expect piety to be rewarded, in the form of material benefits, more frequently well-being is described in emotional or spiritual terms. For example, the young educated devout Sunni Muslims in Blom's (2017, p. 125) study in Pakistan reported that their "rediscovery of Islam [made them] . . . feel 'relaxed', 'good', 'proud' or 'strong'". Surrendering to God, they added, had brought them inner peace (*sukun* in Urdu, as opposed to *pareshani* – worries or confusion), self-respect and a direction for their lives. Similarly, women associated with Al Huda in Islamabad reported that they felt peace (*sukun*) as a result of regular prayer (Ahmad, 2008).

The view that faith in God has to be backed up by effort to achieve well-being is particularly associated with the Pentecostal movement within Christianity and especially those within it who promote the 'prosperity gospel'. However, it is shared by many other churches and their members. The majority of Christians in the SSA and southern Asian countries surveyed by the Pew Forum believe that, in addition to good health and relief from sickness, God will grant material prosperity to those with enough faith (Pew Forum, 2006). Pentecostals are slightly more likely than non-renewalist Christians to see faith in God as a factor in economic success, as (or nearly as) important as hard work and education (Pew Forum, 2006, pp. 29, 53–4). This belief is underpinned by two key ideas: that God is a benevolent deity who wants believers in his son Jesus Christ to be socio-economically prosperous, as well as healthy in body and soul, and that believers can actively claim this-worldly abundance (Sharpe, 2013; see also Gifford, 1998, p. 39). These ideas are backed up by selected Old and New Testament texts that appear to promise rewards to devoted followers of Christ, especially the three-fold blessing promised in 3 John 2 (health, prosperity and salvation), even though for them (unlike fundamentalist Christians) literal interpretations of the Bible are less important than the ecstatic spoken word (Adogame, 2010). The main means to realise the promised rewards are believed to be unwavering faith that God will provide and making financial offerings (tithes) to their churches (10% or more of their gross monthly income), in the expectation that God will in return bless them by fulfilling their wishes. The idea that Christians can seek prosperity in the here and now is a belief shared by the mainstream churches, but while the latter suggest that success and security can only be obtained through secular means (hard work, thrift, wise investment and access to healthcare), the prosperity churches emphasise the role of prayer and godly behaviour and regard prosperity as a reward from God for the faithful (although this does not prevent them also offering practical advice, e.g. the wisdom of avoiding debt, and stressing the values of thrift, sobriety, self-restraint and responsibility) (Jenkins, 2006).

Ethnographic studies of Pentecostal groups reveal the complexity of navigating this terrain in practice. Deacon (2013), for example, argues that the poor Pentecostals he studied in Nairobi use their faith to make sense of their lives in the uncertain, threatening and ambiguous environment in which they live. The idea of personal responsibility for addressing life's challenges through more devout living, he suggests, restores a sense of agency to individuals, who believe that they can improve matters by doing something about their 'mistakes'. It also restores self-esteem and allows people to feel that they can change their situation without implying that the challenges they face will actually be overcome and their lives transformed by achieving good health and increased wealth.

Similarly, a 2008/9 study in Nsofu, a neighbourhood with many Pentecostal churches and fellowships in Zambia's Copperbelt, found that "leaders and laity alike expressed confidence that it was God's will for them to experience 'prosperity', a term usually used in English that they connected not only to economic advancement, but also to physical health, strong family ties, and spiritual development" (Haynes, 2012, p. 126). In the erratic mining economy on which residents depend for their livelihoods, the promised riches elude most. Rather than abandonment of the church, however, this leads to nuancing of the prosperity message, for example by expanding the definition to include intangible blessings, qualifying the speed at which the blessings can be expected to arrive and accepting that some people will advance more than others. This, Haynes suggests, gives rise to a sense of mutual obligation between the better and worse off at both the congregational and national levels, even though it is accepted that inequality will persist and grow.

Concluding comments

In this and the previous chapter, the nature and significance of religious beliefs for adherents of different religious traditions in Sub-Saharan African and southern Asian contexts has been explored by outlining the core teachings in each, presenting quantitative evidence on the prevalence of selected beliefs amongst adherents in contexts where religion continues to be important to most people and using ethnographic studies to illustrate how particular beliefs and teachings are interpreted in a variety of contexts. Inevitably, the outlines in Chapters 3 and 4 do not capture all the variations within faith traditions, whether these are denominational/sectarian or related to the spectrum of adherents' views on the existence and nature of spiritual reality, the meaning they attribute to religion and their perspectives on whether the tenets of a religion should be regarded as literally true or symbolic. However, some of the variations (for example between literalist and interpretative views of the texts, the depth of commitment to core teachings or the extent to which religio-cultural beliefs beyond the core teachings of a tradition/denomination are drawn upon) have started to emerge.

As recognised in the introduction to Part II, beliefs are more central to some religious traditions than others, so focusing on them, especially if they are assumed to be based on official doctrine and teachings, can give rise to a flawed understanding of their nature and role. The analysis has attempted to guard against the potential dangers by recognising the conceptual similarities and differences between the religious traditions and seeking out recent qualitative studies that seek to reveal how individuals and groups articulate and interpret their beliefs and the ways in which these influence day-to-day lives. Clearly, care must be taken in any attempt to generalise from such illustrative case studies, especially as they are only available for some religious traditions and denominations in selected contexts. Nevertheless, they provide an important counterpoint to the official representations of religious beliefs and the interpretations of religious scholars and leaders. They reinforce Lincoln's injunction to pay attention to whose voices are reflected in religious discourse and confirm that interpretations and practices can vary within religious traditions, both in a single and between different contexts.

While based on the view that all contemporary religious ideas are in some sense constructed, the analysis recognises that many beliefs are sincerely held. Most women and men hold ideas and beliefs not just because they have been taught them or grown up in a religious habitus in which they are taken for granted but also because they enable people to make sense of the world in which they live, in which material and spiritual dimensions are intertwined, supernatural forces (the ultimate reality, a supreme being, deities and both benevolent and malevolent spirits) appear to be actively present, and many people have what they interpret as religious experiences. Despite some significant differences between and within the religious traditions, their ideas about well-being, evil, suffering, death and fate have much in common. So does their conviction that people have agency – that personal action (engagement in ritual practices, Lincoln's second domain, in addition to social and political action) matters in achieving prosperity, combating evil, overcoming suffering, coming to terms with death and resisting fate, even if the efficacy of such action varies depending on the circumstances of the individual or group concerned as well as wider social, economic and political forces. For most people in most traditions, as suggested in the analytical framework presented in Chapter 2, religion has social and organisational as well as personal dimensions that interact in complex ways. For them, religion is much more than a series of beliefs – amongst other things, as already revealed in some of the examples given in this chapter, it involves engagement in religious practices, which play significant roles in the lives of many and interact with their beliefs to produce and reinforce their religious habitus. These are explored in the next chapter.

References

Adogame, A. (2010) Pentecostal and charismatic movements in a global perspective, in Turner, B.S. (ed) *The New Blackwell Companion to the Sociology of Religion*, Oxford: Wiley-Blackwell, 498–518.

Ahmad, S. (2008) Identity matters, culture wars: An account of Al-Huda (re)defining identity and reconfiguring culture in Pakistan, *Culture and Religion*, 9, 63–80.

Bhatewara, Z. and Bradley, T. (2013) 'The people know they need religion in order to develop': Religion's capacity to inspire people in Pune's slums, *European Journal of Development Research*, 25, 288–304.

Blom, A. (2017) Emotions and the microfoundations of religious activism: The bitter-sweet experiences of 'born again' Muslims in Pakistan, *The Indian Economic and Social History Review*, 54, 1, 123–45.

Bradley, T. and Ramsay, Z. (2010) *'The people know they need religion in order to develop': The Relationships between Hindu and Buddhist Religious Teachings, Values and Beliefs, and Visions of the Future in Pune, India*, Birmingham: University of Birmingham, Religions and Development WP 51.

Burra, N. (1996) Buddhism, conversion and identity: A case study of village Mahars, in Srinivas, M.N. (ed) *Caste: Its Twentieth Century Avatar*, New Delhi: Penguin, 152–73.

Cantwell, C. and Kawanami, H. (2009) Buddhism, in Woodhead, L., Kawanami, H. and Partridge, C. (eds) *Religions in the Modern World: Traditions and Transformations*, London: Routledge, 67–102.

Davies, D.J. (2006) Death and after life, in Segal, R.A. (ed) *The Blackwell Companion to the Study of Religion*, Oxford: Wiley-Blackwell, 229–37.

Deacon, G. (2012) Pentecostalism and development in Kibera informal settlement, Nairobi, *Development in Practice*, 22, 5–6, 663–74.

Deacon, G. (2013) Satan's snake and political violence in Kenya, in Moksnes, H. and Melin, M. (eds) *Faith in Civil Society: Religious Actors as Drivers of Change*, Uppsala: Centre for Sustainable Development, 158–63.

Devine, J. and Deneulin, S. (2011) Negotiating religion in everyday life: A critical exploration of the relationships between religion, choices and behaviour, *Culture and Religion*, 12, 1, 59–76.

Devine, J. and White, S. (2009) *Religion, Politics and the Everyday Moral Order in Bangladesh*, Birmingham: University of Birmingham, Religions and Development WP 40.

Falk, M.L. (2010) Recovery and Buddhist practices in the aftermath of the tsunami in Southern Thailand, *Religion*, 40, 2, 96–103.

Freeman, D. (2015) Pentecostalism and economic development in Sub-Saharan Africa, in Tomalin, E. (ed) *The Routledge Handbook of Religions and Global Development*, London: Routledge, 114–26.

Gifford, P. (1998) *African Christianity: Its Public Role*, London: Hurst & Co.

Goonasekera, S. (2013) Buddhism in South Asia: Practicing tradition today, in Pechilis, K. and Raj, S.J. (eds) *South Asian Religions: Tradition and Today*, London and New York: Routledge, 101–42.

Haynes, N. (2012) Pentecostalism and the morality of money: Prosperity, inequality and religious sociality on the Zambian Copperbelt, *Journal of the Royal Anthropological Institute*, 18, 123–39.

Jenkins, P. (2006) *The New Faces of Christianity: Believing the Bible in the Global South*, New York: Oxford University Press.

Joakim, E.P. and White, R.S. (2015) Exploring the impact of religious beliefs, leadership and networks on response and recovery of disaster-affected populations: A case study from Indonesia, *Journal of Contemporary Religion*, 30, 2, 193–212.

Kroessin, M.R. (2008) *Concepts of Development in 'Islam': A Review of Contemporary Literature and Practice*, Birmingham: Religions and Development WP 20.

Marranci, G. (2010) Sociology and anthropology of Islam: A critical debate, in Turner, B.S. (ed) *The New Blackwell Companion to the Sociology of Religion*, Oxford: Wiley-Blackwell, 364–87.

Merli, C. (2010) Context-bound Islamic theodicies: The tsunami as supernatural retribution vs natural catastrophe in Southern Thailand, *Religion*, 40, 2, 104–11.

Nadkarni, M.V. (2008) *Hinduism: A Gandhian Perspective*, New Delhi: Ane Books, 2nd ed.

Narayanan, V. (2013) Hinduism: Practicing tradition today, in Pechilis, K. and Raj, S.J. (eds) *South Asian Religions: Tradition and Today*, London and New York: Routledge, 39–72.

Oliver, P. (2010) *World Faiths: An Introduction*, London: Hodder Education, 4th ed.

Pew Forum (2006) *Spirit and Power: A 10-Country Study of Pentecostals*, Washington, DC: Pew Forum, http://pewforum.org/uploadedfiles/Orphan_Migrated_Content/pentocostals-08.pdf

Pew Forum (2010) *Tolerance and Tension: Islam and Christianity in Sub-Saharan Africa*, Washington, DC: Pew Forum, http://pewforum.org/uploadedFiles/Topics/Belief_and_Practices/sub-saharan-africa-full-report.pdf

Pew Research Center (2012) *The World's Muslims: Unity and Diversity*, Washington, DC: Pew Research Center, http://assets.pewresearch.org/wp-content/uploads/sites/11/2012/08/the-worlds-muslims-full-report.pdf

Russell, J.B. (2006) Heaven and hell, in Segal, R.A. (ed) *The Blackwell Companion to the Study of Religion*, Oxford: Wiley-Blackwell, 271–84.

Schlehe, J. (2010) Anthropology of religion: Disasters and the representations of tradition and modernity, *Religion*, 40, 2, 112–20.

Sharpe, M. (2013) Name it and claim it: Prosperity Gospel and the global Pentecostal reformation, in Clarke, M. (ed) *Handbook of Research on Religion and Development*, Cheltenham: Edward Elgar, 164–81.

Simpson, E. (2011) Blame narratives and religious reason in the aftermath of the Gujarat earthquake 2001, *Journal of South Asian Studies*, 34, 3, 421–38.

Smith, D. (2009) Hinduism, in Woodhead, L., Kawanami, H. and Partridge, C. (eds) (2009) *Religions in the Modern World: Traditions and Transformations*, Abingdon: Routledge, 2nd ed, 37–66.

Tibi, B. (2008) *Islam's Predicament with Modernity: Religious Reform and Cultural Change*, London and New York: Routledge.

Tomalin, E. (2006) The Thai Bhikkuni movement and women's empowerment, *Gender and Development*, 14, 3, 385–97.

Tomalin, E. (2007) *Buddhism and Development: A Background Paper*, Birmingham: University of Birmingham, Religions and Development WP 18.

Ukah, A. (2007) *African Christianities: Features, Promises and Problems*, Mainz: Johannes Guttenberg Universität, Institute für Ethnologie und Afrikastudien, WP 79.

White, S.C. (2012) Beyond the paradox: Religion, family and modernity in contemporary Bangladesh, *Modern Asian Studies*, 46, 5, 1429–58.

White, S.C., Devine, J., Jha, S. and Gaines, S. (2011) *Religion, Development and Wellbeing in India*, Birmingham: University of Birmingham, Religions and Development WP 54.

Part III
Religious practices

Beyond the religious discourse, Lincoln suggests, a second domain of practices that embody the religious discourse and render it operational should be examined. While he groups ritual and ethical practices under this heading, I have already suggested that it is analytically helpful to consider them separately. Adherents engage in religious or spiritual practices to express and reinforce their beliefs. In addition, more broadly, their knowledge of and participation in religious practices demonstrates their conception of religion, the religious tradition or group with which they associate themselves, and their understanding of society, the world and their place in it, requiring consideration of all the dimensions of the social field and more than one level of analysis, as suggested in the analytical framework. Participation in religious practices, whether routine or occasional, individual or collective, plays an important part in many people's lives.

Such practices may be categorised in various ways, according to their purpose(s) and characteristics, although many have more than one purpose, and so any such categorisation is simplified. For example, Wuthnow (2001) and Riesebrodt (2006) suggest that it is possible to distinguish between 'spiritual' and 'devotional' practices. Wuthnow defines the former as "those activities in which individuals engage to become more aware of their spirituality or to enrich . . . their spiritual lives" (p. 309). They include participation in rituals such as prayer, meditation, chanting, song and reading religious texts. He defines devotional practices as "activities performed for the purpose of focusing one's attention on the divine . . . usually including efforts to communicate petitions or expressions of worship" (p. 313). Riesebrodt has a particularly instrumental view of devotional practices (prayer, divination and sacrifice), labelling them 'interventive' practices that seek to communicate and exchange gifts and services with a divine power, to manipulate the power or empower the devotee. Both distinguish these purposes from practices that are related to a person's perception of and relationship with the sacred, including those that encourage or enforce compliance with religious rules, norms and moral obligations (with respect to, for example, personal behaviour, providing service or extending hospitality).

Riesebrodt (2006, p. 603) also suggests that it is useful to distinguish between routine practices, which regularly cultivate and affirm adherents' relationships with a superhuman power or ultimate reality; occasional practices linked to life cycle and other events, such as illness, misfortune, disaster or conflict; and virtuoso practices of contemplative or ascetic discipline that seek to tackle the human condition of misery and suffering or to enable the practitioner to come closer to the divine. In this book, I am concerned with the everyday/quotidian, so the latter will not be considered further. Neither writer identifies the ways in which practices enable participants to make sense of their religious and daily experiences (their meaning-making function) or the roles they play in constructing and reinforcing a sense of identity and thus a sense of social cohesion.

Religious practices are underlain by ideas of the sacred, although perceptions differ with respect to whether they are innately/intrinsically sacred because of their association with the divine or supernatural, or *denoted* sacred by adherents. They involve times, objects and places that may be selected *because* they are perceived to be sacred or are *denoted* sacred because of their association with particular religious practices (Kong, 2010). They may also take place in mundane spaces, for example, the widespread practice in Sub-Saharan Africa of opening and closing meetings in secular locations (under trees, in schools and other community buildings, in government offices) with prayer.

They are one way in which religion is embodied. Indeed, it is increasingly recognised that "One cannot adequately understand religious conversion, meanings, rituals, disciplines or communities without attending closely to the handling or behaviors of bodies. . . . Religious beliefs are inscribed into and, in part, govern the behaviors of bodies. And bodies partly incarnate religious beliefs and enact religious practices" (Smith, 2008, p. 1565). All religious practices involve the body: sensations of hearing, sight, touch and taste and bodily postures and movements provide ways of focusing attention, in order to heighten spiritual focus and evoke religious experience. They are also all concerned, to a greater or lesser extent, with moral control of the body and its impulses, through regulating comportment, actions and appearance (e.g. by specifying appropriate dress). Further, they all seek to provide ways of addressing human material concerns, especially health and fertility. All of these are strongly gendered. The notion of embodied religion recognises the mutual constitution of minds, emotions and bodies in religious lives. It is important in understanding religious identity and is a corrective to excessive concentration on religious knowledge and belief.

Often, religious practices are subsumed under the heading of 'ritual', and indeed most of them can be regarded as ritual behaviour. The most common ritual practices are probably prayer and sacrifices/offerings, which may be the basis for other practices, such as religious services, processions or pilgrimages (Collins, 2009). The demands placed on adherents by various rituals differ: some are performed daily, others only on special occasions; some entail minimal amounts of time, others extensive time commitments; some have no financial implications, others require significant expenditure. It is impossible here to provide a comprehensive account of all the significant, regular, life cycle and special rituals associated with the main religious traditions in SSA and southern Asia, let alone their inter-contextual and intra-religious variations. Instead a flavour of those associated with each tradition is provided, with a focus on common and/

or regular religious practices, divided roughly into everyday and occasional practices and events. The former are considered in Chapter 5, which examines the ritual practices associated with each of the main religious traditions and ways in which people seek to communicate with spiritual entities. Some important occasional practices, events and celebrations (fasting, feasting and pilgrimage) are discussed in Chapter 6. However, 'ritual' does not necessarily capture some practices that are important in African and southern Asian religious life: religious giving, socialisation into a religion through learning and education and the transmission of religious traditions, beliefs and practices through proselytism are also considered in this chapter.

Frequency of participation in religious practices has often been used to gauge religiosity or religious commitment, not least because it is relatively easy to obtain the basic data through opinion polls or questionnaire surveys. The available data are used in Chapters 5 and 6. However, polls and surveys tell us little about participants' motives, the ways in which they engage, the meanings they attribute to particular practices or the functions various rituals fulfil for individuals and social groups. The significance of participation will therefore be illustrated by reference to recent qualitative and ethnographic studies in relevant locations.

References

Collins, P. (2009) Religion and ritual: A multi-perspectival approach, in Clarke, P.B. (ed) *The Oxford Handbook of the Sociology of Religion*, Oxford: Oxford University Press, online ed.

Kong, L. (2010) Global shifts, theoretical shifts: changing geographies of religion, *Progress in Human Geography*, 34, 6, 755–76.

Riesebrodt, M. (2006) Religion in global perspective, in Juergensmeyer, M. (ed) *The Oxford Handbook of Global Religions*, Oxford: Oxford University Press, 597–609.

Smith, C. (2008) Future directions in the sociology of religion, *Social Forces*, 86, 4, 1561–89.

Wuthnow, R. (2001) Spirituality and spiritual practice, in Fenn, R.K. (ed) (2001) *The Blackwell Companion to Sociology of Religion*, Oxford: Wiley-Blackwell, 306–20.

5 The role of ritual

Introduction

Religious practices are important both for many adherents and for religious organisations. In this chapter the focus is on rituals, which occupy considerable time in the lives of many people, are a major occupation for religious specialists, often consume significant material resources and have spiritual, doctrinal and emotional significance for both participants and officiants. The emphasis is on rituals typical of the major faith traditions in Sub-Saharan Africa and southern Asia. It briefly describes the ritual requirements of each faith tradition and then reviews some evidence from quantitative and qualitative research to examine the form and significance of some of the key rituals in which adherents participate during the course of their daily lives. Definitions and typologies of ritual abound (for example Coleman, 2010; Collins, 2009; McGuire, 2008; Segal, 2005; Wolfe and Gudorf, 1999). Furseth and Repstad's general definition is as useful as any: "Rituals are standardised interaction patterns with symbolic content that are carried out in specific situations" (2006, p. 128).

No practices are inherently religious – they acquire a religious character by being connected to a discourse that constitutes them as such. However, much ritual has evolved alongside religious discourses, institutions and organisations. In this context, the term is applied to a wide range of actions distinct from mundane practices, which are stipulated by religious texts or specialists and repeated. If repetition leads to religious practices becoming natural and habitual, they may seem to need no further explanation, justification or legitimation, and the key question for participants becomes 'have we got it right?' (Collins, 2009). However, a view of rituals as consisting entirely of a series of stipulated activities implies that participants focus their attention on performance rather than the basis for and meaning of those activities. While critics from within and outside the religious traditions have often dismissed religious rituals on this basis, many observers argue that adherents do have some agency in deciding whether or not to participate in rituals and/or analyse their symbolism and meaning. The latter are generally ambiguous, accommodate different interpretations and may change over time. Symbols condense different associations or meanings in a single object. They are important in terms of the senses, arousing desires and feelings and marking both the links and the differences between the spiritual and mundane. They may also articulate and embody meanings related to the moral and social order, including providing a means of expressing commitment to that order and/or a way of expressing (and handling) rivalry and disagreement (Coleman, 2010). Rituals therefore have expressive, performative and communicative as well as symbolic and physical dimensions. They may be prescribed, formalised and patterned or informal, and they may be public and communal or private, even secretive (Collins, 2009).

Commonly, rituals are seen as fulfilling a variety of functions beyond linking the experience of ordinary life with the sacred. For example, they may provide a means of reiterating religious beliefs or showing devotion to the gods, explaining and dealing with natural and social phenomena and marking key stages in the lives of individuals, families and communities (Tamney, 1998; Wolfe and Gudorf, 1999). The purposes of rituals may include purification; preparation for an important event or task; communicating, petitioning or thanking a spiritual being; achieving union with a spiritual sphere; or the specification and proscription of unacceptable behaviour. Some are primarily concerned with the links between an individual and the divine, for example Hindu *puja* or Muslim prayers, and may be performed at home, especially by women. Others enable encounters and the forging or consolidation of links with co-religionists, for example, participation in church services or mosque prayers. They may be performed to deal with change (Segal, 2005); control unpredictable events (e.g. the weather); honour and seek help and guidance from spiritual powers or beings such as God(s) or ancestors; propitiate or placate antagonistic forces (e.g. evil spirits, vengeful deities); or acknowledge and alleviate guilt (Segal, 2005). The question then becomes 'does it work?' (Collins, 2009; McGuire, 2008).

Rituals generally involve religious specialists, special objects or places, and practices/actions such as recitation, singing, dance, processions, sacrifices and the manipulation of sacred objects (Collins, 2009). They may seek to produce bodily sensations that suggest to participants that the ritual is real and not just imaginary (McGuire, 2008). Rituals associated with proscriptions and taboos may relate to purification, sexuality, fertility, appearance, speech and mourning. Sometimes these taboos are regarded as necessary in preparation for participation in rituals and sometimes they are a means of enforcing behavioural discipline. In SSA and southern Asia, rites of passage to mark births, transitions from puberty to adulthood, marriages and deaths generally have religious significance and content. Typically, many such rituals involve pre-liminal separation from the mundane world, a liminal transition and post-liminal re-integration into society, in which a transformed initiate returns to his/her community with a new status and responsibilities (Collins, 2009; Coleman, 2010; Furseth and Repstad, 2006). Ritual appears to have played a major role in every society historically and continues to do so today, with observers suggesting that, in secularising societies, 'non-religious' rituals have evolved to fill the human need for such practices. The term embraces a very wide, elaborate and varied range of practices, which differ not only between religious traditions and denominations/sects but also between social and cultural contexts. Here, some of those regarded as important in official religious teachings and the lives of adherents in relevant contexts will be considered, although clearly these represent only a small selection of those that are regularly performed around the world. In particular, although life cycle rituals related to birth, coming of age, marriage and death are important in the lives of most individuals and communities, limitations on space preclude full consideration of them here.

Engagement in rituals may differ between people and groups with different socio-economic characteristics, for example level of education or wealth. One of the axes of social difference which has received most attention is gender. In the West, women tend to engage in religious practices more frequently than men, a pattern that is attributed to both social influences and their personal characteristics (Furseth and Repstad, 2006). For example, gender-role socialisation theories argue that women are taught to be more submissive, passive, obedient and nurturing, traits that are supposedly associated with religiousness. The personal psychological characteristics that supposedly foster religiosity may be referred to as a 'feminine orientation'. It is also argued that women's lower workforce participation leaves them with more responsibility and time for childcare, nurturing their families and participation in daily rituals. All these explanations have been dismissed by critics, who focus instead on whether and how religions mobilise female support, for example by proscribing the participation of all or some (e.g. menstruating) women in particular rituals or their presence in religious spaces or, in contrast, offering women equality and the potential for exercising leadership roles. The data presented in Chapter 1 reveal that, although there are differences between male and female religious participation in some countries in SSA and southern Asia, there are few differences where a large majority of people profess religious adherence and claim that religion is important to them. Where the data permit, differences between social groups, particularly men and women, will be considered in the analysis that follows.

Everyday rituals in Islam

Reference has already been made to the Five Pillars of Islam, religious practices essential to Sunni Islam, while Shi'a Muslims subscribe to eight ritual practices that substantially overlap with the Five Pillars. The first is the central tenet of the religion (see Chapter 3), which is repeated in prayer, *salah/salat*,[1] or ritual worship, the Second Pillar, which has to be performed five times a day (before sunrise, at midday, mid-afternoon, at sunset and after nightfall), although Shi'a Muslims permit the noon and afternoon prayers and the evening and night prayers to be joined. Each *salah* is done facing towards the *Kaaba* in Makkah/Mecca, following a specific ritual set by the Prophet and consisting of the recitation of verses from the Qur'an in Arabic, bowing and prostrating. *Salah* is primarily an opportunity

1 In South Asia, more commonly called *namaz* (from the Persian) (Ruffle, 2013).

for Muslims to focus their minds on God, an expression of gratitude and worship. It is not petitionary or intercessory prayer that requests favours from God, which is referred to as *dua*. *Ihsan* refers to a Muslim's responsibility to attain perfection or excellence in worship.

Further guidance on religious practices is provided by the Shar'ia, which deals with two main dimensions of life – rules governing the acts of worship, including ritual purification (*wudu*), worship/prayers (*salah*), religious giving (*zakat*), fasts (*sawm*) and pilgrimage to Makkah (*hajj*) and guidance for behaviour (Oliver, 2010). The third, fourth and fifth of these refer to the Third, Fourth and Fifth Pillars and will be returned to below. *Salah* must be preceded by ablutions so that the worshipper is ritually pure. Performance in a mosque is recommended, especially for the Friday noon congregational worship, during which there is normally a sermon delivered by the *imam/alim* (prayer leader) of the mosque. This is generally related to a few verses from the Qur'an but also provides an opportunity for messages related to other subjects to be conveyed. If prayers are performed outside a mosque, the worshipper generally spreads a prayer mat to ensure ritual purity. If a number of people are worshipping together, one stands in front as leader to give the timing of the specified actions so they are made in unison. The basic content is prescribed, although additional pious expressions may be used. Worshippers are summoned by a muezzin calling from the minaret of a mosque, sometimes amplified, which may become a source of contention in religiously mixed neighbourhoods. Participation in mosque prayer is mandatory for men but not always for women. Often women are required to perform their prayers at home – where they do pray in mosques, they are generally segregated from men, sometimes in a women's section of the building. Recently, some women's mosques have been established, for example in the North East of India and Tamil Nadu (Ruffle, 2013). The proper performance of the specified rituals is important in Islam, often leading to controversy over what is 'correct'. Practice in a West African city is described in Box 5.1.

Box 5.1 Everyday Muslim ritual in Nigeria

In the largely Muslim city of Jos, in northern Nigeria, there are 25 Friday mosques for prayers and the delivery of sermons, attracting congregations of between 50 and 2,000 or more in the three largest. In addition, there are hundreds of other mosques for daily prayers. Since the 1980s, they often have loudspeakers that blare out prayers and recorded sermons (as do hawkers of cassette tapes etc.) (Dugbazah et al., 2009; Hassan, 2010). Although teaching on mosque attendance by women varies, the norm is that they do not attend mosques, with some exceptions. For example, elderly women may attend Friday prayers, female students may pray in college and university mosques and female commuters may pray in public, always in rows behind the men and screened by brick walls or cloth screens. Although a few mosques in Jos have been built by or are on land donated by women, and women make donations towards the cost of construction, they do not currently participate in mosque committees. Many male informants said that they do not object to women praying in mosques, provided men and women are segregated and women screened, although most oppose the idea of women leading prayers (except perhaps in a women-only congregation). In contrast, while accepting the norm that women perform *salah* at home, many Western-educated women respondents advocated a more prominent place, voice and role for women in mosques, citing historic Muslim practices to support this. For example, the then-president of the Plateau State chapter of FOMWAN (Federation of Muslim Women's Associations of Nigeria) said that she believes that women should be allowed reserved areas in mosques and should be represented in mosque committees, quoting the example of a woman who challenged the second caliph, Umar Ibn Khattab, as described in a *hadith*. Some men agree that women could make a contribution as members of committees concerned with mosque construction, improvement and maintenance.

In the early days of Islam in South Asia, many Hindus found the religion appealing because of its message of social equality, and many low-caste Hindus converted. Muslim ritual and devotional practices were often combined with existing texts and practices, forging the Sufi strand of Islam, with its traditions of saint veneration, ecstatic devotional practices and integration of Vaishnava meditative and devotional traditions with Muslim religious practices (Ruffle, 2013). For example, Werbner (2014) provides a case study of a Sufi lodge in Pakistan founded by a contemporary saint, Zindapir (d. 1999), who has followers from across the social spectrum. Similar to many earlier saints, Zindapir's asceticism was seen by his followers "to effect a bodily ritual transformation, suffusing the saint with light and making him a conduit to God's blessings in the world" (Werbner, 2014, p. 73). At the lodge, ritual practices include *zikr* (the reiteration of God's name or Qur'anic verses), prayer, animal sacrifice, food or money offerings, *langar* (blessed food distributed to visitors), healing, exorcism of *jinns* and elaborate rituals during the *'urs'* (annual commemoration of the unification of the saint with God at the moment of his mortal death, including supplication by the saint to God on behalf of the order). Werbner asserts that many of the rituals and practices associated with contemporary Sufism, including the veneration of saints, are in conformity with orthodox Islam (2014, p. 64), although many Muslims would not agree.

Sufi Muslim practices have spread across the world from South Asia and are now widespread in SSA. For example, in Senegal, members of the Mouride Sufi brotherhood are trained to be able to forego the pleasures of worldly life through prayer, fasting and meditation, as well as physical work (Mbacke, 2016). *Marabouts*, spiritual leaders of groups of disciples (*daara*), bind young male followers to them by teaching, order the work they undertake (traditionally agricultural work on the *marabout*'s land) and make amulets which are thought to help solve problems and improve the lives of those who wear them.

In Tanzania, *dhikr* is the recital of praises to the Sufi scholar Abdul Qadir Jilani, normally accompanied by the recitation of *maulid* (verses to commemorate the birth of the Prophet), during the fourth month of the Muslim calendar. Men recite the *maulid*, while women join in the choruses. The former can move into a trance, when they are said to be connected spiritually to Jilani, while women are less likely to do so. The recitations may be accompanied by a sermon similar to a Friday mosque sermon. Fieldwork revealed that in contemporary Tanzania these sermons often contain social messages, including condemnations of extravagant wedding celebrations, criticism of savings and credit associations because they charge interest on the loans they make to members, calls for more cooperation between rival Muslim organisations and the attribution of blame to Western culture for corrupting the morals of young people (Mesaki and Kipacha, 2008).

A minority of Muslims in South Asia are Shi'i, who perform additional rituals. For example, on the tenth day of Muharram, the first month in the Islamic lunar calendar, the Shi'i commemoration of the martyrdom of the grandson of the Prophet and the third Imam Husain at Karbala in Iraq in 680 reaches its peak after a period of mourning lasting two months and eight days. During this period, men and women visit religious buildings containing ritual objects, seek the intercession of saints during visits to the shrines containing their tombs and gather daily to listen to narratives describing the suffering and heroic feats of Imam Husain and his family (Ruffle, 2013). Men and boys perform self-flagellation in mosques and processions as a sign of their loyalty to Islam, Imam Husain and his family. The hagiographic narratives describing the lives and extolling the spirituality and bravery of Imam Husain's female family members hold them up as inspiring examples. Women play a central role in the ritual performance of the devotional mourning poems and have also constructed their own religious buildings for rituals and devotional prayers. By the 1980s in Northwest Pakistan, even single women had started to play a role in these ritual performances (*majales*), which provide spaces for public performances and preaching, contributing to sociability and women's self-confidence. However, they have been curtailed more recently by the increasing influence of Sunni ideologies of female seclusion and submissiveness (Hegland, 2003).

Available data on Muslim populations demonstrate that, in most countries in SSA and southern Asia with significant Muslim populations, the vast majority of Muslims claim to comply with the requirements, stating that they pray at least once a day (75% on average in the 14 relevant countries for which data are available, 3 in southern Asia and 11 in SSA, ranging from as many as 95% in Cameroon and 94% in Ghana to half or less in 4 countries). The vast majority of those who claim that they pray daily stated that they pray five times a day (Pew Research Center, 2012, p. 43). In the countries covered by the Pew Research

Center (2016), men and women are equally likely to claim that they pray regularly (72% of men and 71% of women at least once a day and 56% of men and 57% of women five times a day).[2] The analysis does not discuss their motivation or the significance to them of regular compliance. In Southeast Asia, there are indications that older (35+) people tend to be more observant, but the data do not show this in SSA.

There are, however, gender differences between countries with respect to the proportions of men and women who attend mosque prayers. In most countries, almost all Muslim men (95% or more) claim to attend Friday prayers, and the same is true for women in many African and some Asian countries. However, different local religio-cultural traditions are evident in the lower attendance rates of women in some countries: 77% of women in Bangladesh, 55% in Pakistan, 26% in Senegal, 18% in Cameroon, 16% in Niger and 13% in Chad reported that they never attend Friday prayers (Pew Research Center, 2012, p. 48) (also 12% in Sri Lanka, where Muslims are in a minority, Asia Foundation, 2012, p. 12). This pattern is borne out by the results of surveys in 39 countries where Muslims are the largest religious group: 70% of men and 42% of women report attending weekly prayers, but there is relatively little difference between men and women in some African countries and a much greater difference than the average in others, especially in Southern Asia (Pew Research Center, 2016, p. 24).

The availability and accessibility of places of worship is important in all religions, especially where participation in rituals is expected to be regular and frequent, mobility is limited by the availability and cost of transport and women's movement is limited by their domestic responsibilities and modesty requirements. It may influence people's choice of where to live and changes in the religious composition of residential areas (especially urban neighbourhoods) over time. For example, the largely middle-class residents in Zakir Nagar, a predominantly Muslim migrant *mohalla* (neighbourhood) in south Delhi, which has become more religiously homogeneous over time, value the Muslim character of the area. Their reasons include the presence of religious facilities (mosques and schools); a shared and familiar culture; a sense of community, comfort and belonging; the ability to perform rituals (e.g. relating to a death); and not standing out if they choose to wear Muslim dress in a context where it is reportedly becoming increasingly difficult and sometimes risky to do so. In contrast, poor residents had chosen the area because of the availability of work or relatives who could help them find housing and employment. Kirmani's informants explained that the Muslim ambience of the neighbourhood contributes to their ability to

2 There is evidence from North America that reported church attendance exceeds actual attendance. To assess whether the same might be the case for Muslims, Brenner (2014) compared respondents' claims with respect to daily prayer in the 2007 Pew Global Attitudes sample survey in Pakistan with the results of the Ministry of Finance's 2007 Time Use Survey. The latter is based on diaries, which are generally thought to be more reliable indicators of behaviour than self-reporting. He found a degree of over-reporting in the sample surveys compared to the diary evidence, but a smaller gap than in (largely Christian) North America.

bring their children up as Muslims, makes it easier to celebrate shared rituals and practices (although some residents are marginalised by their regional origins, class or marital status) and provides them with a sense of security (Kirmani, 2008a, 2008b, 2009).

Although the Qur'anic basis for Muslim rituals ensures that their performance is relatively uniform across sects and contexts compared to those associated with other religious traditions, the ways that the specifications are interpreted do vary across sects, and they are also subject to different cultural influences. In addition, they have evolved over time. For example, Fealy (2008) and Howell (2008) describe how in Indonesia, with the advent of television, forms of religious expression such as preaching have increased their reach and influence, developed 'celebrity' characteristics and become commodified.

Christian participation in rituals

Where data are available for both Muslims and Christians, majorities in every SSA country surveyed claim to attend religious services once a week or more (and in most countries, more than three-quarters). These high self-reported attendance rates characterise both Christians (ranging from 88% in Nigeria to 60% in South Africa, median 81%) and Muslims (from 99% in Ghana to 65% in Senegal, median 87%). Similarly, more than two-thirds of both Christian and Muslim respondents in most countries reported that they pray every day. A lower proportion of Christians than Muslims made this claim, although daily prayer is the norm for both (overall median 77%, Christians 72%, Muslims 92%) (Pew Forum, 2010, pp. 27–8). In some SSA countries with a Christian majority or a large Christian population, women are slightly more likely than men to report that they pray daily, but in most such countries there is no significant difference (Pew Research Center, 2016, p. 30).

Christians are expected to participate in prayer and attendance at religious services on a regular basis (daily/weekly), because it is through individual prayer and corporate worship that the promise and hope of life for those who have faith is said to be realised (Pelikan, 1987). However, Christians are generally not subject to rules about practice in the same way as Muslims – the Christian creeds are statements of belief rather than injunctions to engage in religious practices or guides to behaviour (see Chapter 3). Amongst Christians, therefore, frequent attendance and significant time commitment primarily reflect individual religiousness, although social pressure in wider society and within denominations and congregations also plays a significant role. Patterns vary between denominations, related to their history and doctrine. In addition, participation is influenced by the nature of different sources of authority (e.g. the Bible, the Holy Spirit, tradition, the church, the clergy) and the roles played by these sources in mediating between adherents and God, as well as differences in theological teaching. For example, concepts of sin and how to deal with it differ, as does the significance of the sacraments, especially the Eucharist, Holy Communion or Lord's Supper and baptism, which are purely symbolic in some denominations but held in others to demonstrate 'the mystery of faith' by representing invisible divine grace in visible form (Pelikan, 1987, pp. 358–9).

Forms of religious service and prayer associated with the mainstream churches are, to a varying extent, structured and repetitive, orchestrated and delivered by one or more religious specialists (the clergy) rather than actively participatory and didactic and reflective rather than openly emotional. In contrast, Pentecostal Christianity (and similar evangelical and charismatic worship) is a religion of the senses, using oratory and music (backed by technology) to create experiences that are said to enable 'born again' Christians to sense the presence of the Holy Spirit in their bodies and the assembly and to act on those feelings (Obadare, 2016). Outbursts of religious enthusiasm are encouraged as demonstrating the overwhelming of reason by the power of the Holy Spirit, which is thought to capture the self and to be revealed in glossolalia, prophecy, healing and exorcism. Personal religious experience is emphasised over doctrine, producing spectacular and emotionally intense collective religious practices. In this context, the purpose of Bible readings and sermons is not just to teach doctrine and promote values and behaviour but to boost the charge of spiritual power, assisted by song, prayer and testimonies to visions, dreams and miracles (Obadare, 2016; Vásquez, 2009).[3]

The widespread appeal of such informal and ecstatic elements has led to their adoption by parts of the mainstream churches and even, at least in religiously competitive Nigeria, by Muslims. Thus Obadare portrays the country as a 'devotional battleground' in which Muslims have reacted to a "culture whose soundtrack is increasingly Christian charismatic" (2016, p. 77) by offering mosque prayers on Sundays, organising night-time prayer sessions, increasingly emphasising evil, the occult, signs and miracles and developing a new 'dramaturgy' of lengthy, fervent prayers. This Muslim–Christian competition is seen dramatically in the deliberate location in 2001 of the main religious complex of the largest Muslim movement (Nasirul-Lahi-L-Fatih Society of Nigeria, NASFAT, founded in 1995) along the highway between two of Nigeria's largest southern cities (Lagos and Ibadan) adjacent to the complexes (or prayer camps) of major churches, including the Redeemed Christian Church of God and Mountain of Fire and Miracles Ministries (Janson and Akinleye, 2015; Obadare, 2016).

Although the proportion of Christians who attend church services once or more a week varies, it might be expected that revivalist Christians (Pentecostals

3 That the motivations of Christians to attend such services and prayer groups go beyond the devotional and inspirational is revealed by Chong's (2012) study of middle-class South Korean Pentecostal women attending two large churches in Seoul. In addition to surrendering to God and developing a sense of a personal relationship with him, her respondents reported that participating in services (and Bible study groups) provides one of the few opportunities available to them to share problems and obtain help. In addition, their sense of being 'reborn', she observes, transforms their sense of self and builds confidence, helping them to deal with domestic tensions.

Table 5.1 Proportion of Christians who attend church at least once a week

	Proportion of Christians who reported that they attend church at least once a week			*Proportion of Christians who reported that they pray every day*		
	Pentecostals	*Charismatics*	*Other Christians*	*Pentecostals*	*Charismatics*	*Other Christians*
Kenya	92	89	64	82	80	56
Nigeria	83	[a]	89	86	[a]	79
South Africa	75	71	55	77	71	61
India [b]	60	52	51	67	55	52
Philippines	90	65	58	93	80	74

Source: Pew Forum, 2006, pp. 18–19.

Sample sizes varied from 231 in India to 956 in the Philippines. Pentecostal and Charismatic Christians were deliberately over-represented in the samples.

[a] Only 67 respondents, so not included.

[b] The survey was carried out in largely Pentecostal/Charismatic Christian areas in Tamil Nadu, Kerala and Meghalaya, so the figures are not representative of Christians in India as a whole.

and Charismatics), who attach great importance to the manifestations of the gifts of the Holy Spirit that occur during services (as described above), would generally have higher attendance rates. The data show that this is indeed the case in most of the countries for which data are available (Kenya, South Africa, India, the Philippines) but not everywhere (e.g. Nigeria) (see Table 5.1). The same pattern is shown by the data on the proportion who pray daily, suggesting that revivalist Christians are more pious/devout. Nevertheless, in Sri Lanka, where Christians are a small minority and most are Roman Catholic, a high proportion (78%) reported that they attend church at least once a week and 70% that they pray every day (Asia Foundation, 2012, p. 12). Overall, data for SSA countries shows that in most, similar proportions of Christian men and women claim to pray regularly (Pew Research Center, 2016, p. 31).

In India, most Christians converted from Hinduism, and many of the converts were low-caste Hindus. Formerly denied access to many Hindu temples, the construction of prominent church buildings became an important indicator of their new identity. In addition, in contrast to the Hindu tradition of individual expressions of faith (see below), corporate worship is a central feature. Liturgical practices, festivals and celebrations have been influenced not only by the mission churches from which the contemporary Indian churches originated but also by local cultural idioms (language, music etc.), as well as global cultural expressions. Influences beyond those of the mainstream churches also help to explain the emergence and form of home-based rituals alongside church services, including those associated with prayer groups and rites of passage (e.g. puberty, marriages and funerals) (Thangaraj, 2013, pp. 177–8).

Ritual in Hinduism

For Hindus, the performance and practice of rituals are more important than doctrines and belief systems (Narayanan, 2013). In the past not only were literacy rates low but also only male members of the Brahmin, Kshatriya and Vaisya castes (respectively priests and intellectuals; rulers, including warriors and bureaucrats; and traders, artisans and agriculturalists, the twice-born) were permitted to read and study the Vedic texts. Sudras (labourers) and Dalits (formerly untouchables) were not permitted to do so (Sharma, 2010). For most Hindus, celebrating festivals, watching and hearing stories being performed through music and dance, fasting and feasting at appropriate moments in the lunar calendar and pilgrimages were the main forms of religious practice, and this is still the case today, even though literacy is much more widespread. Several days a year may be spent in celebrating the major rituals. In addition, the main ways people learn about their religion are through stories, participation in rituals and visits to temples.

Hindu householders are enjoined to perform the five great sacrifices (*panca mohayajnas*) daily. The first three of these pertain to the worship of Brahman, ancestors and gods, the fourth to the worship of all living things and the fifth to hospitality, the duty to take care of any guest (Sharma, 2010). As outlined in Chapter 3, in addition to Brahman, the ultimate reality, the creator, three principal deities are recognised (Vishnu the preserver, Shiva the destroyer and Devi the goddess), each with more than one avatar or manifestation, as well as innumerable regional and local deities. However, the language and practice of worship is similar. Hindus generally choose one or more of the deities as the object of their love and devotion (*bhakti*). Each deity has a form the worshipper can behold and its own iconography and mythology. Images of deities are displayed in temples, street shrines and the home shrine/altar that every family has. Before every image, a lamp and incense burn. According to Nadkarni (2008), deities and their images are intended to help devotees recognise and worship the ultimate reality behind all forms and images.

Individuals view the deity of their choice as a guest in the domestic setting, who should be honoured with offerings. "*Puja*, the loving offering of light, flowers, and water or food to the divine, is the essential ritual of Hinduism. For the worshipper the divine is visible in the image, and the divinity sees the worshipper. This interaction between human and deity . . . is called *darshan*, seeing" (Smith, 2009, p. 41; see also Sanford, 2013). In return, the deity bestows *prasada* (grace). Typically a woman in the household lights an oil lamp in front of the altar and draws geometric designs in rice flour on the floor. Family members may read and recite prayers from the holy texts and prayer books. Similar rituals take place in front of street and temple shrines. In Sri Lanka, nearly three-quarters of Hindus reported that they do *puja* daily (Asia Foundation, 2012, p. 9). Quantitative data show no statistically significant difference between men and women with respect to their participation in daily prayer (Pew Research Center, 2016, p. 36).

Temples are constructed as sanctuaries for one or more deities – gifts or offerings of water, flowers or fruit are made and may be offered back to devotees once they have been consecrated. At the end of a visit, devotees walk around the central part of the temple and the image within it. Temple priests conduct rituals and may make offerings and engage in devotional acts on behalf of others (Hiltebeitel, 1987). Some specialise in religious narratives, including interpretation of the texts, but they do not perform pastoral roles (Narayanan, 2013). Visitors to temples choose which of the deities displayed in a temple they wish to worship during a particular visit, making offerings, often of food, some of which is taken home once it has been blessed. This is not congregational worship, although in North India, there are sometimes group songs (*bhaja*). Acting, music and dance have traditionally been regarded as possible paths to liberation, through participating or watching. Theoretically, any dance may be divine, but some are devotional in tone and intent, and women's involvement may bypass norms on women's modesty and actions in public (Narayanan, 2013). All Hindus surveyed in Sri Lanka said that they visit a temple, over half more than once a month but a large minority much less frequently or only on special occasions (Asia Foundation, 2012, p. 10).

The Upanishads, the closing section of the Vedas, deal with mystical awareness of the divine – the main theme is "the relationship between Brahman, the divine spiritual force of the universe and the soul of the individual human being" (Oliver, 2010, p. 129). The authors were probably wandering mystics who sought to convey their religious experiences, which were based on meditation and the direct contemplation of God rather than study of the scriptures. Each of the four surviving Upanishads emphasises a different aspect of this theme and provides guidance on religious and ethical practice and daily life. Religious practices emphasise learning from a guru, meditation and yoga, a discipline which links philosophy, ethical (unselfish) behaviour, exercises to control the breath (*pranayama*) and meditation to calm the mind (*dhyana*). Ultimately, through advanced practices of meditation, yoga adepts believe, it is possible to achieve a state of *samadhi* (total peace and union of the human self or *atman* with the universal self) (Nadkarni, 2008; Oliver, 2010). Underlying the practices are notions of auspiciousness and purity, which can be visualised as two intersecting axes (auspicious/inauspicious and pure/polluted). The hierarchical social structure determines the inherent ritual purity of an individual, which varies by caste and gender, influencing what food can be eaten, prepared by whom and with whom (Sanford, 2013). Pollution can also be transitory, through its association with particular life events. Generally, life-promoting rituals and those that promote a person's well-being are considered auspicious: birth is ritually polluting but auspicious, while death brings ritual impurity and is inauspicious. Auspiciousness indicates the potential to bring about good fortune and a good quality of existence. Hindus consider certain people, animals, rituals, foods, places, times and astrological combinations as auspicious and arrange their lives to take advantage of them. They govern the timing of household decisions, life cycle events and festivals, although the details vary enormously between countries, regions, traditions and groups (Narayanan, 2013; Sanford, 2013).

Rituals associated with birth, puberty, pregnancy, birthdays, weddings and death also take place at home (Narayanan, 2013). Every ritual associated with a wedding takes place at a time that is perceived to be auspicious, almost always in front of a sacred fire (*agni*), and many of the mantras chanted are from the Rig Veda. Death is inauspicious but is also the means by which the self can ultimately be liberated from *samsara*. Funerals are considered ritually impure and inauspicious – funeral rituals, during which the body is cremated and the ashes immersed in the ocean or a river, are traditionally performed by men.

Research in Andhra Pradesh, India, records how the adoption of ritual practices, their public performance and the manipulation of 'auspiciousness' is not only intended to propitiate the gods but also to demonstrate the religiousness of family members and help to buttress the middle-class identity of upwardly mobile lower caste families (Säävalä, 2001). Similarly, religious ritual and performance enables returned international migrants in a village in Kerala to transform their material wealth into social and political status, although members of the traditional elite and the new moneyed groups disagree about the correct way of performing participatory rituals (Osella and Osella, 2003).

Research in majority-Hindu but religiously mixed urban and rural communities in Punjab and Orissa, according to Devine and Deneulin (2011), reveals approaches to religious practices and assessments of their purpose and outcomes that appear to be common to Hindus and other religious groups. The study found that people participate in daily practices that include prayer, acts of veneration, ritual austerities, temple service and financial donations, mainly to ensure that they are protected from harm and misfortune but also to seek good favour and good fortune. Participants also believe that prayer and ritual help to reinforce values and strengthen moral and social norms. Believing that the deities have the power to directly influence affairs of this world, many respondents described how they routinely carry out religious practices

> as a preventative measure to avert any potential harm. . . [and] make sure that the influence is benevolent rather than threatening. . . . There are however numerous accounts of people turning to religious practices as a way of overturning or remedying specific harms or dangers that have already occurred. In these cases, the actions taken are specific rather than routine, reactive rather than preventative; and usually entail adopting new, special or simply more religious practices.
>
> (Devine and Deneulin, 2011, p. 68)

The two most common harms are ill health and addiction to alcohol, drugs or gambling, while good fortune relates to concerns that include fertility, birth, finding employment and the education of children. The two categories overlap, since deities can be both protective and threatening. For example, some accounts saw good as coming out of misfortune, or things that could have been worse without divine help. Sometimes, people appeal to deities beyond their own tradition in order to hedge their bets.

However, respondents also expressed concern that participation in religious practices is decreasing. They attribute this to education and modern-day diversions such as television. Although they admit that sometimes traditional religious practices are merely changing, they also perceive a decline which is allowing negative social norms such as greed and selfishness to prevail, threatening society's moral compass and giving rise to social disorder (Devine and Deneulin, 2011).

Ritual practices in Buddhism

For Buddhists, the roots of religious practice reflect both the tradition's Hindu inheritance and the Buddha's teaching, the Noble Eightfold Path, which includes three categories of practice: *sila* (ethical discipline, which is concerned with right speech, right action or behaviour and right livelihood and will be discussed further in Part IV), *prajna* (wisdom), right view (seeing reality as it is i.e. impermanent, thus not becoming attached to anything in the material world) and *samadhi* (meditative contemplation, based on right effort, right mindfulness and right contemplation). Traditions and sects within Buddhism emphasise different aspects of the Eightfold Path (Wolfe and Gudorf, 1999). Here, I am concerned with the last: right effort refers to the requirement to refrain from making value judgements about phenomena and people, to avoid becoming attached to impermanent things; right attention encourages attentiveness to everyday events and bodily actions; and right contemplation encourages meditation to enable the practitioner to see the true nature of the physical world and thus avoid suffering, experience peace and attain equilibrium. Common forms of meditation include *anapanasati*, which involves focusing attention on breathing to calm the mind and prepare it for the next stage of meditation, and *vipassana*, or insight meditation, the process by which the meditator reflects on thoughts that come and go to recognise their impermanence. The practice of venerating the Buddha is not universal, but in Sri Lanka, a 2011 survey found that most Buddhists visit temples (41% more than once a month, 36% once a month, and 23% occasionally or for special festivals) (Asia Foundation, 2012). Fewer observe *sil* (following the routine of a spiritual practitioner for a day) or chant *pirit* on a regular basis (or at all).

In theory, Buddhists reject daily worship and ritual practices as a means of achieving freedom from suffering, but in practice popular Buddhism accommodates many deity cults and seeks to propitiate the gods to secure benefits in mundane affairs. Thus beliefs and rituals may be absorbed from other traditions, although not by all Buddhists everywhere. In southern Asia, for example, Buddhists often believe in deities and demonic beings. The latter are feared, as they are believed to destructively interfere with domestic life, prosperity and health. In addition, Buddhists believe that rebirth occurs and is influenced by *kamma/karma* and have absorbed the astrological tradition to predict and ritually prepare for future events. Theravada Buddhists, for example, have developed a complex ritual life (involving Buddhist and assimilated non-Buddhist practices) and lay and monastic interactions. Many Theravada Buddhist householders perform occasional rituals, including *pirit* (protection ceremonies based on chants to ward off evil influences and bring about good fortune), rituals prior to major life cycle events, rituals associated with *dana* (almsgiving) and calendrical rites (*poya*). Some of the rituals are associated with monasticism, for example *patimokkha* or chanting rules (see Goonasekera, 2013, pp. 121–31).

Box 5.2 Hindus and Buddhists practising their religion in Maharashtra

Research in the slums of Pune, a city in Maharashtra, reveals some of the commonalities and differences in Hindu and Buddhist practices, as expressed in residents' daily lives (Bhatewara and Bradley, 2013). Hindu slum residents generally identified gurus (people with knowledge, wisdom and authority who guide others) and local religious leaders as the main influences on their religious lives, reporting that they select a guru to follow on the basis of feeling a spiritual connection to the person concerned, although life experiences and family influences also contribute to their choice of deities and gurus. Gurus are simultaneously religious teachers and the focus of worship, involving *darsan*, the process by which a devotee gains spiritually by looking at a deity and being in turn observed by that deity, who is believed to be in an image looking back at the devotee. Buddhist converts in the slums position Ambedkar's image next to that of the Buddha in domestic shrines and temples, look to him as their guru and worship him alongside (or at times instead of) the Buddha. While research in rural Maharashtra found that in the households of most Buddhist converts, images of Hindu gods are displayed alongside those of the Buddha and Ambedkar, and most of the informants said that they prayed to both (Burra, 1996), the research in Pune did not find that Buddhist converts continue to display or pray to Hindu gods, although some are still venerated. Slum residents generally attributed their conversion from Hinduism to their desire to overcome caste-based marginalisation (a desire shared by low-caste Hindus) rather than rejection of the Hindu religion and said that their religious lives had not radically changed. In contrast, middle-class respondents were more likely to refer to their personal spiritual journeys, and many educated and politically active informants expressed active hostility to the religion of Hinduism. In the villages, Burra found that public rituals are Buddhist but private affairs remain decidedly Hindu, while in Pune, converts' domestic religious practices are Buddhist, and public rituals tend to be seen as the preserve of the rich and upper castes.

In addition to gurus, who are revered and whose teachings are influential, Pune slum residents described how they seek religious and practical guidance from local religious leaders. Rather than priests or monks, these are authority figures respected for their religious knowledge, prosperity and insight into everyday matters. Most are men and are consulted by both men and women. Their religious knowledge, which has not been gained through formal religious training, is brought to bear on the responses they provide to the day-to-day problems on which they are consulted, such as how to pay back debts, increase household incomes, defuse family tensions, choose a marriage partner, or deal with problems such as domestic violence or alcoholism. The solutions they suggest may include religious duties and rituals. There are also some women leaders, who are consulted by women, particularly on matters related to bodily processes such as pregnancy or reproductive infections. For slum residents in Pune, religious practices are woven into the fabric of everyday life, enabling them to feel that they are building good *karma* or good energy.

Communicating with the spirits

People employ a variety of practices in order to communicate with the spirits, including divination, offerings and sacrifices, life cycle rituals, healing and, if a spirit is thought to possess an individual, exorcism. As discussed in Chapter 3, the idea of God as creator underlies all aspects of religious beliefs, but for many, day-to-day affairs are believed to be governed by other invisible forces, including deities, spirits and the living dead/ancestors, whose relationship with the supreme being may not be explicit. In many folk religions, it is thought that nature spirits inhering in the sky (e.g. the sun, stars, rainbows, storms) or the earth (in hills, boulders, trees, animals or water sources) have an influence on human well-being (Alhassan Alolo and Connell, 2013). Human spirits may include the long dead (ghosts) and the more recently deceased (ancestors). After death, they are recognised, venerated and invoked or feared. The nature and types of spirit vary with geographical location and over time. As discussed in Chapter 3, perceived links between the spirit world, people and their environment are the basis for folk religious traditions, but they also influence the beliefs and practices of those who consider themselves to be adherents of one of the other religions.

In folk traditions, rituals seek to recognise and influence interactions between spirits and people.[4] Although many are common, even universal, the details are

4 The term 'magic' may be used to refer to "the use of ritual to call upon supernatural powers to intervene in the natural world to effect a desired end" (Lawson and Garrod, 2001, p. 295). The aim is to achieve a specific goal. While religion generally requires obedience to the supernatural agency/ies concerned and may require adoption of a particular way or life or adherence to an ethical code, magic is narrower in purpose and time. It is judged by whether or not it works, with failure generally attributed to the ritual not being performed correctly (Bruce and Yearley, 2006).

varied. They seek to maintain or re-establish harmonious relationships with spiritual beings believed to be powerful enough to bring good or harm in order to win their favour or avert their wrath. They often seek to change particular human situations – the ritual is conducted for practical purposes such as curing sickness, calling for rain, ensuring fertility or a good harvest or enabling enemies to be defeated. They are often officiated over by religious specialists and are mostly communal, concerned with relationships between a community and the spirits, with their conduct reaffirming the community by reinforcing moral and social bonds.

Diviners or mediums traditionally perform the task of communicating with the spirits to ascertain the cause of their displeasure, arbitrate between them and humans and offer remedies such as cleansing rituals. A diviner (usually a man) is believed to be an instrument of an invisible power because of his own characteristics or through a spirit which enters him. He may refer to dreams or manipulate physical objects, including live or dead animals. Prayers of thanksgiving or petitions for protection, removal of affliction, deliverance from evil, good fortune, sufficiency and flourishing place an individual or community in the hands of invisible powers. Offerings and sacrifices are made to ask for help or to relay people's requests, preceded by divination to identify what kind of spirit is causing harm and what steps are necessary to obtain a solution or healing. Offerings and sacrifices involve the setting apart of an item for the supernatural powers: offerings tend to be of items used by humans, while sacrifices usually involve an animal (although humans may participate in consuming the meat in order to obtain blessings) (Magesa, 1997). Thomas (2005) identifies four categories: meal and drink offerings, given daily or on special days at a shrine or to a deity; thanks offerings made by individuals; propitiatory sacrifices, generally by a community to avert a crisis or catastrophe; and sacrifices intended to counter a curse or the anger of a divine being who believes that an offence has been committed against him/her. Offerings and sacrifices may be one off or need to be repeated.

In African folk religions, ancestors are seen as ever-watchful beings who can help or harm their living relatives (Mbiti, 1991). They and their descendants are believed to constantly exchange gifts and favours. The ancestors are believed to make their presence and desires known, for example, through divination or appearing in dreams. They are thought to desire to be remembered and honoured, supporting their living kin as long as prayers and offerings are made, visiting calamities if neglected and forcing a response by means of prayers and rituals to restore harmony. Ancestral rituals aim to maintain bonds between the living and the ancestors, repair wrongs that have resulted in misfortune or affliction, and prevent calamities. Affliction is generally regarded as a consequence of human wrongdoing, but misfortune is also attributed to human and nature spirits, who may or may not need provocation, and their visible counterparts, including sorcerers and witches (Magesa, 1997) (see Chapter 3).

People in many parts of Africa and Asia seek contact with the spirits at every stage of their lives, through rites associated with pregnancy, birth, puberty, marriage and death, because these are all periods of transition when it is thought that the spirit world may be of assistance. Rites seek to enhance fertility, to ensure that the life of a family and community continues. Thus pregnancy and birth are considered blessings, indicating that the ancestors are happy and are providing protection from malevolent spirits. If there are problems, various rituals are available to identify the causes, ensure atonement and restore order. Marriage, the primary purpose of which is seen as procreation, is equally important. Death enables ongoing life and communication between the living and the spiritual world, as long as the living refer to the dead by prayer, offerings, sacrifices and other rituals. However, in some cases, it is thought that the spirit of a deceased person can harm the living. For example, in some cultures in Africa, a widow is required to undergo rituals designed to placate the spirit of her deceased spouse. Such rituals may be humiliating and harmful.

Both men and women participate in religious activities and contribute to the spiritual welfare of their lives, families and societies, but women's roles are generally subordinate and are often confined to private rituals in the domestic sphere. The principal intercessor is almost always male, although sometimes female. Even where women are given ritual responsibilities, these are often shared with men. Nevertheless, women play a vital role in personal rituals associated with status transformation (e.g. childbirth, puberty and death), and it is generally women who are thought to be possessed or who act as mediums (Mbiti, 1991; Kilson, 1976). Sexual relations in ritual are interwoven with those in secular life, because they tend to be connected with vitality via reproductive functions – women are regarded as the force of life and crucial for community continuity but also a source of danger. For example, blood associated with menstruation and childbirth is often regarded as polluted, with women being temporarily separated from society and purification rituals performed (Alhassan Alolo and Connell, 2013).

Grillo's detailed study in the cosmopolitan metropolis of Abidjan, Côte d'Ivoire, found that divination and sacrifice are the most common and portable components of ATR/ritual. Many urban residents reported that they seek out diviners to ascertain the causes of problems and misfortunes, which are often attributed to malevolent spirits or witches. Urban diviners ask for (often token) gifts and sacrifices, which are symbolically important, and every divination involves an opening sacrifice and culminates with a prescription for sacrifice, so determining the proper sacrifice is fundamental to divination. Diviners commonly seek the blessings of invisible forces that are said to determine people's destinies and prescribe actions to tame, appease or exorcise the relevant spirits. A wide array of divinatory techniques is on offer, reflecting the cultural diversity of the city. Rather than choosing a diviner on the basis of a common ethnic identity, clients are thus able to choose on the basis of a reputation for efficiency, integrity, accurate diagnosis and the prescription of apparently effective remedies for particular types of problems. Although residents know that divination practices are denigrated, they regard them as a means of harnessing spiritual power and using it to "channel and manipulate the forces that threaten both the individual and the social body" (p. 155). In villages, rituals involve the wider community, but in the city they are more personal, providing individuals with a sense of agency in the face of impoverishment, uncertainty and social alienation (Grillo, 2010).

In a different geographical context, in the southern Ghanaian village she studied, Graveling (2010) describes the use of *juju*. In this village, the spirits are considered to be agents with interests and volition, although they remain largely unseen and do not necessarily conform to functional laws of cause and effect. Interactions with them are therefore uncertain and unpredictable, so residents in the study village seek both to align themselves with and also to influence the spiritual powers. *Juju* includes the use of plants and charms to exploit spiritual power and may be used for good (e.g. healing, divination) or (more often) evil. Christian pastors and Muslim *mallams* tend to perceive *juju* practices as sinister, but people think that some of them can remedy problems, and they may even use some forms of *juju* themselves. Practices incorporate prayers and rituals, although practitioners may also advise and assist the sick to obtain biomedical treatment.

'Tribal religions' in India similarly concern the relationships between nature (land, forest, water), people and the supernatural. God (the supreme being) is regarded as the creator of life, friend and father to the good, guardian of law and morality, and giver of knowledge and wisdom, transcending and controlling all other gods, ghosts and supernatural powers. Spirits may be benevolent (ancestor spirits, village, clan and household spirits), malevolent (those of people who die an unnatural death) or neutral. The latter can be good or evil, so need to be kept tamed. Malicious spirits are thought to bring misfortune, so need to be placated or propitiated. Both religious functionaries and heads of families therefore perform sacrifices during festivals and rites of passage. The former transmit ritual knowledge, perform divination, find witches, make sacrifices to the supreme being and spirits and seek to protect communities and ensure harmony in the universe. They operate occasionally and are considered to be mediators between supernatural beings and communities. The latter operate at household level. Rituals include those that offer thanks to the supreme being, beg for mercy or blessings and seek protection from the evil eye before important undertakings. Offerings, each of which has symbolic value, may include, for example, breaking an egg, offering a plant or drawing a diagram. Rites are performed to purify a mother who has given birth, name a child and incorporate it into the tribe, celebrate marriage and after a death. Often they are performed in sacred spaces (Kujur, 2013).

Ritual offerings and sacrifices clearly play a significant role in folk religions, but they are not confined to these religions. For example, the role of offerings in Hindu ritual was discussed above, and Muslims sacrifice a domestic animal once a year (*qurbani*) during the 12th lunar month of the Islamic calendar, in commemoration of the prophet Ibrahim's sacrifice of a ram in place of his son, symbolically reminding Muslims of their duty to serve God.

As discussed in Chapter 3, closely connected to the belief in spirit possession, which is often associated with evil, misfortune and ill health, are ideas about witchcraft. The identity of supposed witches and spirits that are believed to have possessed a person and ways of delivering people from them and combat their effects involve rituals of exorcism and healing. Traditional approaches to healing embrace a wide range of practices, which do not necessarily have a religious dimension, for example herbalism, or Ayurvedic medicine in South Asia (Nadkarni, 2008), although for many the distinction between 'religious' and 'non-religious' practices is meaningless. While both men and women may be traditional healers, women are often confined to dealing with women's medical needs and treating children, while men treat men, children and only sometimes women.

Spiritual healing practices are commonly associated with folk religions but are not confined to them. For example, beliefs in witchcraft, diabolical forces and spirit possession and their links to ill health are central in many interpretations of Christianity: "For the earliest followers of Jesus – and presumably for Jesus himself [Jenkins suggests] – healing and exorcism were essential components of his proclamation. In his acts of healing, Jesus was not just curing individuals, but trampling diabolical forces underfoot, and the signs and wonders represent visible and material tokens of Christ's victory over very real forces of evil" (Jenkins, 2006, p. 99). In some branches of Christianity, especially Pentecostalism, just as in folk religions, deliverance from evil forces involves miracles and exorcism. The perceived need to confront dark forces, Jenkins suggests, helps to explain the increasingly common practice of all-night services and vigils – not only are these practical (people are not working), they are also symbolically important, because they provide opportunities for "believers [to] proclaim the powers of light and good at times when the forces of darkness are [thought to be] at their strongest and most pernicious" (Jenkins, 2006, p. 110).

In some largely Christian or Muslim contexts, significant proportions of survey respondents report that they turn to traditional healers to cure someone who is ill. The proportion of Muslims who do so ranges in African contexts from 73% in Senegal and 68% in Chad down to 26% in Mozambique, 24% in Uganda and 13% in Kenya (median 44%) and a third to a half in Southern Asian countries (Pakistan 55%, Thailand 48%, Malaysia 40%, Indonesia 38%, Bangladesh 34%) (Pew Research Center, 2012). In earlier surveys in 14 SSA countries, at least three in ten Christians and Muslims said that they sometimes consult traditional healers (half or more in Cameroon, Chad, Guinea Bissau, Mali and Senegal) (Pew Forum, 2010, p. 33). Turning to a traditional healer may, of course, be related to the ease and cost of access to public and private biomedical health services and not just to the prevalence of religious beliefs about the causes and manifestations of ill health.

As noted in Chapter 3, in African countries the volume of allegations about witchcraft always seems to have differed from place to place and fluctuated over time, but in recent years in many places there does appear to have been a resurgence in the number of witchcraft accusations. While Muslims and Christians criticise and even prohibit certain practices, often with beneficial effects (for example, campaigning against humiliating widowhood rituals), they are also conscious that if they do not respond to contemporary concerns, people will continue to resort to traditional intermediaries and practices. Thus many churches (especially Pentecostal churches) practise spiritual healing and exorcism. They draw on the belief that illness results from a combination of physical, mental and spiritual factors, cite biblical references that attribute sickness to sin (and possibly divine punishment), refer to Jesus's healing practices and adopt traditional healing techniques including potent words, touch, prayers and the use of water. In Jenkins's view they see spiritual healing as no worse than traditional healing (and as better if the churches demand less in terms of payment and sacrifices), mostly try to suppress fraudulent claims and bogus miracles and do not prevent their followers using Western medicine where this is available and affordable. Proponents also assert that the churches offer a better chance of success than traditional approaches, because the Christian faith means that people do not need to fear the forces of evil (Jenkins, 2006; see also Spickard and Adogame, 2010).

Rituals may also be associated with secret societies and with conflict. Thus war may be preceded/followed by rituals and the use of medicines, potentially influencing the course and nature of conflict and violence. For example, Jackson's (2010) study of Acholi members of the Lord's Resistance Army in Northern Uganda shows the complex ways in which religion and conflict are entangled. Jackson suggests that the LRA leader's belief system framed the Acholi not only as victims of the Ugandan government but also of the spirit world. He argued that both the traditional Acholi spirits and the Acholi people themselves had failed. Only through purifying acts of extreme violence, he asserted, could the latter be redeemed. His and his followers' "obsession with redemption and purification explains for Jackson not only the range of bizarre quasi-religious battlefield practices of the LRA but also the extraordinary violence perpetrated against the Acholi people themselves" (Alhassan Alolo and Connell, 2013, p. 155).

Thus not only have many beliefs and practices associated with folk religion been absorbed into the beliefs and rituals of other religions, either officially or by informal syncretism, many adherents of these religions continue to hold beliefs and practices associated with folk religion. The varying prevalence of this is revealed by Pew Forum surveys in 18 Sub-Saharan African countries in 2008/9 (Pew Forum, 2010). In four of the countries (Tanzania, Mali, Senegal and South Africa), half or more of all the respondents, both Christians and Muslims, claimed to believe that sacrifices to ancestors or spirits can protect them from harm (although in five countries, fewer than a fifth of respondents said that they believed this) (Pew Forum, 2010, p. 33). In addition, more than a fifth in 12 countries said that they possessed traditional sacred objects (a third or more in South Africa, Senegal and Guinea Bissau, elsewhere mostly between 15% and 30%).

In order to indicate general involvement with ATR amongst avowed Muslims and Christians, the Pew Forum constructed an overall index of African folk religious beliefs and practices. The index includes seven beliefs (the protective power of certain spiritual people, the power of juju and other sacred objects, the evil eye, witchcraft, evil spirits, the protective power of sacrificial offerings to ancestors and reincarnation) and four practices (visiting traditional healers, owning sacred objects, participation in ceremonies to honour the ancestors, participation in traditional puberty rituals). It defines a high level of involvement as

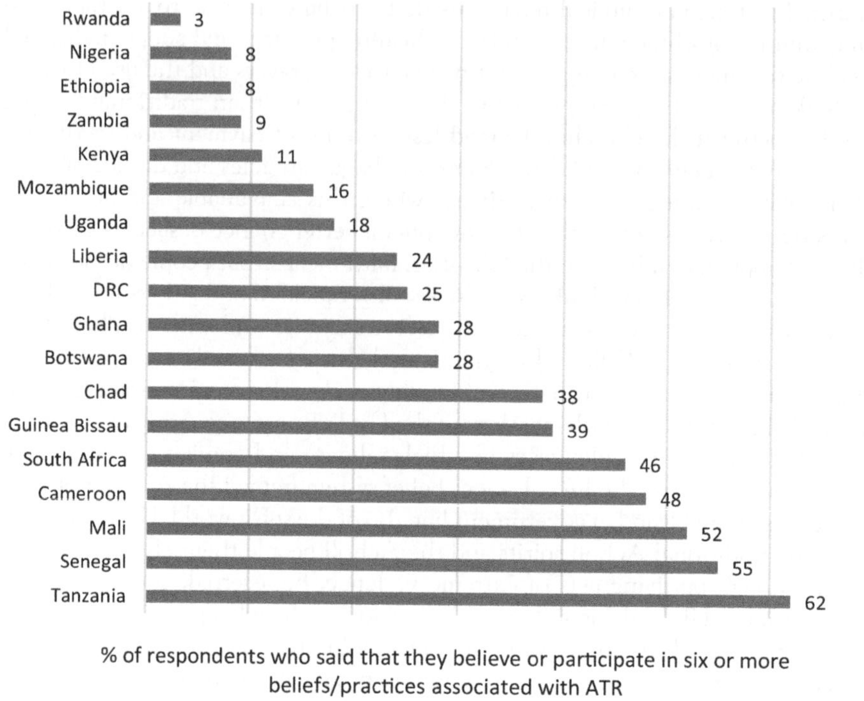

Figure 5.1 Traditional religious beliefs and practices amongst African Muslims and Christians

Source: Pew Forum, 2010, p. 34.

believing or participating in 6 or more of the 11 beliefs and practices. Amongst Christians and Muslims overall, the median proportion with this high level of involvement is 25% (Christians 20%, Muslims 26%), and the countries in which such involvement occurred included some with a Muslim majority, some with a Christian majority and some that are religiously mixed (see Figure 5.1). Although the proportion of both Muslims and Christians with such a high level of involvement varies between countries, from over half of the respondents in three to a quarter or less in half, in general it appears to be higher in West than East African countries. The five countries with the lowest proportion of Muslims and Christians with high levels of traditional beliefs and practices are all predominantly Christian countries, with the exception of Nigeria, where roughly half of the population is Christian and half Muslim.

The findings from the Pew Forum surveys confirm that many Muslims and Christians continue to hold traditional beliefs (sometimes actively encouraged by their church or sect and sometimes against formal religious teaching). However, beyond this, it is difficult to know what to make of the results, for a variety of reasons. First, the same selection of beliefs and practices was used in every country, although there are many variations; second, the question posed and the construction of the index were both quite simplistic; third, observation and anecdotal evidence throw doubt on some of the findings; and finally, the responses to the questions may reflect the perceived social acceptability of such beliefs and practices as much as actual beliefs and behaviour. Moreover, as has been repeatedly noted, like other similar organisations, the Pew Forum does not offer explanations of the patterns identified.

Concluding comments

In the framework adopted at the outset of this analysis, practices are identified as constituting the second domain of religion. They can, I suggested, be subdivided into religious and social or ethical practices – this chapter has focused on the former, illustrating some of the ways informal and formal institutions associated with the religious traditions and denominations reproduce and sometimes modify religious practices, while asserting their transcendent value and eternal validity. Like religious narratives, Lincoln (2014) suggests, ritual can best be understood as an authoritative mode of symbolic discourse, which has gestural and dramatic rather than solely verbal characteristics. Religious specialists and leaders claim the authority to preserve and interpret practices and assume responsibility for supervising them. However, without recognising the motives and experiences of participants, it is impossible to understand why participation in religious practices continues to be so widespread across religious traditions and different social contexts in Sub-Saharan Africa and southern Asia, especially as knowledge of many basic rituals (such as daily prayer in Hinduism or Islam) is passed on within families, and their performance is individual and private.

For many adherents, not only is their religious tradition syncretic, they also practise bricolage in their own religious practices. For example, informants in White et al.'s study in India, like the African survey respondents described above, reported drawing on insights from and engaging in religious practices associated with different religious traditions: people describe alternating visits to churches, mosques, *gurdwaras* and temples, learning from other religions and visiting healers from traditions other than their own. Their religious practices are thus subject to a range of influences and the religious identity they develop can be quite fluid (White et al., 2011).

Engagement in religious practices has important gender dimensions. In Bangladesh, for example, most women, both Muslim and Hindu, share a sense of ritual responsibility for their families, and the observance of religious practices is part of their everyday duties, although the frequency of participation (for example in worship services such as Friday prayers in Islam) may differ between men and women (Devine and White, 2009; White, 2010, 2012). Across South Asia as a whole, the view that women's self-fulfilment lies in carrying out their gendered roles of serving other household and family members is sustained not only through cultural narratives of women's sacrifice but also through religious practices, including daily prayer and fasting. Fasting and related practices will be considered in Chapter 6, followed by those intended to teach children and adults about their religious tradition, instil the habit of regular engagement in religious practices and, in some religious traditions, encourage adherents to persuade those affiliated with other traditions or considered to be mistaken in their beliefs or lax in their practices to become more observant.

References

Alhassan Alolo, N. and Connell, J.A. (2013) Indigenous religions and development: African traditional religion, in Clarke, M. (ed) *Handbook of Research on Development and Religion*, Cheltenham: Edward Elgar, 138–63.

Asia Foundation (2012) *National Values Survey, Sri Lanka 2011*, Colombo: The Asia Foundation.

Bhatewara, Z. and Bradley, T. (2013) 'The people know they need religion in order to develop': Religion's capacity to inspire people in Pune's Slums, *European Journal of Development Research*, 25, 288–304.

Brenner, P.S. (2014) Testing the veracity of self-reported religious practice in the Muslim world, *Social Forces*, 92, 3, 1009–37.

Bruce, S. and Yearley, S. (2006) *The SAGE Dictionary of Sociology*, London: Sage (electronic book).

Burra, N. (1996) Buddhism, conversion and identity: A case study of village Mahars, in Srinivas, M.N. (ed) *Caste: Its Twentieth Century Avatar*, New Delhi: Penguin, 152–73.

Chong, K.H. (2012) Revisiting religious power: The Korean evangelical church as a disciplinary institution, in Bender, C., Cadge, W., Levitt, P. and Smilde, D. (eds) *Religion on the Edge: De-centering and Re-centering the Sociology of Religion*, Oxford: Oxford University Press (online 2013).

Coleman, S. (2010) Recent developments in the anthropology of religion, in Turner, B.S. (ed) *The New Blackwell Companion to the Sociology of Religion*, Oxford: Wiley-Blackwell, 103–21.

Collins, P. (2009) Religion and ritual: A multi-perspectival approach, in Clarke, P.B. (ed) *The Oxford Handbook of the Sociology of Religion*, Oxford: Oxford University Press, online ed.

Devine, J. and Deneulin, S. (2011) Negotiating religion in everyday life: A critical exploration of the relationships between religion, choices and behaviour, *Culture and Religion*, 12, 1, 59–76.

Devine, J. and White, S. (2009) *Religion, Politics and the Everyday Moral Order in Bangladesh*, Birmingham: University of Birmingham, Religions and Development WP 40.

Dugbazah, J., Labeodan, H. and Dadughun, S (2009) *The Relationship between Religious Teachings, Values and Beliefs and Concepts of Development in Nigeria*, Birmingham: Religions and Development Research Programme, unpublished research notes.

Fealy, G. (2008) Consuming Islam: Commodified religion and aspirational pietism in contemporary Indonesia, in Fealy, G. and White, S. (eds) *Expressing Islam: Religious Life and Politics in Indonesia*, Singapore: Institute of Southeast Asian Studies, 15–39.

Furseth, I. and Repstad, P. (2006) *An Introduction to the Sociology of Religion: Classical and Contemporary Perspectives*, Aldershot: Ashgate.

Goonasekera, S. (2013) Buddhism in South Asia: Practicing tradition today, in Pechilis, K. and Raj, S.J. (eds) *South Asian Religions: Tradition and Today*, London and New York: Routledge, 101–42.

Graveling, E. (2010) 'That is not religion, that is the gods': Ways of conceiving religious practices in rural Ghana, *Culture and Religion*, 11, 1, 31–50.
Grillo, L. (2010) 'When you make sacrifice, no one is a stranger': Divination, sacrifice and identity among translocals in the West African urban diaspora, in Adogame, A. and Spickard, J.V. (eds) *Religion Crossing Boundaries: Transnational Religion and Social Dynamics in Africa and the New African Diaspora*, Leiden: Brill, 143–64.
Hassan, I. (2010) *Muslim Women in Development: Perceptions from Muslims in Jos, Nigeria*, Birmingham: Religions and Development Research Programme, unpublished research notes.
Hegland, M.E. (2003) Shi'a women's rituals in northwest Pakistan: The shortcomings and significance of resistance, *Anthropological Quarterly*, 76, 411–42.
Hiltebeitel, A. (1987) Hinduism, in Eliade, M. and Adams, C.J. (eds) *The Encyclopedia of Religion*, New York: Macmillan, 336–60.
Howell, J.D. (2008) Modulations of active piety: Professors and televangelists as promoters of Indonesian '*Sufisme*', in Fealy, G. and White, S. (eds) *Expressing Islam: Religious Life and Politics in Indonesia*, Singapore: Institute of Southeast Asian Studies, 40–62.
Jackson, P. (2010) *Politics, Religion and the Lord's Resistance Army in Northern Uganda*, Birmingham: University of Birmingham, Religions and Development WP 43.
Janson, M. and Akinleye, A. (2015) The spiritual highway: Religious world making in megacity Lagos (Nigeria), *Material Religion: The Journal of Objects, Art and Belief*, 11, 4, 550–62.
Jenkins, P. (2006) *The New Faces of Christianity: Believing the Bible in the Global South*, New York: Oxford University Press.
Kilson, M. (1976) Women in African traditional religions, *Journal of Religion in Africa*, 8, 2, 133–43.
Kirmani, N. (2008a) Competing constructions of 'Muslim-ness- in the South Delhi neighbourhood of Zakir Nagar, *Journal of Muslim Minority Affairs*, 28, 3, 355–70.
Kirmani, N. (2008b) Constructing 'the other': Narrating religious boundaries in Zakir Nagar, *Contemporary South Asia*, 16, 4, 397–412.
Kirmani, N. (2009) Deconstructing and reconstructing 'Muslim women' through women's narratives, *Journal of Gender Studies*, 18, 1, 47–62.
Kujur, J.M. (2013) Tribal religions in India: Practicing tradition today, in Pechilis, K. and Raj, S.J. (eds) *South Asian Religions: Tradition and Today*, London and New York: Routledge, 101–42.
Lawson, T. and Garrod, J. (2001) *Dictionary of Sociology*, Chicago, IL: Fitzroy Dearborn, 2nd ed (electronic version).
Lincoln, B. (2014) *Discourse and the Construction of Society: Comparative Studies of Myth, Ritual and Classification*, New York: Oxford University Press.
Magesa, L. (1997) *African Religion: The Moral Traditions of Abundant Life*, Nairobi: Paulines Publications Africa.
Mbacke, S. (2016) *The Mouride Order*, Washington, DC: Georgetown University, Berkeley Centre for Religion, Peace and World Affairs and World Faiths Development Dialogue.
Mbiti, J. (1991) *Introduction to African Religion*, London: Heinemann.
McGuire, M. (2008) *Lived Religion: Faith and Practice in Everyday Life*, Oxford: Oxford University Press.
Mesaki, S. and Kipacha, A. (2008) *The Relationship between Religious Teachings, Values and Beliefs and Concepts of Development in Tanzania*, Birmingham: Religions and Development Research Programme, unpublished research notes.
Nadkarni, M.V. (2008) *Hinduism: A Gandhian Perspective*, New Delhi: Ane Books, 2nd ed.

Narayanan, V. (2013) Hinduism: Practicing tradition today, in Pechilis, K. and Raj, S.J. (eds) *South Asian Religions: Tradition and Today*, London and New York: Routledge, 39–72.

Obadare, E. (2016) The Muslim response to the Pentecostal surge in Nigeria: Prayer and the rise of charismatic Islam, *Journal of Religious and Political Practice*, 2, 1, 75–91.

Oliver, P. (2010) *World Faiths: An Introduction*, London: Hodder Education, 4th ed.

Osella, F. and Osella, C. (2003) Migration and commoditisation of ritual: Sacrifice, spectacle and contestations in Kerala, India, in Gardner, K. and Osella, F. (eds) *Migration, Modernity and Social Transformation in South Asia: An Overview*, New Delhi: Sage.

Pelikan, J. (1987) Christianity: An overview, in Eliade, M. and Adams, C.J. (eds) *The Encyclopedia of Religion*, New York: Macmillan, Vol. 3, 348–62.

Pew Forum (2006) *Spirit and Power: A 10-Country Study of Pentecostals*, Washington, DC: Pew Forum, http://pewforum.org/uploadedfiles/Orphan_Migrated_Content/pentocostals-08.pdf

Pew Forum (2010) *Tolerance and Tension: Islam and Christianity in Sub-Saharan Africa*, Washington, DC: Pew Forum, http://pewforum.org/uploadedFiles/Topics/Belief_and_Practices/sub-saharan-africa-full-report.pdf

Pew Research Center (2012) *The World's Muslims: Unity and Diversity*, Washington, DC: Pew Research Center, http://assets.pewresearch.org/wp-content/uploads/sites/11/2012/08/the-worlds-muslims-full-report.pdf

Pew Research Center (2016) *The Gender Gap in Religion Around the World*, Washington, DC: Pew Research Center, http://assets.pewresearch.org/wp-content/uploads/sites/11/2016/03/Religion-and-Gender-Full-Report.pdf

Ruffle, K.G. (2013) Islam in South Asia: Practising Islam today, in Pechilis, K. and Raj, S.J. (eds) *South Asian Religions: Tradition and Today*, London and New York: Routledge, 192–221.

Säävalä, M. (2001) Low-caste but middle-class: Some religious strategies for middle-class identification in Hyderabad, *Contributions to Indian Sociology*, 35, 3, 293–318.

Sanford, A.W. (2013) Hinduism and development, in Clarke, M. (ed) *Handbook of Research on Development and Religion*, Cheltenham: Edward Elgar, 79–96.

Segal, R.A. (2005) Theories of religion, in Hinnell, J.R. (ed) *The Routledge Companion to the Study of Religion*, London: Routledge, 49–60.

Sharma, A. (2010) Hinduism and poverty, in Galston, W.A. and Hoffenberg, P.H. (eds) *Poverty and Morality: Religious and Secular Perspectives*, Cambridge: Cambridge University Press, 160–79.

Smith, D. (2009) Hinduism, in Woodhead, L., Kawanami, H. and Partridge, C. (eds) *Religions in the Modern World: Traditions and Transformations*, Abingdon: Routledge, 2nd ed, 37–66.

Spickard, J.V. and Adogame, A. (2010) Introduction: Africa, the new African diaspora and religious transnationalism in a global world, in Adogame, A. and Spickard, J.V. (eds) *Religion Crossing Boundaries: Transnational Religious and Social Dynamics in Africa and the New African Diaspora*, Leiden: Brill, 1–29.

Tamney, J.B. (1998) Animism, in Swatos, W.H. Jnr (ed) *Encylopedia of Religion and Society*, London and Altamira: Sage, 22–3.

Thangaraj, M.T. (2013) South Asian Christianity: practicing tradition today, in Pechilis, K. and Raj, S.J. (eds) *South Asian Religions: Tradition and Today*, London: Routledge, 161–192.

Thomas, D.E. (2005) *African Traditional Religion in the Modern World*, Jefferson, NC: McFarland & Co.

Vásquez, M.A. (2009) The global portability of pneumatic Christianity: Comparing African and Latin American Pentecostalisms, *African Studies*, 68, 2, 273–86.
Werbner, P. (2014) Reform Sufism in South Asia, in Osella, F. and Osella, C. (eds) *Islamic Reform in South Asia*, Cambridge: Cambridge University Press, 51–78.
White, S.C. (2010) Domains of contestation: Women's empowerment and Islam in Bangladesh, *Women's Studies International Forum*, 33, 4, 334–44.
White, S.C. (2012) Beyond the paradox: Religion, family and modernity in contemporary Bangladesh, *Modern Asian Studies*, 46, 5, 1429–1458.
White, S.C., Devine, J., Jha, S. and Gaines, S. (2011) *Development, Religion and Wellbeing in India*, Birmingham: University of Birmingham, Religions and Development WP 54.
Wolfe, R.W. and Gudorf, C.E. (1999) Introduction, in Wolfe, R.W. and Gudorf, C.E. (eds) *Ethics and World Religions: Cross-Cultural Studies*, Maryknoll NY: Orbis Books, 1–20.

6 Discipline, joy and learning

As noted in the introduction to the previous chapter, it is useful to distinguish between routine and occasional practices. The discussion so far has concentrated on the former, with an emphasis on regular everyday participation in prayer and rituals, although some occasional practices linked to key points in the life cycle and others intended to propitiate or win the favour of deities and spirits were also mentioned. This theme is continued in the first part of this chapter, which examines fasting, feasting, pilgrimage and religious giving, which are common in many religious traditions. Fasting, feasting and pilgrimage may be an expression of penitence, a means of enforcing religious discipline or a way of maintaining religious identity, commitment and fervour through celebration. They may be individual or collective practices, bringing co-religionists together at a family, congregational, denominational, national or international level. Giving for religious purposes is not only critical for the maintenance of religious organisations but also plays important roles in the expression and demonstration of religiosity.

The analysis then moves on to explore ways in which adherents are socialised into a religious tradition, as both children and adults, through various modes of learning. Finally, for many, adherence to a religious tradition, sect or denomination involves an obligation to proselytise. Some aspects of this commitment, some of the forms it takes in contemporary Sub-Saharan Africa and southern Asia and some of its implications are explored.

Fasting and feasting

The practice of fasting, or abstention from eating and other activities, such as drinking alcohol or smoking, is part of most religious traditions. Frequently, the end of a fasting period is marked by a feast, which has both religious and social significance. Typically, 80% or more of respondents in Pew Forum surveys in a number of African countries said that they fasted (Pew Research Center, 2012, p. 54).

In every country with a significant Christian population included in the organisation's surveys in 2008/9, more than half of Christians said that they fast during holy times, especially during Lent, the period prior to the annual commemoration of Jesus's crucifixion and celebration of his resurrection at Easter. The fasting lasts for 40 days and entails giving up certain pleasures or foods. 80% or more of respondents in Chad, Nigeria, Ethiopia and Ghana said that they fast during this period. Although only a third reported fasting in South Africa, the overall median is 69%. In a majority of the countries surveyed, equal proportions of Protestants and Catholics reported that they fast (Pew Forum, 2010, p. 29). In addition, a higher proportion of Pentecostals than non-renewalist Christians reported that they fast, the majority several times a year (except in South Africa) (Pew Forum, 2006). In Sri Lanka, nearly two-thirds of Catholics claim to refrain from eating meat during Lent (Asia Foundation, 2012, p. 9).

The obligation to fast during Ramadan (*sawm*), the ninth month of the Muslim calendar, when the Qur'an is believed to have been sent down to Muhammad, is the fourth pillar of Islam. Detailed in *shari'a*, it entails refraining from eating, drinking, smoking and sexual activity between sunrise and sunset for 30 days and stricter-than-usual observance of prayer and attendance at mosques to participate in reading the Qur'an. It is even more commonly observed amongst Muslims than Lent fasting among Christians, ranging from 100% of Muslims in Cameroon, 90%+ in Chad, Senegal, Ghana, Ethiopia, Kenya, Nigeria, Tanzania and Mali, down to 69% in DRC (median 93%) (Pew Forum, 2010). Some Muslims are excused: those who are ill or on a journey can fast for a similar number of days at another time of year, while pregnant and menstruating women, the elderly and children are also exempt. At sunset, a snack is eaten to break the fast and then a substantial family meal later, with the first meal of the day eaten before sunrise. Towards the end of Ramadan, various events are celebrated: on the 27th day, Muslims remember the first night during which Muhammad received a revelation, while Eid ul-Fitr, involving a family gathering, feast and the exchange of gifts, is the celebration that ends Ramadan (Oliver, 2010; Ruffle, 2013; Watt, 1996).

Other festivals are celebrated in some Islamic traditions, especially in Sufi variants (Bop, 2005; Sieveking, 2007). For example, members of the Tiyanniyya *tariqa* in Senegal celebrate the birth of the Prophet, members of the Muridiyya the return of the order's founder from exile by the colonial powers and members of the Layennes the call of the founder. Such convivial occasions, in which women play an important role, involve visits to the shrine of the founder, religious meetings and chanting. In these rituals, the public manifestation of women's spirituality is not perceived as being in conflict with Muslim tradition.

Fasting is also practised by Hindus. For example, 21% of Hindus in Sri Lanka claim that they fast at least once a week and 22% between one and three times a month (although 18% say that they do not fast) (Asia Foundation, 2012, p. 11). Every lunar cycle is marked by days of fasting and feasting at home, depending on individuals' choice of deities that they worship. There are a variety of fasting practices, typically involving abstaining from certain foods. Some are specific to women: they honour particular deities and have different purposes, for example ensuring conjugal harmony and the welfare of the family or attaining a goal (for example, to conceive or to end a child's illness). Often food cooked daily is ritually offered at the home altar and then consumed, and there are many rules and regulations related to its preparation, offering and consumption, although observance of practices related to food and eating is generally more relaxed today than in the past. The guidance provided is mostly either oral or pamphlets available at temples and pilgrimage centres rather than the classic texts. Practices vary geographically and over time. The Hindu calendar is also filled with festivals at different scales and with many regional variations. They sometimes commemorate events in the epics, through rituals that interpret these events literally or symbolically, and may include dance, bathing, lighting lamps, giving gifts/offerings, praying and feasting (Narayanan, 2013; Sanford, 2013).

Pilgrimage

Pilgrimage to places perceived as sacred, mentioned above with reference to Sufi shrines, is a feature of many religious traditions. It is regarded as an expression of devotion, a means of self-discipline and a way of obtaining a response to illness or a problem.

During the early period of Buddhism in India, *stupas*, "memorial monuments replete with cosmological and associated royal symbolism that represented the Buddha and were, in most cases, believed to contain a portion of his relics" as well as images, were constructed (Reynolds and Hallisey, 1987, p. 339). Over the centuries, these have been associated with many ritual practices and have been the destination for pilgrimages by adherents of many Buddhist sects. Often they were associated with places such as mountain peaks, whose sacred associations predated the sacred sites associated with the Buddha's life. These pilgrimage practices assimilated deities and practices associated with local religious traditions, which became imbued with Buddhist connotations. Buddhism penetrated not only the sacred topography of the areas into which it expanded but also their calendrical rites. The latter include both large-scale ritual occasions and the lesser rituals associated with daily life, such as life cycle rites, especially those associated with death and initiation into adulthood.

In Hinduism, pilgrimage to sacred sites or personages is an important part of the tradition. Because it is believed that divinity may be found in human form, as described in many of the Hindu texts, some of the most inspirational gurus are held to be incarnations of a deity, enabling them to teach and sometimes perform miracles; some Brahmins claim the same for themselves, although this is often contested (and some saintly figures come from lower caste groups); and practitioners of yoga may claim divine powers. There are therefore a large number of Hindu pilgrimage sites on the Indian subcontinent, where sacred geography is important, because particular sites are linked to different deities and events in the epics. The sacred rivers, the Yamuna and the Ganges, are particularly important – not only are they important suppliers of water to much of North India, they are also theologically important because they are thought to be the incarnation of the Goddess. Bathing in one of the rivers is believed to wash away impurities and transgressions. Often pilgrims bring back images or pictures of deities for their household shrines or Ganges water to be used in rituals on special occasions.

Some pilgrimages are annual, some are once in a lifetime (Sanford, 2013; Smith, 2009).

Perhaps the best-known pilgrimage globally is the *hajj*, the fifth pillar of Islam. Every Muslim has a duty to perform the *hajj* at least once in his/her lifetime if possible. The main pilgrimage is held in the 12th month of the Islamic calendar, although it can be done at any time (the lesser pilgrimage or '*umra*'). In 632 the Prophet Muhammad completed a *hajj* to Mecca, and it was on this occasion that most of the ritual was specified, so the requirements are detailed in *shari'a*. Since that first pilgrimage, all adults, provided they are well, able to afford the journey and able to make arrangements for the care of their families in their absence, are expected to make the pilgrimage. Wearing simple white garments (*ihram*) signifying a state of purity, it involves pilgrims making a series of predetermined visits to and ceremonies in locations in Mecca and its surroundings and concludes with the festival of Eid ul-Adha. Some pilgrims then go to Medina to visit the place where Muhammad is buried (for further details, see, for example, Oliver, 2010, p. 104; Ruffle, 2013, p. 212; Watt, 1996, pp. 84–5). Many governments of countries with significant numbers of Muslim citizens provide financial assistance. As the number of Muslims has increased and international travel has become easier, the number of Muslims making the pilgrimage has grown, leading the Saudi authorities to spend large amounts to enable them to accommodate the growing number of visitors but also to impose quotas for the maximum number of pilgrims permitted from any one country. On their return, those who have made the journey are entitled to use the honorific *hajji*, so that as well as being a rewarding experience for many pilgrims, making the pilgrimage confers status.

In some Islamic traditions, pilgrimages to the shrines of saints are also popular – for example, increasingly so in Indonesia, according to Quinn (2008). Pilgrims to such sites are seeking out places associated with people who are considered to have succeeded in getting close to God in the hope of themselves attaining blessings. Because all good fortune is thought to come from God, it is considered natural for those who prosper after praying at such a site to repay their good fortune in kind. Quinn suggests that such practices are likely to continue, despite the disapproval of other Muslims, because the intertwining of religion and business at such pilgrimage destinations benefits local communities.

Religious giving

In this book, a distinction will be drawn between giving for religious purposes, which is regarded as a religious duty, and altruism or philanthropy, which may be seen as a moral imperative or obligation, although not everyone would draw this distinction, the dividing line between the two is fuzzy, and donors' motives are complex. Religious giving can refer either to the religious activity of giving (altruism motivated by religious belief) or to giving for explicitly religious enterprises (Vacek, 2010). The latter may be narrowly or broadly defined. A narrow definition refers to the establishment and operation of religious institutions and organisations and activities to promote their growth, typically by proselytising.

A broader definition may include support for activities undertaken by religious organisations. Philanthropy, based on a value promoted by many religions, is discussed in the next chapter, which considers values and ethics and their intended and actual impacts on behaviour. The former is examined here.

Religious giving as a religious duty is often expected to yield religious rewards or merit. Smith (2010), for example, identifies three ways of understanding the relationship between God and giving in Christianity, Islam and Judaism. First, God commands obedience, in this instance to the duty to give. Second, God as redeemer rescues people from guilt, punishment for sin and fear of death – giving by believers is regarded as an expression of gratitude. Third, God as love encourages the believer to participate in the life and mission of God, which may be seen as enabling the life and expansion of the church or realising the love of God for other people and the natural world. This is most commonly perceived as a duty but may also be regarded as voluntary – something all those who love God and have been transformed will want to do. Such dutiful giving may be expected in turn to bring spiritual rewards for the donor.

In Christianity, religious giving traditionally and commonly takes the form of tithing, generally an expectation that adherents will give 10% of their incomes to the church of which they are members, to support the clergy, build and maintain churches and related buildings and enable proselytising activity, including preaching and church planting. For example, in Pentecostal churches, the obligation to give is emphasised in every worship service. When congregants give, they are not only providing sustenance for the pastor, they are also enabling the renewal of the whole church community and justifying their own striving to accumulate wealth. In Africa, what is often a public display of generosity is recognised and valued. It builds on a traditional expectation that the wealthy (for example, chiefs) will redistribute part of their wealth for the benefit of the community. However, unlike a person's obligations to kin, the obligation to give to the church is not limitless and so for the most part is said to enable members to retain sufficient resources to achieve individual social mobility and success (Swidler, 2013). The majority of Christians in all the countries where the Pew Forum carried out surveys reported that they practise tithing (Pew Forum, 2010, p. 29).

Islam distinguishes between *zakat* and voluntary almsgiving. The third pillar of Islam, *zakat* is the obligatory payment by Muslims of a proportion of their wealth for the benefit of the poor, because, it is thought, extremes of wealth and poverty distract from the moral, God-conscious life for both the wealthy and the destitute. It is considered to be an important means to "counter the human urge toward greed, miserliness and hoarding, (all strongly condemned in Q.3:180, 9:34, 64:16)" (Hashmi, 2010, p. 185). Because it is a religious duty, paying it is regarded as an act of worship (*ibadah*). Normally paid once a year at the end of Ramadan, it is associated with various rituals (Ruffle, 2013, p. 213). The Arabic verb *zakā* means to increase and grow as well as to purify. *Zakat* worship is thus supposed to help safeguard, purify and multiply wealth. Retsikas (2014) summarises the teaching as follows (see also Hashmi, 2010; Ruffle, 2013):

> Conducting zakat worship brings not only the faithful closer to Allah but also serves to purify the hearts of believers from unwanted desires such as greed, self-indulgence, hedonism and decadence. It also renders the remainder of wealth clean (*bersih*) for one to enjoy and dispose of. Moreover . . . handing over zakat to its rightful owners, i.e. the eight categories of people as enumerated in the Qur'an, opens the door to more rewards, material and otherwise, being received from Allah. In addition, zakat protects the wealth that remains from being stolen, misplaced, lost or simply scattered around. When zakat is not handed over, it is said to bring about misfortune. . . . Those who refuse to conduct zakat are sometimes said to have become kafir [unbelievers] and on Judgement Day, they [are expected to] . . . come face to face with the wealth they withheld, for not transferring the zakat due is [seen as] a major sin.
>
> (Retsikas, 2014, p. 343)

Zakat is a tax on wealth (beyond that necessary to provide the necessities for living), providing this is capable of increase. Such wealth is customarily conceived of as livestock, agricultural produce and savings, for example gold. *Zakat* is not payable on fields, which Muslims are encouraged to use for productive purposes, in order to achieve prosperity. Nor is it payable on one's own home or jewellery, which are in daily use. Details of how the amount due is calculated, as well as to whom it should be allocated and through what channels, are set out in *shari'a* (Watt, 1996). Today, the *zakat* due is commonly equated by Sunni Muslims to 2.5% of their disposable wealth (the income beyond what is required to provide the necessities of life) at the end of a year. *Zakat* has implications both for understandings of property rights and for the status of those eligible for assistance. First, the possessor of wealth is held to have rights of possession and disposal only after the share of property and/or income to which others have rights under *zakat* (mainly the poor and destitute) has been disbursed. In addition, those who receive *zakat* are not regarded as 'recipients' but as people entitled to *zakat* (so that the non-payment of *zakat* is considered to be theft). Moreover, because the payment of *zakat* is expected to ensure Allah's protection for the remainder of one's wealth, failing to pay is regarded not just as avoiding submission in worship and so akin to unbelief and an affront to Allah but also as rendering the non-payer susceptible to punishment by misfortune in this life (Retsikas, 2014).

The majority of Muslims in every country included in the Pew Research Center's surveys claimed that they adhere to the injunction to pay *zakat* to help the poor and needy, although in countries where the representative samples include a large proportion of poor people, it is not clear what these claims mean – perhaps that those who do not consider themselves to have sufficient wealth to sustain their families exempt themselves from the obligation. In the Asian countries included in the surveys, 90% or more of Muslims reported that they pay *zakat*, while the proportion varies more between African countries, ranging from 87% in Kenya to 58% in Mozambique, median 77% (Pew Research Center, 2012, p. 52). However, other evidence, for example from Nigeria and Tanzania, shows that although many Muslims pay voluntary alms (see below), the payment of *zakat* is not widespread, in part because many Muslims are poor. In Tanga, a largely Muslim coastal city in northern Tanzania (population 244,000 in 2002), for example, declining sisal production, post-liberalisation de-industrialisation and reduced port and railway employment have resulted in a depressed economy and widespread poverty. Relatively few people are sufficiently well off to be liable for payment of *zakat* – according to a local religious leader, "*ufukara umeingia kiasi kwanba kuamba hata kuabud u ni taabu*" (there is so much poverty that even fulfilling religious obligations is a problem) (Mesaki and Kipacha, 2010, p. 16).

In the last 40 years or so, in Indonesia (and elsewhere), *zakat* has been repositioned from an annual obligatory act of worship to a means of achieving poverty reduction, ensuring people's welfare and securing socio-economic justice in the face of changing circumstances (urbanisation, the emergence of a sizeable salaried/waged population associated with industrialisation and economic liberalisation, but also stubborn poverty, impoverishment and increased inequality) (Retsikas, 2014). This attempt to redefine the use of *zakat* from providing charitable assistance to various categories of poor and needy people to a possible tool for socio-economic transformation has a variety of implications for the organisational framework for its collection and disbursement, which will not be examined here. However, it is worth noting that some of the original categories of people identified in the Qur'an as eligible for *zakat* have been reconsidered. For example, Dehlvi (2015) draws attention to the category of *Fi Sabeelillah* (in the cause of God). The original purpose of this was to meet the costs of defensive expeditions that Muslims had to launch not to spread their faith by force but for their own survival and safety. Today, Dehlvi notes, this has been construed by contemporary jihadists as including all methods of promoting Islam, including attempts by militant individuals and organisations to spread the faith and establish Islamic states.

In Hinduism, there is a strong tradition of renouncing the things of the world in order to seek spiritual fulfilment and *moksa* (liberation from *samsara*). Most individuals choose to pursue the first two stages of life (*ashramas*): student and householder. However, once children are grown up and householder duties fulfilled, a man (and his wife) may retire to a secluded place (traditionally a forest) to devote himself to religion. Finally, (or indeed at any stage) a man may, if he wishes, give up his wife, possessions and material comforts and become a wandering ascetic, offering moral and spiritual guidance to those he encounters, subsisting by begging and so dependent on the generosity of others (*dana*). Many renunciants wander alone (*sadhus*), although some may become members of monastic orders. The tradition also has a trained and ordained Brahmin priesthood, including *pujaris*, who perform temple rituals, and *purohits*, who perform rituals in other locations, especially life cycle rituals for families (although many do both).[1] Priests are paid for the services they perform and may also engage in agriculture or trade. Regarded as the protectors of religious learning, many

1 Today, there are some non-Brahmin and women priests.

earn a living as teachers, which is considered preferable to depending on gifts or alms. Spiritual guides and teachers of religious practices include *swamis* or *gurus*, who may be ordained priests or self-styled, men or women. They may be venerated, even worshipped, attracting large numbers of followers, maintaining retreats (*ashrams*) and initiating welfare or educational activities and organisations. Temples, which fulfil educational, welfare and medical roles in addition to their religious ones, are regarded as the property of the presiding deity. They are often managed by committees, with or without a degree of state involvement. For the land on which they are built, construction and day-to-day costs, temples and ashrams depend on voluntary donations from landowners and devotees, who earn religious merit by their giving.

In Buddhism, *dana* or almsgiving is considered part of the *dharma* (duty) of individuals and expected to increase their religious merit (Tomalin, 2009). The monastic orders are more important in Buddhism than in Hinduism, and alms are given primarily to monks (and nuns), for whom they are often the sole source of subsistence. Theravada Buddhists, for example in Sri Lanka, may give anywhere, but often ritual offerings are made following *pirit* ceremonies, which are held to seek protection, good fortune and merit prior to many events and activities, including major life cycle events such as marriage, housewarming, during mortuary rituals, in commemorations of the death of a relative or when embarking on a business activity. Alms are given after offerings have been made to the Buddha and appropriate verses have been chanted, by placing part of the alms (which may include alms bowls, robes, towels, books or umbrellas) before a picture or statue. Laypeople may take their alms to a monastery or invite a group of *bhikkhus* to their houses. Often, in Sri Lanka, a meal is shared after an all-night *pirit* ceremony. "The almsgiving ritual ends with a transference of merit obtained from the act of giving to all in the gathering as well as to the gods and the deceased kindred who are awaiting rebirth or are in inauspicious rebirths" (Goonasekera, 2013, p. 122). The religious merit gained from giving alms is influenced by the spiritual status of the latter, with more karmic merit gained from giving food to a monk than to a low-status nun and least to a starving homeless person, who may be regarded as experiencing the consequences of poor *karma* (Loy, 2010, pp. 52, 59). As is the case with respect to religiously mandated giving in many traditions, the act of apparent generosity is often regarded as more beneficial to the donor than to the recipient.

Socialisation, learning and proselytising

Religious traditions, sects/denominations and congregations have generally developed elaborate formal and informal mechanisms to ensure that adherents are familiarised with and socialised into their religious tradition to secure its continuation and, often, expansion. Socialisation or enculturation refers to assimilation of the contents, practices and values of a culture or institution by individuals (who may influence that culture or institution, as well as being its products). Some of the absorption is unconscious – individuals are conditioned, independently of their own will, by the beliefs, values and rituals of the society and religious group of which they are part. These provide them with conceptions of life and death and with an understanding of religious interpretations of the material and spiritual dimensions of life. They may be transmitted through everyday activities, symbols, myths and/or participation in rituals. The latter, as has been argued in Chapter 5, permit encounters with co-religionists, reinforce belonging and deepen values, as well as making affiliation visible. The ensemble of beliefs and practices forms a value-laden habitus into which people are socialised. Even if many people in a society do not actively participate in religious activities, Cipriani (2001) suggests, mass religious socialisation results in long-term conditioning and the diffusion of beliefs and values throughout a society, especially where a particular religious tradition is hegemonic. In addition to such a process of enculturation, most traditions have explicit mechanisms and systems to ensure that both those born into a religious and converts learn about the beliefs and practices associated with the religious group.

Religious education and learning

Important in the socialisation process is religious education for children, which may be provided by religious organisations, as part of the state education system or within families, in combinations which differ between the faith traditions, depend on the legal and political relationships between states and religious groups and are related to social characteristics such as parents' level of education.

Many religious groups and parents seek to socialise children through specific activities, particularly educational activities associated with a church/mosque (Sunday schools, *madrasas*), in schools run by the religious group concerned or in state schools whose curriculum contains religious content. The role played by schools and further education institutions run by religious organisations varies between the faiths, depending on their history and relations with successive governments. For example, for the Christian churches, the provision of education has played a central role in missionary activities in Sub-Saharan Africa, partly because of the importance attached to adherents reading the Bible for themselves and partly because during the colonial era, these educational institutions were the main source of indigenous staff for the administrations of the colonial powers. Similarly, in Islam, most mosques run *madrasas* in which children (especially boys) learn to recite and read the Qur'an in its original language, while advanced *madrasa* education provided administrators for Muslim states and empires, although not usually for colonial administrations, a source of political tension with repercussions since independence in many countries. The role of religious organisations in providing secular education will not be explored further here but will be taken up in the linked volume.

Religious content in school curricula may be permitted, promoted or reinforced by state action, depending on the political and legal relationships between a state and religious groups. For example, research amongst Nigerian Christians in 2008/9 found that they consider the purpose of religious education to be "to bring people into a right relationship with God and help them to develop their potential gifts to the glory of God", equipping them to serve the church and society. This, some think, can be achieved by integrating Christian values in school curricula alongside secular subjects (Interview with Rev. E. Ali, of the Evangelical Church of West Africa Education Department, 8/10/09). While both religious and secular educationalists emphasise the need to impart knowledge and skills, in Nigeria (and elsewhere), many of the former believe that it is the role of religious education to impart values.

The influence of religion on the content of school curricula is a contentious matter – conflicts between religious and secular educationalists (and between religious educationalists from different traditions or points on the conservative–liberal spectrum) are quite common, for example with respect to values related to human rights, freedom of the individual and the prioritisation of personal advancement. In Nigeria, for example, some religious educationalists are reportedly happy with the latter (for example, believers in the prosperity gospel), but others prefer to stress the values of modesty and restraint. Differing doctrinal teachings also have implications for curriculum content, for example with respect to creation and evolution. In addition, the operation of schools by religious organisations or the incorporation of curriculum content influenced more by one tradition than another is thought to be socially divisive, potentially exacerbating inter-religious conflict and hindering social cohesion. For example, in Sri Lanka, four-fifths of respondents in the 2011 survey referred to above indicated that they (or other members of their family) had attended state or religious schools where religious education was part of the curriculum (or were attending such schools at the time of the survey). The receipt of religious education in school was almost universal for Buddhists (and lowest among Hindus), and nearly all the respondents believed that religious education had become more important during the five years prior to the survey (Asia Foundation, 2012). In a religiously mixed and politically volatile country, the dangers for resolution of post-conflict tensions and social cohesion are clear.

Arguably, religious education within the family is as, if not more, important than that provided by religious organisations or in state schools. A large majority of respondents in most of the 12 relevant countries surveyed by the World Values Survey in 2000 said that it is important that children learn about their religious faith at home: 93% in Indonesia, 86% in Pakistan, 75% in Tanzania, 74% in Zimbabwe, 70% in Bangladesh, 68% in Nigeria and Uganda, 61% in South Africa and 59% in the Philippines. Exceptions were Singapore (30%) and Vietnam (8%), the latter presumably a legacy of the clampdown on religiosity during the communist period. In India, only 29% of parents surveyed in 1990 said that they thought it important for children to receive religious education at home, but 47% responded positively to a similar question in a 2000 survey (Halman et al., 2007, p. 31). As discussed in Chapter 3, although the nature and status of religious texts varies between the faith traditions, in many they are accorded special status and so play central roles in religious education. For example, the majority of Christian and Muslim parents in nearly all the countries surveyed by the Pew Forum in

2008/9 said that they pray or read the scriptures with their children (Pew Forum, 2010, p. 30). Although this question is deliberately phrased not to differentiate between parents with differing degrees of literacy, their level of education (as well as their work and other commitments) inevitably affects the type of religious education they can offer to their children. As will be seen below, this is also an important influence on trends in and channels for adult religious learning.

Many religious groups recognise that adults, whether long-standing adherents or newcomers to a faith, need to learn about the teachings of their religion on an ongoing basis. Sometimes this is tackled through the content of traditional channels, for example Sunday or Friday sermons, which remain popular, and potentially reach large audiences. For example, Millie (2008) examines the practice of a preacher in West Java with a large following, noting how his sermons, which encourage self-reflection, use examples from village life and demonstrate his religious credentials by reference to the texts and the use of Arabic alongside the local language, remain popular amongst lower- and middle-class rural Indonesians.

However, the channels through which religious educational activities are delivered have proliferated in recent years as a result of the increased availability of new technologies. The packaging of religious messages into products accessible to individuals has increased the frequency of practices such as reading the scriptures or listening to recorded sermons and enabled adherents to increase their religious knowledge (Guest, 2009). For example, large majorities of both Christians and Muslims in all the SSA countries surveyed by the Pew Forum said that they read religious pamphlets, books and magazines, and in most countries large numbers also said that they listen to or watch religious radio and television (Pew Forum, 2010). In Kenya and Nigeria, in the mid-2000s, about three-quarters (and in South Africa nearly two-thirds) of Christians reported using these religious media at least weekly, and an even higher proportion of Pentecostalists than non-renewalist Christians did so, especially in Kenya. The proportion of Christians who reported using these print and electronic media was markedly lower in both India and the Philippines (42% and 46% respectively). However, as in the African countries, renewalists were more likely to do so than Christians associated with the mainstream churches (Pew Forum, 2006, p. 20). While some of the production of such materials is done by individuals, much packaging and transmission of knowledge involves the collective structures (churches and other organisations), helping to influence interpretations of religious teachings and the reproduction of religious identities.

The religious links of some states influence their policies and the media and messages by which they promote their aims. For example, in Malaysia, there are close links between the state and conservative Muslim (and Christian) forces, which both play a role in the national conversation about family values. A campaign to inculcate appropriate social values was launched in 2002 by the newly established Ministry of Women and Family Development and its National Population and Family Development Board. This promotes Islamic/family values (including homemaking and childcare as women's most important tasks) through parenting advice, programmes, events and partnerships with conservative national and international religious groups (Stivens, 2006).

In addition to the above, many religious groups advocate and organise continued learning for adults. In most if not all faith traditions, reading religious texts, both in public and in private, is an important religious practice. The results of the Pew Forum's surveys in 19 SSA countries show that, while reading sacred texts is less common than attending religious services or praying, it is still widespread: a majority of respondents in seven of the countries and more than 40% in nine additional countries reported that they read the scriptures at least once a week (Pew Forum, 2010). Of the Muslim respondents in its surveys in five SE Asian countries, between a third and a half claimed that they read the Qur'an once a day, while none in three of the countries (southern Thailand, Indonesia and Malaysia) and only 14% in Pakistan and 5% in Bangladesh admitted to never reading it. Only in Pakistan did significantly more women than men (46% compared to 26%) report that they read the Qur'an every day (Pew Research Center, 2012). Amongst Christians, while less common than daily prayer, 40% of Pentecostalists, a larger proportion than amongst charismatic or non-renewalist Christians, claim to read the Bible daily (Pew Forum, 2006, p. 19).

Often, religious texts are read alongside other activities in prayer and other groups. Amongst respondents to Pew Forum surveys in SSA, a large majority of Pentecostalist Christians reported that they participated in prayer groups, Bible study and other activities at least weekly, varying from 62% in South Africa to 82% in Nigeria. The proportion of charismatic Christians in the mainstream denominations who reported such engagement was invariably lower, although it varied from country to country, while fewer than half of non-renewalist Christians reported participating in such activities (Pew Forum, 2006, p. 22).

Islamic teaching and socialisation are concerned above all with teaching Muslims how to worship correctly, as well as how to behave and interact with others, including segregation of the sexes, respect for elders and the need for modest dress. The range of education and learning activities available to Muslims in a typical African city is illustrated by research findings from Jos in northern Nigeria (Box 6.1).

Box 6.1 Islamic education in Jos, Nigeria

In Jos, like most places with a Muslim population, religious education for young children has traditionally been provided in Qur'anic schools, while specific channels for inculcating Muslim beliefs and values and increasing religious knowledge amongst women have evolved more recently. Especially since the 1970s, in addition to preaching sessions inside and outside mosques, the number of Qur'anic classes for women and special preaching sessions has grown. These are almost always delivered by men, who provide oral exegesis (*tafsir*) of the Qur'an, especially during Ramadan. Books produced by senior scholars contain chapters on women's issues, and teachers often attempt to explain to women how they can be both 'Western' in their mode of life and also devoted Muslims.

Religious education and socialisation also occur through the media. The oldest Hausa-language newspaper in Jos was started in 1939. Today there are a variety of newspapers and journals, with varying religious content. Radio and TV are popular media (all except one station are government owned) and are widely used to disseminate religious messages. All the radio and TV stations make available free and paid-for airtime for both Muslim and Christian preaching. Jama'atu Nasril Islam administers the free airtime allocated to Muslims, dividing it amongst different (sectarian) organisations. Every day, airtime on the one independent radio station is bought by Muslim organisations and individuals for Qur'anic exegesis (daily during Ramadan), with Izala, a Wahhabi-influenced organisation, dominating. The richer Pentecostal churches dominate the commercial airtime available to Christian organisations. In addition, broadcasts from radio stations in neighbouring states reach Jos.

As well as the texts themselves, commentaries are important because they offer influential and potentially conflicting interpretations and guidance. In Islam, for example, both individual scholars and Muslim organisations issue *fatawa* (authoritative opinions) on aspects of *shari'a*. For example, in Indonesia, in addition to those issued by individual scholars, they are produced by the three main Muslim organisations, the mass movements of Muhammadiyah and Nahdlatul Ulama and the *Majelis Ulama Indonesia* (Indonesian Council of Islamic Scholars). In the past and conventionally, these have been issued in paper form, but today, increased access to the internet (especially in urban areas) has widened the range of sources available to Muslims, made it quicker to obtain a response and enabled enquirers to pick a preferred interpretation (Hosen, 2008). While online "fatwa shopping" is increasing, it is controversial and, for the foreseeable future, in Hosen's view, is likely to complement rather than replace traditional sources of guidance.

The frequency with which Christians and Muslims read religious texts is clearly associated with the status and standing accorded to the Qur'an and Bible in these two religious traditions. In contrast, as discussed in Chapter 3, other religious traditions may not have a single founder and central written text(s). In Hinduism, for example, study of the foundational Sanskrit texts, as noted in Chapter 5, was restricted to Brahman boys, who could choose to do this all their lives rather than moving on to the second stage in life as householders, with responsibilities for their families and communities. Moreover, the "Vedas, the oldest and most authoritative texts, are less relevant to daily devotional practices than are [others]. . . . Today, . . . later works such as Hindu epics or treatises dedicated to specific deities play far greater roles in Hindus' lives" (Sanford, 2013, p. 80). Hindus emphasise orality and sound over the written texts, many of which lend themselves to being read aloud and listened to rather than read silently. Oral narratives (for example of the epics and the Bhagavad Gita) and performance provide stories that can inform devotional practices and ethical choices and behaviour while allowing flexibility and improvisation.

All the religious traditions use both textual sources and oral communication, but they are located in different (and overlapping) places along a broad oral–textual spectrum. While Hinduism reflects a more oral tradition than either Christianity or Islam, folk religions are located even closer to the oral end of the spectrum. In contrast, the growth of revivalist movements in Islam and Christianity is, as discussed in Part II, associated with literalist interpretations of the texts. It is marked by laypeople's desire to read the texts for themselves, related to their wish to develop a better understanding of their faith as part of their quest to forge more pious selves, increasing levels of education and the feeling amongst some population groups that they have been excluded in the past from opportunities to receive a religious education. While Bible study has always been important in Christianity and has been boosted by the growth in revivalist Christianity and the proliferation of Bible study groups for adults that has accompanied it, the trend is most noticeable in Islam.

Muslim women, in particular, feel that they have had many fewer opportunities to read the texts and learn about their religion than men. In gender-segregated societies, assisted by financial flows from the Middle East, recent years have seen increased numbers of *madrasas* for girls (as well as boys) in many countries, at every level of education from children's classes in local mosques to post-secondary education. Since the 1980s, as described in Chapter 3, study groups have also proliferated, especially for women. These may be associated with mosques and draw on the knowledge of male teachers, but many are organised by women who have had the opportunity to acquire some religious learning themselves. They are particularly associated with urban women who are at least literate and are often well educated in both secular and religious subjects. Women meet to read the religious texts (especially the Qur'an and *ahadith*), listen to commentaries and sermons, learn Arabic and pray together. Often using Arabic alongside the local language, they aim to increase participants' knowledge of Islam, especially its ritual requirements and rules for daily life (Chao, 2014; Gomez-Perez, 2016; Schultz, 2008a, 2008b, 2011). Participants' self-stated motives include a desire to understand the written sources for themselves, acceptance of individual responsibility for developing a close relationship with God, and the desire to cultivate a religious disposition, understood as modesty, submissiveness and patience. The groups emphasise learning and understanding the texts rather than interpretation of them through theological study or reinterpretation. They also stress the role of learning as a conduit to moral renewal of self and society, because it enables adherents to defend their conception of a proper Islamic way of life. The scope for discussion and debate varies – in some groups those present listen passively to readings, commentaries and sermons, while in others there is much discussion of correct religious and social conduct in the face of the challenges of everyday life in communities, work situations and family relationships. In addition to their religious study functions, groups may provide their members with social support and financial assistance (for instance, through rotating savings and credit arrangements) and forge social links across religious/sectarian, ethnic and other social boundaries.

In recent years there have been detailed studies of the phenomenon in various Muslim countries, some of which are referred to in Box 3.1 (in Bangladesh, Sri Lanka and Mali). In addition, Chao (2014) provides an account of *pengajian* (sermon groups) in Indonesia, which are fostered by the well-established mass Muslim movements, Muhammadiyah and NU, as an alternative to more extremist elements of the piety movement such as the religious political party PKS (Partai Keadilan Sejahtera). All over Indonesia, women seek religious instruction, read religious texts and engage in prayer in *majelis taklim* (study groups), some of which are independent and others organised into networks. A 2010 survey showed that three-quarters or more of Muhammadiyah and NU members often participate in these types of community-based religious activities, including *majelis taklim*, *pengajian* and local celebratory rituals (Bush, 2014, p. 13). Gomez-Perez (2016) has studied similar groups in Burkina Faso.

Often, religious educational opportunities and learning activities are sex segregated and the religious knowledge conveyed is gendered, with ambiguous implications for gender relations (see also Chapter 9). As noted above, the increased opportunities for women to engage in learning activities are closely associated with revivalist movements in Christianity and Islam. Women's groups may serve to reinforce and perpetuate patriarchal power relations within the family, religious group and wider society (see, for example, Deacon and Parsitau, 2017, on Pentecostal churches in Nairobi). However, they may also enable members to discuss issues considered too subversive or indecent for general discussion, on the grounds that they are referred to in the Bible. For example, Gecaga (2006, in Jenkins, 2006) records how the story of Tamar, who is raped by her brother, provides women in South Africa with a way of discussing female sexuality, rape and incest. The Muslim study groups described above are generally religiously and socially conservative, but participation in them may lead to increased participation by women in public rituals and the affairs of mosques and other Muslim organisations. Chong's findings are similarly contradictory: women's organisations, cell meetings, Bible study groups and seminars in the South Korean Pentecostal churches she studied enable women to experience self-transformation and personal fulfilment, but above all to maintain existing family and gender relationships, by stressing women's obedience (especially to their husbands) and endurance (to ensure that a marriage lasts). These, she argues, are enforced by group disciplinary techniques such as confession, self-criticism and learning appropriate attitudes and behaviour (Chong, 2012). Questions of gender and sexuality are discussed further in Chapters 9 and 10.

Proselytisation and conversion

Beyond the socialisation and continuing education of adherents, those religions that consider themselves the one true faith believe that they have a duty to transmit this message through proselytising or *daw'ah* (invitation to Islam), either to convert non-believers or to increase the level of awareness and observance of non-practising fellow adherents.

Proselytisation and conversion, or what the Pew Research Center has taken to calling religious switching, have been the subject of detailed study. The complexities of conversion are discussed by, for example, Freston (2001), Pew Research Center (2015) and Woods (2012). Freston and Woods note that the forms and meanings of 'conversion' vary over space, time and tradition. Analyses have tended to concentrate on successful Christian and Muslim proselytisation and to neglect non-conversion (failure) and non-proselytising groups (e.g. Buddhists). Human agency perspectives seek to track and explain the conversion trajectories of individuals but are said by critics to lead to a psychological bias that emphasises the fulfilment of unmet emotional needs and treats converts as autonomous of the social fields in which they live. In addition to understanding individual motives for conversion it is, therefore, necessary to examine social, economic and political drivers, such as social relationships at the family, local and national levels, the results of modernisation and changes in the socio-political context, especially those that result in social disorganisation and the marginalisation of certain groups. Woods is also concerned with how contextual factors may foster or constrain efforts to proselytise and influence individual and group responses, noting that conversion implies not only a (re)-definition of self but also changed perceptions of the 'others' with whom converts have social relationships, possibly leading to tensions at both the political and individual levels. The questions posed in surveys of conversion tend to oversimplify affiliation, assuming that people 'belong' to one religious tradition, sect or denomination, whereas in practice, as has been discussed above, much religious belief and practice is syncretic, and some processes of conversion allow converts to combine a new religion with existing beliefs and practices.

A substantial proportion of respondents in the Pew Forum's surveys of Christians and Muslims confirmed that they regard it as their duty to seek to convert others (Pew Forum, 2010, p. 30) (see Table 6.1). For Christians, this perceived duty is consistent with a belief that accepting Jesus Christ is the only way of being saved from eternal damnation. In some countries, a majority of Pentecostal Christians said that they share their faith with non-believers at least once a week (59% in Kenya, 58% in South Africa, 56% in the Philippines, 41% in Nigeria and 37% in India) (Pew Forum, 2006, p. 21). This was a larger proportion than of charismatic Christians (except in Kenya and India) or non-renewalist Christians. Relatively few Pentecostalists, but a larger proportion of other Christians, said that this is something they never do (Pew Forum, 2006).

Table 6.1 Proportion of Christians who believe that they have a duty to evangelise

	% of all Christians	*% of Pentecostals*	*% of charismatics*	*% of other Christians*
Kenya	91	91	91	87
Nigeria	91	96	a	93
South Africa	58	70	64	56
India	32	64	57	34
Philippines	58	77	58	57

Source: Pew Forum, 2006, p. 27.

[a]In Nigeria, the survey included only 67 respondents who self-identified as charismatics, too few for them to be analysed separately.

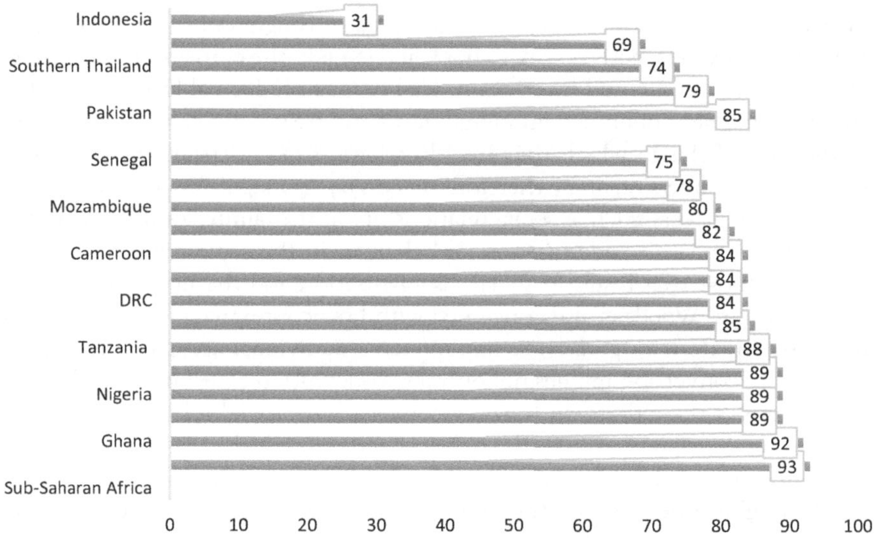

Figure 6.1 Proportion of Muslims who believe that it is their religious duty to convert others to Islam

Source: Pew Research Center, 2013, p. 112.

In most countries, the vast majority of Muslims believe that theirs is the one true faith and that it is their religious duty to convert others to Islam (see Figure 6.1). Figures are available for 15 SSA and five southern Asian countries, amongst which Indonesia is unusual. Generally more devout Muslims are more likely to believe that they have an obligation to try and convert non-Muslims. However, how extensively they engage in *da'wah* in practice varies, depending on the degree of religious pluralism, inter-religious relations and whether the number of adherents is a sensitive issue for political or policy reasons.

In the past, much conversion to Islam or Christianity was from those engaged in folk religious traditions. Today, in practice, almost all Muslims in the 39 countries in which the Pew Forum conducted surveys in 2011 reported that they had been born and raised as Muslims (except in some SSA countries, where a minority said that they had converted from Christianity, and in Thailand, where a few had grown up as Buddhists) (Pew Research Center, 2012). Susceptibility to conversion varies between adherents of a particular religious tradition, as do the motivations of converts, both individual and socio-political. Sometimes marriage may be accompanied by conversion of one of the spouses (usually the woman to the religion of the man).

In India, the desire to escape caste discrimination has motivated conversion from Hinduism, with lower-caste or Dalit communities converting to Islam, Christianity, Sikhism or Buddhism (White et al., 2011). Some Hindus do not see the caste system as part of the religion, and indeed Hinduism in some parts of the country is less caste stratified than elsewhere, partly because of the influence of other religions such as Sikhism or Islam, for example in Punjab. In other cases, Hindus have turned to different strands within the religion in which the Sanskritic/Brahmanical tradition is less dominant. For example, Gooptu (1996) describes how untouchable groups amongst the urban poor in early-twentieth-century North India turned to Bhakti devotion, which emphasises the direct relationship between a devotee and God and de-emphasises the ritual role of Brahmin priests. However, for others conversion has seemed to offer an escape to a new less-stigmatised status. White et al. identify a number of earlier studies of this process. For example, Searle-Chatterjee (2003) describes the number of Muslims amongst the poor in Uttar Pradesh who were previously lower-caste or Dalit Hindus, and Robinson (2003) notes that over half of Indian Christians are drawn from the former untouchable castes. In the Indian context, White et al. (2011) suggest, there is a tendency for upper-caste adherents to "be more squarely located in the centre of their own religious tradition. . . [whereas] *all* the cases of people changing their religion, wishing to change, or becoming involved with a sect or *ashram* [that they came across in their own study] . . . involved people of OBC [Other Backward Caste] or Dalit identity" (p. 656). As shown by Bhatewara and Bradley (2013), conversion (in this instance of Hindus to Buddhism in Maharashtra) has increased converts' (and their informants' predecessors') access to educational opportunities and their sense of their own worth and dignity. Informants reported that these have resulted in some improvements in their daily lives but also that petty discrimination, abuse and poverty continue. These they attribute to caste-based discrimination in the wider society, exacerbated by gender inequality within families, workplaces and communities. Even within non-Hindu religious groups, as a result, the stigmatisation and marginalisation of lower-caste groups may be replicated, as reported by some of White et al.'s informants. Similarly, Arun (2007) found that the material status of the untouchable Paraiyar community in Tamil Nadu had not changed in spite of their mass conversion to Christianity, because they continue to be exploited by upper-caste Hindu patrons, and Robinson (2003) describes how Christian Pulayas and Paraya in Kerala are trapped in occupations as landless labourers for upper-caste Hindu and Syrian Christian landlords.

Switching between Christian denominations is somewhat more common than conversion between religious traditions, and some switch their denominational allegiance and attendance quite frequently. Within Christianity, the piety movement has been an important driver of this. Thus a varying but sometimes significant proportion of Pentecostal Christians (42% in Kenya, 36% in Nigeria, 18% in South Africa, 10% in India and 74% in the predominantly Roman Catholic Philippines) said that they had switched from mainstream Christian churches, especially the Catholic Church (Pew Forum, 2006, p. 33).

Within Islam, rather than switching between sects, Muslims are inspired by various revivalist movements to change their own practices and bring unobservant Muslims into greater awareness and observance. One of the largest and best-known apolitical piety movements is Tablighi Jama'at, although it is one of many that combine religious learning with outreach to other Muslims. For example, ABIM (Angkatan Belia Islam Malaysia or the Islamic Youth Movement) grew in the 1970s as young men and women alienated from the established Malay elites and influenced by international *da'wah* movements sought to awaken religious consciousness and create a society and state infused by religious values, including restoration of a gender-stratified society, by enacting religious practices borrowed from Muslim history and Middle Eastern societies, reinforced by the Malaysian state since the early 1980s (Ong, 1995). Others include Al-Huda in Islamabad, Pakistan, which provides educational programmes for women (Ahmad, 2008; Levitt, 2012); Al-Muslimaat in Sri Lanka (Haniffa, 2008); Jamaat-i-Islami in Bangladesh (Huq and Rashid, 2008; Riaz, 2013; Rosario, 2006) and similar movements in West Africa, for example in Burkina Faso (Gomez-Perez, 2016). Many involve and reach out to women, not least because of the increased numbers of women who have gained a religious education, some of whom have become preachers.

The Tablighi Jama'at movement originated in India, from where it spread initially to Bangladesh and Pakistan, then to a number of southeast Asian countries and then to West Africa and beyond. In South Asia, travelling groups of preachers head for a Muslim locality (Noor, 2010; Reetz, 2004). Usually based in a local mosque, participants do door-to-door preaching, inviting people to attend the mosque for prayer and a religious education session (a sermon, *bayan* or inspirational talk, recitations or teaching based on the core and Tablighi texts). The movement has local centres and organises regional, national and international congregations (sometimes for students and young people), during which prayers are offered, inspirational talks and readings delivered and calls made for more volunteers to come forward. The annual gatherings attract up to 2 million people in Bangladesh and between 1 and 1.5 million in India and Pakistan. The Southeast Asian assemblies in, for example, Indonesia and Thailand, are now nearly as large. Increasingly, the gatherings are used to organise mass marriages, originally to avoid the financial burdens of dowry and lavish weddings. Such marriages are regarded as particularly blessed (Horstmann, 2007; Reetz, 2004, 2006). The TJ's six articles of faith are familiar to conventional Muslims: the confession of faith (*kalima*), praying regularly and correctly (*salat*), acquiring religious knowledge and remembering God daily (*'ilm, zhikr*), respecting fellow Muslims (*ikram*), reforming oneself through pure intentions (*niyya*), rituals and pious deeds and, importantly, going out in the way of God (*da'wah*) (Reetz, 2004, 2006). Their activities include learning groups or circles and preaching tours or camps.

In South Asia, TJ groups are based in local centres, with women doing *da'wah* locally and men doing preaching tours. In southern Thailand, TJ members run mosques, in which they organise study circles, and schools (Horstmann, 2007, Liow, 2011). In Indonesia, attendees at some *pesantren* (religious boarding schools) do *da'wah* locally (Nisa, 2014). In some West African countries, for example the Gambia, married women may join preaching tours, teaching women in their host's home and assembling groups of married women to hear a sermon, which is delivered by a male member of the group in Arabic and Mandinka from behind a curtain (Janson, 2005, 2008, 2014). Preparation for participation in preaching tours includes learning the principles of Islam as set out in the Qur'an, adopting a reformed lifestyle based on religious discipline and hard work and building social relationships based on trust, reciprocity and solidarity (at least within the group). Members who participate in preaching tours agree that their participation furthers their knowledge of Islam, especially how to live their lives properly. However, in India and elsewhere, there are tensions between members and their families over the former's commitment to religion, with TJ membership requirements sometimes being felt to be at the cost of fulfilling family responsibilities (White, 2010, 2012; Devine and White, 2009).

Conclusion

That religious practices are important to people is demonstrated by widespread participation and the ways in which they are integrated into everyday life, as well as involving special occasions, such as pilgrimages, commemorations and festivals. Also most life cycle transitions are marked by rites with religious dimensions. McGuire likens ritual to "a chain of . . . embodied practices, each link having the potential to activate deep emotions and a sense of social connectedness, as well as spiritual meanings [She suggests that ritual] produces a real effect, sometimes privately for the individual practicing it and sometimes for a whole social group engaged in ritual practice" (McGuire, 2008, pp. 100–2). Performed through words, music and actions, rituals reflect and construct the meaning of spaces, including both apparently mundane venues (homes, offices, squares, stadiums) and places endowed with a sacred quality, which may be demonstrated through the construction of a religious building (church, mosque, shrine, temple, tomb) or associated with a natural feature (water, a rock, a tree). Regular engagement in religious practices takes effort and commitment.

Accounts reveal that many adherents participate because they find that regular participation renews and reinforces their awareness of a dimension they perceive as spiritual, divine or sacred. In addition to finding participation rewarding, it is also thought by many to be necessary to stave off misfortune, protect against evil, tackle illness and ensure good fortune and good health. Religious practices are, for many, embedded in everyday social relationships and are an important part of their daily, weekly, monthly and annual routines. In addition to providing ways of responding to uncertainty, they play a role in constructing and (re-)affirming people's selfhood, worldviews and religious commitment. They are expressions of a particular religious identity and habitus (Riesebrodt, 2006). As with religious beliefs, crises may threaten existing practices and lead to modifications.

For analytical purposes in this volume, a distinction has been made between religious beliefs and practices, and religious practices have been categorised into various types. However, for many these are not meaningful distinctions. Rather, the content and meaning of beliefs is expressed through and reinforced by religious practices, while many practices overlap in their performance, place and significance.

The widespread, regular and intense engagement in religious practices apparent in most of the countries relevant to this analysis should not, of course, be romanticised: some participation occurs because it is socially expected (and this may also be associated with over-reporting by informants), some may be essentially precautionary ('better safe than sorry'), some may be used for enforcement purposes (self- or externally imposed discipline), some may be associated with hatred, power and violence, and some may be superficial. Thus rituals may be referred to as 'lifeless', generally meaning that they are understood in purely cognitive terms (relating to thoughts or beliefs) rather than meeting emotional needs or being related to a person's spiritual life, everyday experiences and social ties.

Religious narratives and practices, Lincoln (2014) suggests, play roles in the construction of societies, by evoking sentiments that create and sustain affinity between co-religionists or define boundaries by emphasising differences, leading to estrangement and separation. The capacity to evoke such sentiments, which play important roles in the shaping and reshaping of societies, is not confined to religion, but religion plays a particularly powerful and persuasive role because of its claim to authority based on transcendent truths. The complex ways in which practices, along with beliefs, help to construct individual and social identities and to build, challenge and renew social relationships are illustrated by Roszko's detailed study of a coastal settlement in central Vietnam. This reveals the ways how, over time, religious practices were the outcome of complex interactions between beliefs and actors associated with local folk religion and Buddhism, local people and the state (changing central political regimes and policies, as interpreted and implemented by the security services and local officials). In this instance, she concludes, "religious practice is a coproduction of the state – through its diverse agents and agencies – and religious reformers through their purifying discipline, as well as of various categories of villagers who use indiscipline as a local tactic when acting on behalf of their gods" (Roszko, 2018, p. 20). This example demonstrates some of ways religious domains, dimensions of social life and levels of analysis interact with each other, as suggested in the analytical framework outlined in Chapter 2.

Religious practices also provide ways of conveying, elaborating the implications of and enforcing values, ethics and morals, the subject of the following chapters, which seek to explore the implications of adherents' religious affiliation, beliefs and practices for their values, behaviour and social practices, within their families, kinship networks, local communities and beyond.

References

Ahmad, S. (2008) Identity matters, culture wars: An account of Al-Huda (re)defining identity and reconfiguring culture in Pakistan, *Culture and Religion*, 9, 63–80.

Arun, C.J. (2007) From stigma to self-assertion: Paraiyars and the symbolism of the Parai drum, *Contributions to Indian Sociology*, 41, 4, 81–104.

Asia Foundation (2012) *National Values Survey, Sri Lanka 2011*, Colombo: The Asia Foundation.

Bhatewara, Z. and Bradley, T. (2013) 'The people know they need religion in order to develop': Religion's capacity to inspire people in Pune's Slums, *European Journal of Development Research*, 25, 288–304.

Bop, C. (2005) Roles and the position of women in Sufi brotherhoods in Senegal, *Journal of the American Academy of Religion*, 73, 4, 1099–19.

Bush, R. (2014) *A Snapshot of Muhammadiyah: Social Change and Shifting Markers of Identity and Values*, Singapore: Asia Research Institute WP 221.

Chao, En-Chieh (2014) 'Not fanatical': The evolution of sociable piety and the dialogic subject in multi-religious Indonesia, *The Asia Pacific Journal of Anthropology*, 15, 3, 242–64.

Chong, K.H. (2012) Revisiting religious power: The Korean evangelical church as a disciplinary institution, in Bender, C., Cadge, W., Levitt, P. and Smilde, D. (eds) *Religion on the Edge: De-centering and Re-centering the Sociology of Religion*, Oxford: Oxford University Press (online 2013).

Cipriani, R. (2001) Religion as diffusion of values. 'Diffused religion' in the context of a dominant religious institution: The Italian case, in Fenn, R.K. (ed) (2001) *The Blackwell Companion to Sociology of Religion*, Oxford: Wiley-Blackwell, 292–305.

Deacon, G. and Parsitau, D. (2017) Empowered to submit: Pentecostal women in Nairobi, *Journal of Religion and Society*, 19, 1–17 (online)

Dehlvi, G.R. (2015) The institution of zakat in Islam and its contemporary applications, *New Age Islam*, 13 July, www.newageislam.com/islamic-ideology/ghulam-rasool-dehlvi

Devine, J. and White, S. (2009) *Religion, Politics and the Everyday Moral Order in Bangladesh*, Birmingham: University of Birmingham, Religions and Development WP 40.

Freston, P. (2001) *Evangelicals and Politics in Asia, Africa and Latin America*, Cambridge: Cambridge University Press.

Gecaga, M. (2006) Rape as a tool of violence against women, in Wamue, G. and Getni, M. (eds) *Violence Against Women*, Nairobi: Acton.

Gomez-Perez, M. (2016) Women's Islamic activism in Burkina Faso: Toward renegotiated social norms? *Canadian Journal of African Studies/Revue canadienne des études africaines*, 50, 1, 45–63.

Goonasekera, S. (2013) Buddhism in South Asia: Practicing tradition today, in Pechilis, K. and Raj, S.J. (eds) *South Asian Religions: Tradition and Today*, London and New York: Routledge, 101–42.

Gooptu, N. (1996) Urban poverty and the politics of caste and religion in early 20th century North India: Implications for development practice and research, *Oxford Development Studies*, 24, 3, 221–40.

Guest, M. (2009) The reproduction and transmission of religion, in Clarke, P.B. (ed) *The Oxford Handbook of the Sociology of Religion*, Oxford: Wiley-Blackwell, online ed.

Halman, L., Inglehart, R., Dioz-Medrano, J., Luijkx, R., Moreno, A. and Basanez, M. (2007) *Changing Beliefs and Values in 85 Countries: Trends from the Values Surveys from 1981 to 2004*, Leiden: Brill.

Haniffa, F. (2008) Piety as politics amongst Muslim women in contemporary Sri Lanka, *Modern Asian Studies*, 42, 2–3, 347–75.

Hashmi, S.H. (2010) The problem of poverty in Islamic ethics, in Galston, W.A. and Hoffenberg, P.H. (eds) *Poverty and Morality: Religious and Secular Perspectives*, Cambridge: Cambridge University Press, 180–203.

Horstmann, A. (2007) The Tablighi Jama'at, transnational Islam, and the transformation of the self between Southern Thailand and South Asia, *Comparative Studies of South Asia, Africa and the Middle East*, 27, 1, 26–40.

Hosen, N. (2008) Online fatwa in Indonesia: From fatwa shopping to googling a *kiai*, in Fealy, G. and White, S. (eds) *Expressing Islam: Religious Life and Politics in Indonesia*, Singapore: Institute of South East Asian Studies, 159–73.

Huq, S. and Rashid, S.F. (2008) Refashioning Islam: Elite women and piety in Bangladesh, *Contemporary Islam*, 2, 7–22.

Janson, M. (2005) Roaming about for God's sake: The upsurge of *Tablighi Jama'at* in the Gambia, *Journal of Religion in Africa*, 35, 4, 450–81.

Janson, M. (2008) Renegotiating gender: Changing moral practice in the Tablighi Jama'at in the Gambia, in Janson, M. and Schulz, D. (eds) *Reconfiguring Gender Relations in Muslim Africa, a special issue of the Journal for Islamic Studies*, 28, 9–36.

Janson, M. (2014) *Islam, Youth and Modernity in the Gambia: The Tablighi Jama'at*, New York: Cambridge University Press.

Jenkins, P. (2006) *The New Faces of Christianity: Believing the Bible in the Global South*, New York: Oxford University Press.

Levitt, P. (2012) Religion on the move: Mapping global cultural production and consumption, in Bender, C, Cadge, W., Levitt, P. and Smilde, D. (eds) *Religion on the Edge: De-centring and Re-centering the Sociology of Religion*, Oxford: Oxford University Press (Online version published 2013).

Lincoln, B. (2014) *Discourse and the Construction of Society: Comparative Studies of Myth, Ritual and Classification*, New York: Oxford University Press.

Liow, J.C. (2011) Muslim identity, local networks and transnational Islam in Thailand's Southern border provinces, *Modern Asian Studies*, 45, 6, 1383–1421.

Loy, D.R. (2010) The karma of poverty: A Buddhist perspective, in Galston, W.A. and Hoffenberg, P.H. (eds) *Poverty and Morality: Religious and Secular Perspectives*, Cambridge: Cambridge University Press, 44–61.

McGuire, M. (2008) *Lived Religion: Faith and Practice in Everyday Life*, Oxford: Oxford University Press.

Mesaki, S. and Kipacha, A. (2010) *The Relationship between Religious Teachings, Values and Beliefs and Concepts of Development in Tanzania*, Birmingham: University of Birmingham, Religions and Development, draft research report, unpublished.

Millie, J. (2008) 'Spiritual Meal' or ongoing project? The dilemma of Dakwah oratory, in Fealy, G. and White, S. (eds) *Expressing Islam: Religious Life and Politics in Indonesia*, Singapore: Institute of Southeast Asian Studies, 80–94.

Narayanan, V. (2013) Hinduism: Practicing tradition today, in Pechilis, K. and Raj, S.J. (eds) *South Asian Religions: Tradition and Today*, London and New York: Routledge, 39–72.

Nisa, E.F. (2014) Insights into the lives of Indonesian female Tablighi Jama'at, *Modern Asian Studies*, 48, 2, 468–91.

Noor, F.A. (2010) On the permanent *Hajj*: The *Tablighi Jama'at* in South East Asia, *South East Asia Research*, 18, 4, 707–34.

Oliver, P. (2010) *World Faiths: An Introduction*, London: Hodder Education, 4th ed.

Ong, A. (1995) State vs Islam – Malay families, women's bodies and the body politic in Malaysia, in Ong, A. and Peletz, M. (eds) *Bewitching Women, Pious Men: Gender and Body Politics in Southeast Asia*, Berkeley: University of California Press.

Pew Forum (2006) *Spirit and Power: A 10-Country Study of Pentecostals*, Washington, DC: Pew Forum, http://pewforum.org/uploadedfiles/Orphan_Migrated_Content/pentocostals-08.pdf

Pew Forum (2010) *Tolerance and Tension: Islam and Christianity in Sub-Saharan Africa*, Washington, DC: Pew Forum, http://pewforum.org/uploadedFiles/Topics/Belief_and_Practices/sub-saharan-africa-full-report.pdf

Pew Research Center (2012) *The World's Muslims: Unity and Diversity*, Washington, DC: Pew Research Center, http://assets.pewresearch.org/wp-content/uploads/sites/11/2012/08/the-worlds-muslims-full-report.pdf

Pew Research Center (2013) *The World's Muslims: Religion, Politics and Society*, Washington, DC: Pew Research Center, http://assets.pewresearch.org/wp-content/uploads/sites/11/2013/04/worlds-muslims-religion-politics-society-full-report.pdf

Pew Research Center (2015) *The Future of World Religions: Population Growth Projections, 2010–2050*, Washington, DC: Pew Research Center, http://assets.pewresearch.org/wp-content/uploads/sites/11/2015/03/PF_15.04.02_ProjectionsFullReport.pdf

Quinn, G. (2008) Throwing money at the holy door: Commercial aspects of popular pilgrimage in Java, in Fealy, G. and White, S. (eds) *Expressing Islam: Religious Life and Politics in Indonesia*, Singapore: Institute of Southeast Asian Studies, 63–79.

Reetz, D. (2004) Keeping busy on the path of Allah: The self-organisation (Initizām) of the Tablīghī Jamā'at, *Oriente Moderno*, 23, 84, 1, 295–305.

Reetz, D. (2006) Sufi spirituality fires reformist zeal: The Tablighi Jama'at in today's India and Pakistan, *Archives de Sciences Sociales des Religions*, 51, 135, 33–51.

Retsikas, K. (2014) Reconceptualising zakat in Indonesia: Worship, philanthropy and rights, *Indonesia and the Malay World*, 42, 124, 337–57.

Reynolds, F.E. and Hallisey, C. (1987) Buddhism, in Eliade, M. and Adams, C.J. (eds) *The Encyclopedia of Religion*, New York: Macmillan, Vol. 2, 334–51.

Riaz, A. (2013) The new Islamist public sphere in Bangladesh, *Global Change, Peace and Security*, 25, 3, 299–312.

Riesebrodt, M. (2006) Religion in global perspective, in Juergensmeyer, M. (ed) *The Oxford Handbook of Global Religions*, Oxford: Oxford University Press, 597–609.

Robinson, R. (2003) Fluid boundaries: Christian communities in India, in Chakrabarty, B. (ed) *Communal Identity in India*, New Delhi: Oxford University Press.

Rosario, S. (2006) The new burqa in Bangladesh: Empowerment or violation of women's rights? *Women's Studies International Forum*, 29, 368–80.

Roszko, E. (2018) A triad of confrontation: State discipline, Buddhist purification, and indiscipline as a local strategy in Central Vietnam, *Journal of Contemporary Ethnography*, 1–26 online. Doi: 10.1177/0891241618758854

Ruffle, K.G. (2013) Islam in South Asia: Practising Islam today, in Pechilis, K. and Raj, S.J. (eds) *South Asian Religions: Tradition and Today*, London and New York: Routledge, 192–221.

Sanford, A.W. (2013) Hinduism and development, in Clarke, M. (ed) *Handbook of Research on Development and Religion*, Cheltenham: Edward Elgar, 79–96.

Schultz, D.E. (2008a) Piety's manifold embodiments: Muslim women's quest for moral renewal in Mali, *Journal for Islamic Studies*, 28, 66–93.

Schultz, D.E. (2008b) (Re)turning to proper Muslim practice: Islamic moral renewal and women's conflicting assertions of Sunni identity in urban Mali, *Africa Today*, 54, 4, 21–43.

Schultz, D.E. (2011) *Muslims and the New Media in West Africa: Pathways to God*, Bloomington, IN: Indiana University Press.

Searle-Chatterjee, M. (2003) Caste, religion and other identities, in Searle-Chatterjee, M. and Sharma, U. (eds) *Contextualising Caste: Post-Dumontian Approaches*, New Delhi: Rawat.

Sieveking, N. (2007) 'We don't want equality, we want to be given our rights': Muslim women negotiating global development concepts in Senegal, *Africa Spectrum*, 42, 1, 29–48.

Smith, D. (2009) Hinduism, in Woodhead, L., Kawanami, H. and Partridge, C. (eds) *Religions in the Modern World: Traditions and Transformations*, Abingdon: Routledge, 2nd ed, 37–66.

Smith, D.H. (2010) One man's extrapolations: Conclusions after two years of listening, in Smith, D.H. (ed) *Religious Giving: For the Love of God*, Bloomington, IN: Indiana University Press, 185–94.

Stivens, M. (2006) 'Family values' and Islamic revival: Gender, rights and state moral projects in Malaysia, *Women's Studies International Forum*, 29, 354–67.

Swidler, A. (2013) African affirmations: The religion of modernity and the modernity of religion, *International Sociology*, 28, 6, 680–96.

Tomalin, E. (2009) *Hinduism and International Development: Religions and Development Background Paper*, Birmingham: Religions and Development WP 19.

Vacek, E. (2010) A Catholic theology of philanthropy, in Smith, D.H. (ed) *Religious Giving: For the Love of God*, Bloomington, IN: Indiana University Press, 99–124.

Watt, W.M. (1996) *Islam: A Short History*, Oxford: One World.

White, S.C. (2010) Domains of Contestation: Women's empowerment and Islam in Bangladesh, *Women's Studies International Forum*, 33, 4, 334–44.

White, S.C. (2012) Beyond the paradox: Religion, family and modernity in contemporary Bangladesh, *Modern Asian Studies*, 46, 5, 1429–1458.

White, S.C., Devine, J., Jha, S. and Gaines, S. (2011) *Development, Religion and Wellbeing in India*, Birmingham: University of Birmingham, Religions and Development WP 54.

Woods, O. (2012) The geographies of religious conversion, *Progress in Human Geography*, 36, 4, 440–56.

Part IV

Values, ethics and everyday lives

'Religion' is said to provide a repository of values, an ethical framework, moral norms and processes of moral reasoning, especially in societies in which it plays a central role in the lives of most people. It is assumed that these are linked to behaviour and social practices. However, links between religious adherence, values and morals on the one hand and everyday practices and behaviour on the other cannot be taken for granted.

The concepts of 'values', 'ethics' and 'morality' and their relationships with religion have long been debated, and there is no consensus on the meaning of the terms and the differences between them. For the purposes of this discussion, they can be conceived of in the following ways.

Values can be defined as broad preferences regarding, for example, justice, freedom, respect, compassion, care, solidarity and equality. It is often assumed that values are universal, at least across a social group and/or within a religious tradition, that they are enduring and that they should not be compromised. However, 'religion' and values are both socially constructed and so are influenced by socio-cultural, historical and economic factors that change over time. In addition, unless values are embodied in norms and practices, they are abstract – they do not have meaning and may not be acted on or reproduced unless they are expressed as social or personal ethical and moral norms.

Ethics will here be taken to refer to the standards of right and wrong that provide a basis for prescribing what human beings ought to do, in terms of rights, obligations, virtues such as loyalty, compassion and honesty and ensuring justice. For ethical standards to be realised, they need to be made explicit, for example, through ideas about what is good, beneficial, useful, important etc., and people need to be enabled to make appropriate choices (Gasper, 2004; see also Blasi, 1998). The basis for ethics may or may not be religion – it may instead be a political ideology or socio-cultural worldview that promotes, for example, individualism, wealth or stability.

Morality differentiates between intentions, decisions and actions that are good/right and bad/wrong – it concerns the "practical matter of acting rightly" (Blasi, 1998, p. 168). It is often suggested that a distinction needs to be made between social and personal moral issues – social norms prescribe actions beneficial to and proscribe those considered harmful to a social and/or religious group and tend to be shared, while personal matters are considered to be those of less significance to the social or religious group as a whole. This involves processing messages and making decisions, taking into account social norms and/or religious prescriptions in evaluating alternatives, influenced by an individual's level of cognitive development, experiences and/or familiarity with religious reasoning (Duriez and Soenens, 2006; Marquette, 2013). The distinction between public and private or social and individual ethical and moral matters varies between societies and religions. The moral reasoning through which individuals and social groups address conflicts between values and ethics may be influenced by religion, including an individual's level of commitment, but it is also shaped by other factors, including age, educational level, gender and the nature of the moral issue itself. The process of reasoning is complex: for some a moral norm may be regarded as a matter of personal choice while for others it is considered to be generally mandated; one religious or social group may prioritise personal striving to realise a value such as integrity, while another prioritises strengthening the community; and within any religious or social group there are likely to be socially liberal and conservative wings (Marquette, 2013). Bochow et al. (2017, p. 447) suggest using the concept of 'ordinary ethics', entailing "people's convictions, value judgements and sentiments on how to live a morally good and/or just life", to examine how ethical ideals are translated into the practicalities of daily lives and social relationships, reinforcing or challenging social boundaries and norms in the process.

The conception of a value system, ethical framework or moral code implies norms that are shared by the people associated with a particular social or cultural group, who identify with the values and norms, even if they do not share all of them. It is often assumed that a religion provides such a coherent value system, because its teachings are based on revelations and sacred texts or because it has an organisational framework for specifying, teaching and enforcing norms, but in practice the importance and influence of religion varies widely between traditions, individuals and contexts. This is illustrated by Noy's research on the social visions of a range of people concerned with development in 80 religious and 74 secular organisations across Africa, Latin America and Asia. In practice, the specific religious group to which his respondents were affiliated (and its doctrines) were less important in explaining their vision than their interpretation of the implications of their religion for social change, with an important distinction between the socially and religiously conservative and liberal wings within each tradition (Noy, 2009). Moreover, while religions often do identify norms of social behaviour, explain why they must be obeyed and encourage/enforce compliance, moral behaviour does not necessarily bear much resemblance to their formal teachings, because many situations pose ethical dilemmas, responses to which differ; societal challenges lead to changes in the contours and characteristics of ethical fields; and non-religious institutions and organisations are also influential (Bochow et al., 2017; Reeves, 1998).

This part of the book is concerned with the ways in which the values and ethical stances of religious adherents in Sub-Saharan Africa and southern Asia are (or are not) influenced by their religious affiliation and commitment and reflected in their lives and social relationships. Religious teachings are outlined and benchmarked against internationally accepted norms, the discussion illustrated where possible with evidence from empirical, especially ethnographic, studies. The analysis could consider a much wider range of values, but because space is limited, the selection focuses on those that seem particularly relevant to understanding the links between religion and day-to-day behaviour. As with beliefs and practices, the values considered do not operate independently. Rather, they influence each other and interact in the ways they influence morality and behaviour. The values or ethical norms considered, selected for their relevance in understanding social change, are freedom, rights, justice and tolerance; well-being, poverty, equality and altruism; and gender equality and sexuality, considered in Chapters 7, 8 and 9 respectively.

References

Blasi, A.J. (1998) Ethics, in Swatos, W.H. Jnr (ed) *Encyclopedia of Religion and Society*, London: Sage, Altamira, 168–70.

Bochow, A., Kirsch, T. and van Dijk, R. (2017) Introduction: New ethical fields and the implicitness/explicitness of ethics in Africa, *Africa*, 87, 3, 447–61.

Duriez, B. and Soenens, B. (2006) Religiosity, moral attitudes and moral competence: A critical investigation of the religiosity-morality relation, *International Journal of Behavioral Development*, 30, 1, 76–83.

Gasper, D. (2004) *The Ethics of Development: From Economism to Human Development*, Edinburgh: Edinburgh University Press.

Marquette, H. (2013) Corruption, religion and moral development, in Clarke, M. (ed) *Handbook of Research on Development and Religion*, Cheltenham: Edward Elgar, 220–37.

Noy, D. (2009) Material and spiritual conceptions of development: A framework of ideal types, *Journal of Developing Societies*, 25, 3, 275–307.

Reeves, A.B. (1998) Morality, in Swatos, W.H. Jnr (ed) *Encyclopedia of Religion and Society*, London: Sage, Altamira, 309–10.

7 Freedom, rights and justice

Introduction

For human beings to have control over their lives, it is often argued that both freedom and free will are required. Free will refers to the capacity of rational creatures to think and deliberate for themselves, while freedom implies that individuals can think or act in accordance with their deliberations. Negative freedom refers to the absence of external constraints on an individual, who is thus 'at liberty' to act as s/he wishes – freedom of choice. Positive freedom relates to the achievement of some identifiable goal or benefit. Any freedom has both dimensions. For example, freedom from ignorance means being free to obtain an education. For many, freedom suggests the ability to satisfy one's own interests or, more generally, to realise one's potential, i.e. freedom is regarded as the basis for happiness and well-being (Heywood, 2000, p. 129).

Freedom is inextricably tied to the idea of rights, which underpins internationally accepted values and norms, especially those set out in the Universal Declaration of Human Rights (UDHR). A right is to have an entitlement, as a matter of justice. Rights may be held by entities (e.g. nations, states, firms) or individuals, who may include all humans, by virtue of their humanity, or some categories of people, such as citizens or parties to a contract. Universal human rights are regarded as inalienable – they may be denied or violated, but a person's entitlement to them cannot be removed. The rights of a group depend on the status accorded to the category of people concerned. For example, in some societies, civil rights may apply only to citizens. While political and civil rights are long established, getting recognition for social and economic rights has been a struggle: many have only been widely recognised in the last 40 years, and they are still controversial.

Rights may be justified on the basis that our duty to respect people's dignity requires us to respect their rights or because particular rights have desirable outcomes. Their conceptualisation, definition and realisation raise a number of difficult issues (An-Na'im, 1995; Chanock, 2002; Ensor, 2005; Heywood, 2000). These include, first, the individualist bias of international conceptions of rights, the promotion of which is criticised by some as a form of cultural imperialism. Chanock (2002) offers a critique of this view. A second issue concerns the difficulty of balancing individual with group rights, especially freedom of thought, conscience and religion with rights to economic, social and cultural development, or between the rights of different individuals (e.g. the debate on abortion is concerned with the rights of the unborn child as well as those of the mother). Thirdly, it is generally recognised that limits to untrammelled freedom and rights are needed, to ensure justice, balance rights with responsibilities and sustain social harmony. Such limits entail duties that may be negative (e.g. not to rape, kill, steal or lie), positive (requiring people to perform certain actions, such as feeding the hungry or healing the sick) or both (for example, the right to freedom of speech implies a negative duty – not to silence people – and a positive one – to protect those trying to exercise freedom of speech in the face of threats). The limits, for example, on practising a religion, may seek to protect public safety, order, health, morals or the rights and freedoms of others (see Art. 12(3) of the International Covenant on Civil and Political Rights, 1976, which conveys freedom of thought, conscience and religion). The last of these caveats is intended to outlaw discrimination justified on religious grounds, for example, against women (see Chapter 9). Fourthly, social change and definitions of rights are inter-related. Ensor (2005), for example, argues that freedom of choice is necessarily exercised in the context of shared cultural meanings and social practices, so that an individual's well-being or interest is influenced by the context and varies across social forms and cultures (see also Bochow et al., 2017). In addition, An-Na'im (1995) suggests that social change may lead to reinterpretations of religious teachings, with implications for the recognition of rights once sufficient support has developed amongst scholars and the community of adherents. Finally, the notion of rights and entitlements entails a concept of justice, which is applicable to the distribution of any 'goods', including freedom and rights. Because of the need to hold those responsible for protecting rights and ensuring that obligations are fulfilled accountable for their actions, at both the individual and governmental levels, the values of honesty and integrity are also considered. The debate over justice most often focuses on the distribution of material benefits and burdens (especially income/wealth) and is returned to in Chapter 8, while religious perspectives affecting rights related to gender and sexuality are analysed in Chapter 9.

As in the earlier chapters, the discourses employed by both religious authorities and adherents are considered in order to, first, develop an understanding of how they conceive of relevant values and norms and, second, how they seek to reflect such values in ethical codes and social practices (Lincoln's first and second domains of religion). The extent to which these codes and practices (in addition to core beliefs and religious practices) are common to individual religious traditions, contributing to a shared identity and worldview, or vary between and within them is explored to throw further light on the nature of religious communities (Lincoln's third domain) and the ways in which religious institutions and organisations interpret a religion's official teachings and enforce its ethics (the fourth domain). The discussion of religious perspectives on freedom and rights in this and the succeeding chapters will benchmark the values promoted by official religious teachings against internationally accepted norms and also explore why and how religious teachings place limits on these norms.

While Lincoln recognises that the practice domain includes ethical practices, he does not elaborate on the links between discourses (which express beliefs and are embodied in practices) and action. As suggested in Chapter 2, the links between discourses related to religiously ordained values and ethics and action/behaviour have personal, social and organisational dimensions. Despite the difficulties of doing so, these all need to be teased out to assess what we know about the values and ethics of religious organisations and individuals in Sub-Saharan Africa and southern Asia, how they are interpreted in a changing world and whether and how they influence behaviour with respect to interactions between individuals, within primary groups and in wider societies (the other crosscutting dimensions of the analytical framework set out in Chapter 2).

Psychological research, almost all of which has been conducted in North America and Europe (i.e. largely Christian settings), has sought to understand the relationships between religion/religiosity and morality by evaluating individuals' approaches to religion, how people process religious content and how these are affected by other factors such as education or age (Duriez and Soenens, 2006; Wulff, 2001). Some of this research indicates that whether religious texts are interpreted literally or symbolically influences people's moral reasoning more than whether they are religious adherents. It suggests that literalists are more likely to respect convention and prioritise conforming with and maintaining the socio-religious order. In contrast, those who interpret texts symbolically are more likely to refer to principles in their moral reasoning. Such research does not appear to have been replicated in other global regions or among subjects from non-Christian religious traditions.

Such individualist approaches have been criticised for failing to recognise the importance of socialisation and social interaction in forming a moral community which reinforces norms by overt or covert disciplinary action. In societies with shared values, by and large individuals endorse the moral code by which affairs are supposed to be conducted, even if they themselves do not all or always behave morally. In addition, they expect the individual behaviour and social relationships of others to conform to these stipulations, even if they are prepared (or at least tempted) to exempt themselves. Insofar as religion prescribes or endorses the moral code, it can function as the arbiter of proper moral conduct. Thus in contexts in which religious affiliation is widespread, religion pervades other social institutions, and mass religious socialisation transmits and diffuses religious values throughout society, where they form the accepted basis for daily and community life, social action and individual choices. In such societies, compliance is so widely expected that injunctions may be adhered to even by those who are not religiously committed (Devine and Deneulin, 2011). Even if religiosity and participation in rituals decline, the values and moral expectations associated with a previously hegemonic religion may have become established social norms, reducing the scope for conscious moral choices and helping to maintain social cohesion.

In contrast, economic and social change may be perceived as a threat to the existing social order and its moral underpinnings. For example, in Europe, industrialisation and the emergence of capitalism undermined the ability of the Roman Catholic Church to ensure that adherents maintained their loyalty and fulfilled their obligations to the Church, reducing its dominance. Instead, a moral system centred on diligence, sobriety and responsibility emerged, i.e. an individualised ethic related primarily to work and reinforced by the teachings and institutions of Protestantism (Wilson, 2001).

Similar fears that the prevailing social order, already undermined by colonialism, is threatened by ongoing economic and political changes and the spread of cultural globalisation are often expressed today in SSA and southern Asia. For many, these fears have given rise to a sense of moral decline: at least one in every two respondents in surveys of Christians in five countries for which data are available consider moral decline to be a major concern (89% in India, 73% in Kenya, 70% in the Philippines, 68% in Nigeria and 53% in South Africa) (Pew

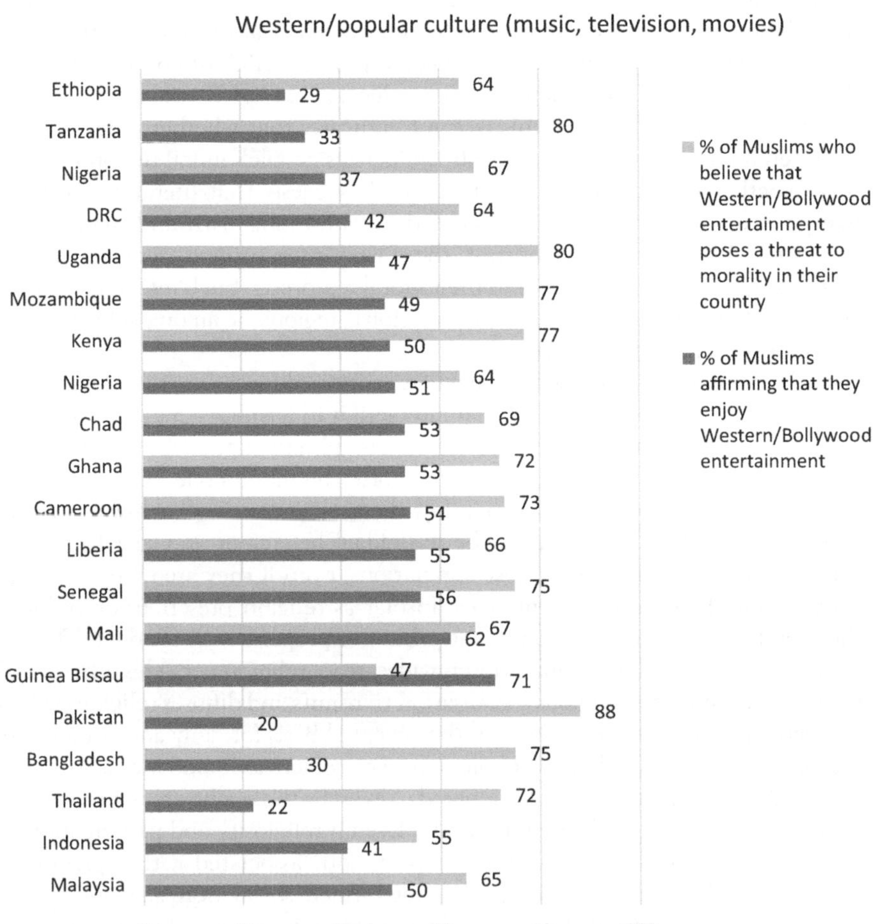

Figure 7.1 Muslim perceptions of the influence of cultural change on morality

Source: Pew Research Center, 2013, data selected from pp. 134, 136.

Forum, 2006, p. 65). Such decline is sometimes associated with the growing influence of foreign culture. For example, between a third and two-thirds of Muslim respondents in 15 SSA countries and between a fifth and half in five southern Asian countries reported that they enjoy Western or Bollywood entertainment. However, a large majority (two-thirds or more in 18 of these 20 countries, in both SSA and southern Asia) believe that Western popular culture in the form of music, television and the movies poses a threat to morality in their countries, with older and more devout Muslims more likely to see it as damaging (Pew Research Center, 2013) (see Figure 7.1). This sense of moral decline has been one of the factors fuelling religious change, especially the emergence of revivalist and radical movements.

This chapter concentrates on values that are central to how people live in the world – freedom (especially with respect to religious beliefs and practices), tolerance (which is a precondition for exercising freedom and expressing difference), rights (and how these are to be realised and protected) and honesty and integrity (to sustain social relationships based on trust). The main purpose of the chapter is to examine the evidence on prevalent discourses, attitudes and actions with respect to freedom, tolerance, rights and justice, as well as honesty and integrity. As in earlier chapters, an introduction is provided to the basic ethical frameworks of the main religious traditions to provide a common starting point for the discussion.

Perspectives on values and ethics from the faith traditions

Islamic values and laws

In Islam, the most basic moral injunction is 'to enjoin the good and abjure evil'. Values and principles are derived from the Qur'an[1] and *sunnah* and set out as a set of clearly articulated laws in *shari'a*, although all have been extensively (re-)interpreted in subsequent centuries. By the tenth century, five schools of jurisprudence had emerged, all of which recognise the validity of the others. Of the four Sunni schools, the Hanafi is prevalent in the Indian subcontinent and the Shafi'i school in Southeast Asia, while Shi'a Muslims mostly adhere to the Jaafari school (Wolfe and Gudorf, 1999). Islam is thus different from other religious systems because it does not distinguish between Islamic values and moral considerations and their expression in law (Rahman, 1987). By the late nineteenth century, however, codified state law started to replace the role of scholarly legal opinion in the Muslim world, and the colonial powers sometimes forced Muslim states to change their laws, leading to different interpretations of *shari'a*. During the twentieth century, the views of those who regard the practices of early

1 The view that values and morals are based on the doctrines and laws of God/gods/the divine are commonly labelled 'divine command theories'. In such religious communities, one might expect to find a well-defined set of moral rules which simply require compliance. However, divine command theories have been challenged on both theoretical and practical grounds.

Muslims as authentic and correct (such as Wahhabi/Salafi Muslims) diverged from those of other Muslims. For the latter, while core Islamic values remain valid, the law can and should be reinterpreted to adapt it to modern life, including acceptance of the idea (at least in principle) of universal human rights and recognition that in many parts of the world Islam itself has a distinctly syncretic character (Rahman, 1987).

Shari'a governs most aspects of human behaviour and interaction, requiring the individual to act or not to act in certain ways and regulating those requirements, although it only offers a model for human society if the arrangements promoted by the Prophet and adopted by early Muslims are considered to be appropriate for today. Rather than a right, freedom implies the ability to recognise no authority other than God and to personally surrender to the divine will. However, according to Traer (n.d.), reporting on a 1980 seminar on human rights in Islam for jurists, Muslims gain certain freedoms and rights if they fulfil their obligations to God, nature and other humans, i.e. freedoms and rights are linked to duties rather than being unlimited. This view is shared by many Muslims. For example, the scholar Maududi holds that the concept of a person having inalienable rights by virtue of his/her humanity alone does not exist in Islam, which regards all rights as God given and transmitted by divine revelation (Maududi, 1976).

The contemporary human rights discourse in Islam reflects the ongoing intellectual rivalry between modernists, traditionalists and fundamentalists (Kroessin, 2008; Mayer, 1995). Some Muslims affirm the universality of human rights, arguing that their roots are to be found in Islamic doctrine and respect for them is enshrined in *shari'a*, although they acknowledge difficulties with some, especially those related to religious freedom, gender and marriage. However, critics such as Tibi (2008) assert that the rights specified in the UDHR are based on cultural concepts (especially individualism) that have been enshrined in European and international law, which is biased towards furthering the interests of the West and inconsistent with both Islamic law, in which rights must be accompanied by duties (a view not confined to Islam), and the Muslim worldview, which is based on collective rights, in which the individual is considered to be merely a limb of the *ummah* (the community of Islam) (Tibi, 2008). Moreover, critics note that the fundamental distinctions in Islamic teaching and law between the rights of Muslims and non-Muslims, and those of Muslim men and women violate the principles of equality and non-discrimination on which universal human rights are premised (An-Na'im, 1995, 1996; Twining, 2009). Further, fundamentalists may regard any comparison between the secular concept of human rights and Islamic law as blasphemous in itself (Abraham, 1998, p. 316). In their view, the only meaningful human rights are those supported by Islamic teachings, which can only be universal when Islam is universally recognised as representing the will of God (Traer, n.d.).

The 1990 Cairo Declaration of Human Rights in Islam, produced by the Organisation of the Islamic Conference (OIC), is perhaps the most prominent attempt to develop an alternative to the UDHR, although in practice it has been less influential than the latter. It affirms *shari'a* as the sole source of human rights. For example, Article 22 restricts the UNDHR's commitment to freedom of speech to those expressions that are not in contravention of *shari'a*. In addition, Article 5 prohibits the imposition of any restrictions on marriage stemming from 'race, colour or nationality', notably excluding religion from the list and requiring that Muslim women marry only Muslim men (Hardy, 1987, p. 402). In practice, of course, the principle is implemented in different ways in different contexts. For example in southern Asia, marriages are also influenced by the caste system. Further, in Article 6, women are guaranteed equal dignity but not equality in other matters (for example, responsibility for maintaining the welfare of a family is allocated to husbands while no such obligation is placed on wives).

Tibi (2008) is critical of all the efforts to develop an Islamic alternative to the UDHR, on the grounds that they fail to recognise the cultural obstacles to universal individual human rights in Muslim countries. In his view, the only solution is to reform *shari'a* to recognise human rights as entitlements (pp. 130–46). An-Na'im (1995, 1996) agrees that Qur'anic principles need to be reinterpreted in the contemporary context. This implies accepting the necessity and validity of *ijtihad*, based on discussion between Islamists and their opponents within Islam. However, such discussions are problematic because the former claim a monopoly of Islamic authenticity and authority, while the latter are currently on the defensive, perhaps because in many Muslim-majority countries they are associated with the discredited (liberal and socialist) post-independence project (Asad and An-Na'im, 2009; Twining, 2009).

Muslim views on tolerance can be traced back to the Qur'anic verse that says 'let there be no compulsion in religion', implying that people should be free to choose and practise a religion. However, interpretations vary: one view is that religion cannot be compelled because it is only God who grants or withholds the gift of faith, but an alternative perspective is that, according to Qur'anic teaching, the will of God can only be known through the texts. It follows that only people of 'the book' are allowed freedom of religious practice, because only they have access to God's guidance through the relevant scriptures. There is, however, a tension in the Qur'an: on the one hand it insists on the unity of the monotheistic religions, but on the other it recognises that God has given different guidance to Jews, Christians and Muslims and teaches that non-Muslim 'people of the book' should be tolerated (*tasamuh*, toleration). Thus the Prophet, in his last years, gave religious and cultural autonomy to Jews and Christians but sought to forcibly convert Arab pagans to Islam. However, the former were tolerated as subdued minorities rather than equals (Rahman, 1987). Moreover, while adherents of non-monotheistic religions were classified as *kafirun* (infidels, unbelievers), in contemporary political Islam, this label is also applied to Christians and Jews, who are seen as being allied against Islam (Tibi, 2009, pp. 214–17). As Tibi points out, Muslim teaching and behaviour often fall short of true tolerance, which should be based on an acknowledgement that Islam and other religions are equal and respect for the human rights of all.

Christianity, ethics and social responsibility

Jesus and his followers lived by the moral code of the Hebrew bible, centred on the Ten Commandments, which Jesus summarised in a double commandment to love God and neighbour, including the outcast, and the 'golden rule', 'Do unto others as you would have them do to you' (Kim, 2007, p. 6). As well as a set of (often contested) beliefs, Christianity presented itself as a way of life, based on obedience to God's will and imitation of the human life of Jesus, as portrayed in the New Testament. Just what this imitation implies is complex, Pelikan comments, because the man Christians are invited to imitate "was not married, did not hold public office, and was not supported chiefly from a trade or profession" (Pelikan, 1987, p. 360). Also, not only were the life, teachings, death and resurrection of Jesus recorded later, they have also subsequently been subject to a variety of interpretations. Nevertheless, all Christians accept that they imply, first, seeking to live individual lives that reflect underlying values, especially faith, hope and love, which need to be interpreted in particular social contexts rather than requiring strict observance of a legal code (Curran, 2005; Pelikan, 1987); and second, creating a new social order, although there is no single Christian vision of a desirable society. While all the Christian denominations have a sense that God is involved in the world and desires peace and justice, and all are committed to improving the human condition, they differ on the best way of achieving this (Kim, 2007).

The Catholic Church, which has the largest number of adherents among the denominations, has developed perhaps the most coherent interpretation, known as Catholic social teaching (Curran, 2002; Deneulin and Bano, 2009; Plant and Weiss, 2013). This is based on the values of freedom, dignity, responsibility and solidarity. Freedom of conscience is regarded as a right. Human beings have equal dignity because they are made in the image of God. Human well-being is conceptualised as having both material and spiritual dimensions (integral human development), the realisation of which should be the aim of all economic, social and political processes. However, it is believed that individual well-being can only be achieved if the common good is realised, so some freedoms may need to be restricted. Since the 1960s, the Church has begun to champion human rights, especially of the poor, and to urge the duty of all to become engaged in political, economic and social affairs to build a more just social order (Deneulin and Bano, 2009, pp. 136–42). Its social teaching deals firstly with economics, out of a concern for the poor. Secondly, it is concerned with the political order necessary to bring about a just society, which it believes to be democracy. The Church therefore supports economic and social as well as political and civil rights, insofar as they uphold human dignity, provided that they are related to duties and considered alongside other values such as truth, justice and charity (Curran, 2002, pp. 215–22).

So-called social Christianity emerged within the liberal Protestant tradition in the mid-late nineteenth century (Kim, 2007). It focuses on Jesus's ethical teaching to love God and neighbour, rather than doctrine, downplays sin and judgement and emphasises the immanence (presence) of God in the world, leading to a Christian responsibility to help build the kingdom of God on earth. Critics of the mainstream missionary churches consider that their focus on compassion for the poor and disadvantaged is unlikely to result in changes to the structural drivers of injustice, inequality and poverty. However, increasingly they and their successors in the global south recognise the extent to which the Bible addresses issues of injustice and assert that it can be achieved by establishing democratic and restraining market forces. This led them to become active in democratisation movements in many countries in the 1980s. In addition, they have cooperated on international affairs, including developing the UDHR, and in 1973 the interdenominational body of which these churches are members, the World Council of Churches, adopted a 'comprehensive' understanding of salvation (or development) as liberation in four inter-related dimensions: economic justice, dignity, solidarity and 'hope in personal life'. On the basis of this formulation, the WCC began to lend its support to movements that were challenging the structures of injustice, followed in the 1980s by adopting the 'option for the poor' of liberation theology. This had emerged in Latin America in the 1960s, when some Roman Catholic priests brought together Catholic social teaching, political theology, anti-colonial sentiment and Marxist analytical tools, even though this brought them into conflict with the higher rungs of the Catholic hierarchy (Plant and Weiss, 2013; see also Deneulin and Bano, 2009, pp. 142–8). Liberation theology has had wide influence beyond Latin America. For example, it has influenced Dalits' struggle against caste oppression in India.

In contrast, the evangelical churches, including Pentecostal denominations and the charismatic movement within the mainstream churches, emphasise conversion (the need for an individual to renounce sin – being 'born again') and evangelism. For much of the twentieth century, they withdrew from social and political involvement. However, in recent years, they have started to re-engage, arguing that evangelism should not be separated from social concern. This is driven by a view of "society as being in a state of [moral] crisis to which spiritual activities such as prayer and worship are the appropriate response" (Yamane, 1998, p. 81). It is also motivated by compassion, which is interpreted as participation in the transforming action of God in the world, although the purpose of this is evangelism rather than promoting rights and justice. Their perspective is based on individual responsibility, acceptance of the capitalist system and a belief that religion and the state should be separate. Social issues are thus to be dealt with at the personal and family levels, emphasising the responsibility of individuals to lead moral lives and foregrounding family life. Their social conservatism is commonly associated with opposition to some internationally accepted human rights, such as gender and sexual equality (see Chapter 9). Moreover, their involvement in the political sphere has been limited, although since numbers have grown, especially in the south, engagement in politics has become more common. However, the churches are generally fragmented and fissiparous, and many lack the organisational structures and specialised wings that might take a lead on developing stances on human rights and political engagement.

Hindu ideas about right social ordering

As discussed in earlier chapters, Hinduism is centrally concerned with right social ordering. Hindu philosophy and religious practice do not provide a succinct ethical framework for individual and social life,[2] but ideas about the basis for social order underlie the values and ethics associated with the tradition. The underlying principle is *dharma*, a complex concept which refers to religion, the underlying cosmo-moral order and the ethical principles applicable to all human behaviour (including truth, non-violence, compassion and regard for others and their welfare according to the same standard that one would apply to oneself) (Nadkarni, 2008; Tomalin, 2009). Ethical conduct based on these principles will, it is believed, ultimately produce a stable moral order that will ensure the security and welfare of all. Higher priority is given to *achara* (one's conduct according to *dharma*) than either *vichara* (beliefs, philosophy) or ritual, although the latter, as discussed in Part III, is also important. *Dharma* can be specific to one's caste (*jati*), class (*varna*) and stage in life (*ashrama*). Fulfilment of one's ritual and moral duty is considered necessary both to achieve release from the cycle of rebirths and to uphold the cosmic order (*rita*). *Karma* (action) seems individualistic in the sense that rebirth is attributed to individual behaviour in a previous life, but it is also considered unacceptable to fail to help someone in need on the basis that his/her suffering can be attributed to his/her *karma*.

In practice, Hindu values and ethics are based more on tradition and social consensus than on the authority of the texts, implying that there is scope for change as society changes. While some values are considered to be universal, others may be appropriate in particular circumstances or are specific to social roles. Texts such as the Bhagavad Gita recognise the moral dilemmas human beings face, advocating an approach that concentrates on motivations: rather than specific actions being approved or prohibited, what is considered important is dedication to the deity/ies and having unselfish motives (Oliver, 2010, pp. 14–15). Freedom of thought is regarded as necessary for the pursuit of truth and scientific knowledge and for individual spiritual development, based on the view that revelation is ongoing. Particular texts or individual teachers may be emphasised, with disagreements dealt with by the formation of a new tradition (*sampradaya*), way or road (*pantha*) or school of thought (*mata*), which nevertheless draws on a common pool of scriptural and cultural resources (Nadkarni, 2008).

Rights (*adhikāra*), for example to property, prosperity and a livelihood, imply duties and vice versa. For example, Gandhi uses the concept of trusteeship, implying that while it is acceptable to acquire wealth, it should be used to benefit others. Similarly, historically a king had rights over his subjects but a duty to protect them, establish law and order and care for the destitute. If a ruler fulfils these responsibilities, his subjects in turn have a duty to cooperate. However, equality of rights is more problematic. Some texts in the Hindu scriptures uphold the principle that all human beings are equal before God, but others have been used

2 Although see the discussion of the Manusmriti in Chapter 3.

to provide an ideological justification for the caste system, and some rituals reflect and reinforce caste differences (Nadkarni, 2008). In practice, the links between Hinduism and the caste system are complex, contested and contradictory. Today, Mosse (2018) argues, caste is treated by policy makers as "an archaic cultural and ritual phenomenon and source of historical disadvantage" (p. 423) that can be tackled by remedial provisions, even though it (alongside class, religion and gender) provides a basis for social, economic and political advantage and discrimination, exacerbating inequality.

The pre-eminence of the spiritual path in Buddhism

In Buddhism, morality plays a subordinate role in the spiritual path that leads to liberation, which involves realisation that the sense of separation between the self and the world is a delusion (see Chapter 3). Once this non-duality is recognised (wisdom), a responsibility to the world follows, based on the central values of compassion and the non-harming of living beings (Loy, 2010). Three of the eight elements of the Buddha's path are concerned with behaviour – the injunctions to adopt right speech, conduct and livelihood. Right speech means to reflect before speaking and not to use speech to harm others; right conduct prohibits killing or harming living creatures, sexual misconduct and stealing; and a right livelihood is one that does not involve the harming of other living things (*ahimsa*) (Loy, 2010; Oliver, 2010). Monastics are expected to adhere to many additional precepts, including celibacy, although specification of these precepts and how they should be applied differs between the traditions (Wolfe and Gudorf, 1999). Compliance, through a combination of good religious practices and good actions, it is believed, will reduce one's *dukkha/duhkha* (suffering) and improve one's *kamma/karma*. Compassion is stressed rather than justice – practising compassion is regarded as essential to the spiritual progress of the practitioner (Loy, 2010).

Justice is only a Buddhist concept in the sense that *kamma/kharma* implies that justice is built into the way the cosmos functions. Most individuals' motives for adopting the five precepts of Buddhism are to reduce their own *dukkha/duhkha* and improve their *kamma/karma*, not to make the wider social order more just, even though, almost as a by-product, certain good deeds, such as accepting responsibility or considering consequences, can have beneficial social effects as well as making merit for individuals (Clarke, 2011; Strong, 2005). The notion of human rights appears to be inconsistent with the Buddhist understanding of the impermanence of the individual self. Instead, the tradition, like Hinduism, foregrounds duty, with an emphasis on the duty not to harm other living beings. Like Hinduism, this provides a basis for negative rights, whereas there is weaker support for positive rights or entitlements, which are seen as something that is good for others to provide rather than as inalienable rights.

Apart from its general ethical guidelines, Buddhism "has little to say concerning 'correct' family or social relationships – [regarding these as] 'worldly' matters, outside the proper scope of the religious tradition" (Cantwell and Kawanami,

2009, p. 77). Although recent socio-economic and political change has encouraged some rethinking of the traditional approach, including that associated with socially engaged Buddhism, Cantwell and Kawanami think that it is unlikely that normative Buddhist teachings will emerge, partly because of the differences between Buddhist societies and partly because the flexibility associated with the religious tradition and its scriptures is helpful in a rapidly changing world.

Ethics in folk religions

For most adherents of folk religions, ethical values are manifest in norms that regulate relationships between individual members of a society and their social groups (Alhassan Alolo and Connell, 2013). In the absence of religious texts, norms are illustrated in folklore, proverbs and myths, and values are transmitted through oral media (song, dramatic performances), ritual and symbolic objects and performances associated with shrines and their adjacent spaces. As noted in Chapter 3, in folk belief systems the high god/supreme being is seen as being distant from human affairs, so the relationship between god and humans may or may not have a moral content governing personal and social conduct. In practice, most emphasis is placed on the latter: what is morally right is generally regarded as that which is conducive to the well-being of individuals when, and only when, this is in harmony with the well-being of society (Wiredu, 2011). Thus the central ethical imperative is to realise sociability, binding together a community's living members, remembered dead and members yet to be born (Mageza, 1997). The basic social unit is generally the familial or lineage group, and morality is conceived of as being primarily concerned with duties towards this group, with the ancestral spirits seeking to ensure that these duties are performed by living members of the group and expecting repentance and propitiation if they are disappointed. The ultimate aim of punishments and rewards for individual im/moral behaviour is regarded as being to ensure the continuity and stability of the group rather than to punish the individual, although individual or group misfortune, as discussed in Part II, may also be attributed to the ill will of a spiritually empowered being (witch or sorcerer), who may be motivated by envy, spite or resentment. In practice, many adherents of folk religions conceive of values and norms as being based on culture and custom rather than 'religion'. In addition, folk (like other) traditions have invariably been influenced by socio-economic change and political debates about freedoms, rights and responsibilities, complicating the interactions between values and norms, practices and individual and communal interests.

Religious freedom

The UNDHR was drawn up in 1948 and has subsequently been elaborated into various statements of political, civil, economic and social rights and freedoms. Freedom has both positive and negative dimensions: on the one hand rights to free speech, belief and action, on the other freedom from oppression. Political and civil rights depend on freedom of speech and association, while social rights are often interpreted as including cultural freedom. For example, the UNDP's 2004 Human Development Report, on cultural diversity, which was strongly influenced by Amartya Sen's liberal individualist thinking, embraces religion as part of culture and regards cultural liberty and rights as compatible, although it draws attention to the danger that powerful movements or individuals may promote their own interpretations of cultural freedom and attempt to dictate how lives should be lived (for example, Hindutva political ideology in India or Islamic political parties) (Deneulin and Bano, 2009). Chanock (2002) has a more sceptical view of 'culture' and its relationship with rights, while even committed liberals disagree over whether individual freedoms sufficiently recognise that people are social beings, that community may be valued over and above individualism and that there should be limits to individual freedoms, for example with respect to families.

For adherents of all religious traditions, religious freedom is particularly important. It was enshrined in Article 18 of the UNDHR in 1948: "everyone has the right to freedom of thought, conscience and religion; this right includes freedom to change his or her religion or belief, and freedom, either alone or in community with others and in public or private, to manifest his [sic] religion or belief in teaching, practice, worship and observance".[3] This implies freedom both between and within religions. It implies freedom of belief and conscience, including the freedom to dissent, freedom to comply or not and tolerance of those with different religious identities and beliefs (An-Na'im, 1996). While it may appear to be the concern of remote and bureaucratic organisations and international laws, people's freedom to practise their religion has major implications for their everyday lives.

Evidence on religious freedom around the world is collected by, for example, Aid to the Church in Need, an international Catholic organisation (bi-annually).[4] In 2014, it noted that in 81 of the 196 countries for which it had produced reports, religious freedom was impaired or had declined since 2012. Only in 80 had there been no regular or systematic religious freedom violations. Between 2012 and 2014, changes in the extent of religious discrimination or persecution occurred in 61 countries: only in 6 had the position of religious minorities changed for the better, while in 55 their situation had deteriorated. In 14 of the 20 countries with high levels of restrictions on religious freedom, ACN linked the persecution of non-Muslim minority groups and certain sects within Islam to 'extremist' Islam. Perpetrators included the state, through legal or constitutional discrimination, including the imposition or implementation of *shari'a* (relating, for example, to blasphemy or apostasy), or crackdowns on all religions by authoritarian states such as China. They also included international and national non-state organisations, most obviously Daesh and related

3 www.un.org/Overview/rights.html.

4 Also the US Department of State (annually). Both produce country reports that are available online www.state.gov/j/drl/rls/irf/; www.religious-freedom-report.org.

movements. In most African countries religious freedom is a right, exercised against a background culture of tolerance and mutual respect, although Islamic fundamentalism has grown in some countries (notably through the activities of Boko Haram in Nigeria and surrounding areas). Violations of religious freedom are, ACN found, more widespread in Asian countries, affecting Christians, various Muslim groups and occasionally Hindus and Sikhs.[5]

The organisation's 2016 report concentrates on the growth of what it calls Islamist hyper-extremism, characterised by an extremist creed, the desire for radical changes to law and government, systematic attempts to annihilate or drive out groups which do not conform to its outlook (including moderate co-religionists), cruel treatment of victims and the use of social media to recruit followers and intimidate opponents. In 38 countries, there was evidence of significant violations of religious freedom, severe in 23 (persecution) and moderate in 15 (discrimination). The situation had worsened in 14 of these since 2014 and only improved in 3. Many (including Bangladesh, Indonesia, Kenya, Niger, Pakistan, Sudan and Tanzania) had experienced Islamist violence, contributing to increased numbers of refugees. Persecution of religious minorities by militant nationalist Hindus (in India) and Buddhists (in Burma/Myanmar) had also occurred. In some countries where one religion is identified with the state, such as India and Pakistan, ACN noted that governments had taken steps to defend the rights of that faith rather than those of individual believers, resulting in increased state rhetoric against religious pluralism; restrictions on minority faith groups, evangelisation and conversion; and increased sanctions against 'blasphemy' (ACN International, 2016).[6] In ten countries, including Vietnam, the state either tightly regulates all religious institutions or is actively hostile to religion (ACN, 2016).

The Pew Research Center has followed up the issue of government restrictions on religious freedom by examining the extent to which national governments favour specific religions over others, officially or unofficially. Of the 199 countries analysed in 2015, 43 had an official state religion (most commonly Islam) and 40 a preferred religion (most often Christianity), compared to 106 with no official or preferred religion – where religions are treated more or less equally and their governments have a neutral relationship with religion, broadly maintaining a separation of religion and the state (an issue that will be taken up in the linked volume). In SSA, three-quarters of countries have no official or favoured state religion: of the five that have an official religion, four are Muslim and only Zambia is Christian. Even if a particular religion is not designated the state religion and religious freedom is constitutionally guaranteed, favoured status may bring tangible advantages in terms of legal or tax status, ownership of property and financial support, sometimes for particular activities such as education (Pew Research Center, 2017a).

5 https://combonianum.org/2014/11/06/religious-freedom-world-report-2014/ accessed 20/11/2017
6 The report does not examine whether similar restrictions have emerged in explicitly Christian states such as Zambia.

Adoption of an official religion is often accompanied by stricter regulation of religious practice, sometimes, but not only, by government, including restrictions or bans on minority religious groups. In Malaysia, for example, a 1996 *fatwa* required that Sunni teachings be followed, with other Muslim sects being banned and not allowed to assemble, worship or speak freely about their faith. In 2015, more than three-quarters of the governments of countries with an official or preferred religion had interfered with worship by particular religious groups, compared with fewer than half the governments of countries with no official or preferred religion. The 23 countries with extensive government restrictions on religion include Malaysia, Indonesia and Vietnam (Pew Research Center, 2017a, 2017b, p. 15). In Buddhist-majority Sri Lanka, a recent survey found that, while the vast majority feel free to express their religious opinions in their local area, some Hindus and Muslims feel restricted, except amongst their friends and relatives. Members of minority religious groups are also more likely to oppose the erection of religious statues or places of worship and the display of religious symbols and signs in public places, because they perceive this as being potentially divisive (Asia Foundation, 2012). In Sri Lanka and elsewhere, majority/minority status and a history of conflict appear to have greater implications for freedom to express one's views and identity than religious beliefs and practices per se.

Tolerance

The principle of tolerance (forbearance, a willingness to accept views or practices with which one disagrees or of which one disapproves) accompanies but can be distinguished from freedom and liberty, because attempts to enforce religion can never succeed, the state's ability to discern people's individual good is limited, it is arguably desirable to respect people as autonomous agents even if we disapprove of their decisions, and tolerance helps people to live together, thus enhancing social stability.[7] It is based on two moral judgements: disapproval of a form of behaviour or set of beliefs, and refusal to impose one's views on others, although influence may be exerted through example or persuasion. A distinction can be made between positive and negative toleration, with diversity celebrated in the former and passively accepted in the latter (Heywood, 2000). As with other values, some think that there should be limits to tolerance to protect others from damage or offence and individuals or social groups from self-harm. In addition, some believe that a state should go beyond toleration to protect and even promote things of which it approves, although opponents think this is asking too much of those who disapprove of those things. Homosexuality is one issue that comes to mind. In addition, some suggest that freedom should include, as

7 It is not the same as permissiveness, a social attitude that allows people to act as they choose, reflecting either moral indifference (the actions in question cannot be judged in moral terms) or moral relativism (the belief that moral judgements can only be made from the perspective of the person(s) concerned) (Heywood, 2000).

mentioned above, freedom to respond to perceived divine will and freedom to take responsibility.

Evidence on the extent to which religious people value and practise tolerance is scattered and there is no unanimity on what questions might produce the most reliable and relevant evidence. In addition, some aspects of the issue may be considered too sensitive to explore, at least in opinion poll type surveys. Nevertheless, some recent studies have attempted to assess tolerance.

For example, in a 2010 study of Nahdlatul Ulama, the largest Muslim organisation in Indonesia, and Muhammadiyah, the second largest, members were asked about their tolerance of other faiths (Bush, 2014). About 70% of the respondents indicated that they would be tolerant of non-Muslims playing most roles in their local community, but somewhat contradictorily, more than 60% said that they would object to a non-Muslim place of worship being built in their neighbourhood (75% of Muhammadiyah and 63% of NU members). The lower tolerance of generally modernist/reformist Muhammadiyah than traditionalist NU members may, Bush suggests, be linked to their concentration in urban areas, where most churches are built, religious rivalry is more pronounced and there have, in recent years, been outbreaks of violence against religious minorities, including Ahmadis, Shia Muslims and Christians. The Indonesian population is democratically inclined and open-minded on many social issues. However, this study of two mainstream mass organisations shows that there are limits to Indonesian Muslims' tolerance (Bush, 2014).

In a Pew Research Center survey of Muslims in SSA and southern Asia, many respondents, especially in SSA, said that they know little about other faiths (Christianity, in southern Thailand Buddhism), but those who claimed familiarity with another religion are more likely to think that the faiths have a lot in common. Even in countries where a substantial proportion of people are non-Muslims, almost all Muslims in the southern Asian countries covered reported that all or a majority of their friends are also Muslim, and only a small proportion said that they would be comfortable with their son or daughter marrying a Christian or Buddhist. In SSA, few said that they would be comfortable with this, even though a substantial proportion in many countries had Christians in their immediate families (Pew Research Center, 2013, pp. 119, 125).

Most recently, a 2014/15 Afrobarometer poll in 31 SSA and two North African countries identified five aspects of social difference and asked respondents whether they would like, dislike or be indifferent to having people with these characteristics as neighbours. Respondents in most of the countries expressed their acceptance of people from different religions (87%), different ethnic groups (91%) and immigrants or foreign workers (81%). Acceptance of people living with HIV/AIDS (PLWHA) (68%) and homosexuals (21%) was much less common, indicating that HIV/AIDS is still stigmatised and homosexuality highly controversial (see Chapter 9). However, the levels of tolerance vary by region and country. Thus while 80% or more respondents in 28 of the 33 countries expressed tolerance for people of other religions, this varied from 98% in Côte d'Ivoire to 51% in Niger (Dulani et al., 2016, p. 8).

In addition, if the indicators are grouped into a single index, the analysis shows that levels of tolerance vary by age, educational level, exposure to news media, urban/rural residence and religion.[8] Overall, those who express tolerance are younger, male, more educated, more exposed to news media and live in urban areas: 91% of urban and 85% of rural residents expressed tolerance of having people professing another religion as neighbours. Christians are somewhat more likely than Muslims to express tolerant views, scoring 3.19 on the combined tolerance index compared to 3.02. The authors do not explore possible explanations for this difference (Dulani et al., 2016, pp. 4, 15). In addition, there are, as expected, variations within the religious traditions, with Christians appearing to be more tolerant if they live in religiously diverse contexts and some Muslims less tolerant than others if they are older and live in rural areas. Differences between countries suggest that pluralism, proximity and frequent contact between groups contribute to tolerance, so people living in countries that are religiously mixed and urbanised, with better-educated populations, tend to be more tolerant.

Linked to freedom of religion and tolerance is whether or not adherents to a particular faith or denomination seek to convert others. As discussed in Chapter 6, Muslims and Christians regard their religion (and sometimes the beliefs associated with their particular sect or denomination) as the true religion and seek to proselytise amongst adherents of other mono- and polytheistic religious traditions (or none). How extensively and aggressively they engage in proselytising varies, as does the precise purpose and nature of such activities. Today, fear that proselytising can threaten religious freedom and spark conflict has led some states to curtail or forbid attempts to convert others, especially to what may be perceived as a foreign religion, although others regard this as interfering with religious freedom.

Honesty and integrity

All religions, more or less explicitly, condemn dishonest, corrupt, fraudulent, deceitful, insincere and untrustworthy behaviour. Where religious adherence is widespread, as noted above, it seems reasonable to assume that people derive their moral framework from their religion and that even if they are non-adherents, their moral system is influenced by the religion(s) that form(s) part of their cultural heritage, affecting individual behaviour, social relations and dealings with political and bureaucratic structures. To a differing extent, the religious traditions also offer models for government and public administration that complies with religious injunctions (see, for example, Tomalin, 2009, on Hinduism). If moral standards are derived primarily from religious norms, it might be expected that religious people are less likely to engage in dishonest or corrupt behaviour, because it entails theft and mistreatment of others (Beets, 2007) and also because

8 Overall the two Muslim North African countries (Morocco and Tunisia) show the lowest levels of tolerance with respect to all the indicators except homosexuality, of which they appear to be more tolerant than Muslims elsewhere in the continent (Dulani et al., 2016).

they may think that such behaviour demonstrates their own religiosity. We might anticipate, therefore, that high levels of religious adherence at the national level will be associated not only with social and economic relationships based on trust but also with low levels of corruption. However, many of the most corrupt countries (according to the Transparency International Perceptions of Corruption Index) rank highly in terms of religiosity and, in practice, dishonest and corrupt behaviour is widespread, so the influence of religion is not clear.

There are many definitions and types of dishonesty and corruption. It may be defined as spiritual or moral impurity or deviation from an ideal, demonstrated in dishonest business deals, cheating, deception and even the abuse of religious or spiritual power and authority. It may include political corruption (the use of legislative powers by politicians to gain political office or government officials for illegal private gain), embezzlement (the theft of entrusted funds, misappropriation of property) and bribery (a 'gift' that is intended to alter the conduct or behaviour of the recipient by transferring money, goods or property to him/her in return for preferment, being given an advantage or action taken by a person in his/her official capacity). In the political science literature, it is commonly defined as the abuse of public office for private gain and a broad distinction is made between 'grand corruption' (e.g. political corruption, state capture) and 'petty corruption' (the everyday practices involving street-level public-sector employees, in which most people are involved). A further distinction can be drawn between public and private corruption: the former is regarded as harmful to the social group, while what is perceived as corruption in the private sphere is regarded as hurtful to individuals. In practice, it may be more appropriate to conceive of these scales and types of corruption as being a spectrum rather than separate categories, since the distinctions between them are not clear. All these forms of corruption are widely condemned by both religious and non-religious people (Marquette, 2012).

Attempts to understand how religious teachings on integrity influence ethical attitudes and behaviour are scarce. In much of the literature, corruption is regarded as an institutional failure, and the potential links with religious teachings are either not made explicit or not dealt with systematically. Where 'religion' *is* considered, much of the work has been concerned to prove or disprove the existence of a relationship between levels of 'religious adherence' and perceptions of the extent of 'corruption' at the national level (for example, Beets, 2007; Mensah, 2014).[9] In practice, the data and methods used in such studies have been inadequate to 'prove' or 'disprove' such a relationship. Such studies cannot tell us how people's attitudes to corruption are formed, the impact of religion (and other socio-cultural factors) on attitude formation or individuals'

9 Variations on this theme occur in studies of differences in ethical attitudes (how far people view certain actions as always justified, never justified or in between) between countries with different dominant religions, using World Values Survey data (for example, Norris and Inglehart, 2004). WVS data were also used in a study that showed that while family values and reported church attendance are associated with higher aversion to corruption, there is no significant difference in attitudes between the major religions (Gatti et al., 2003).

grounds for condemning/justifying corrupt behaviour. Further, they cannot reveal the links between moral values and actual behaviour (Marquette, 2012, 2013). Given the wide range of what may be regarded as dishonest or corrupt behaviour, the diversity and syncretic nature of religious values and moral teachings, the difficulty of disentangling religion and culture and the complexity of links between values and behaviour, this is hardly surprising.

De Sardan (1999) takes up the question of whether and how 'religion' can be disentangled from 'culture' and 'governance' (the latter generally focusing on political efficacy and legitimacy), advancing the idea of a 'corruption complex', to capture the intertwined nature of corruption, governance, culture and religion. He identifies six cultural logics engrained in social life in West Africa that help to explain the ubiquity and persistence of generalised corruption, which pervades every social domain, including religion, despite religious teachings and efforts by different types of political regime to combat it.

To begin to develop an understanding of whether and how religiously based ethics shape attitudes to corruption, an interpretive approach is needed, based on constructions of both religion and corruption as lived experiences, within which morality is constructed and evolves. This was attempted in a study of religion and attitudes to corruption in India and Nigeria, in both of which corruption at a variety of scales is perceived to be widespread.

Box 7.1 Religion and attitudes to corruption in India and Nigeria

Research was conducted among Hindus in Hyderabad, Andhra Pradesh, and Sikhs in Chandigarh and Amritsar in India and amongst Christians, Muslims and adherents of traditional religion in four locations in Nigeria (Kano in the north west, where Islam is dominant, Abuja, the Federal capital, which is religiously mixed, Owerri in the south east, where Christians and adherents of traditional religion live, and Ibadan in the south west, where Christians and Muslims live alongside each other). The study sought to understand actors' own interpretations of relevant concepts, their experiences and their own and others' responses to these experiences.[10] Semi-structured interviews were carried out with leaders and ordinary members of religious organisations, policy makers and staff in selected public- and corporate-sector organisations, members of NGOs, young people, media persons and those involved in anti-corruption policy design and

10 The research was also driven by a recognition that recent anti-corruption initiatives in both countries (and elsewhere) appear to have had little success in reducing the scale and scope of corruption and to evaluate whether ethical values rooted in religion and religious organisations could be harnessed to tackle the problem, although these questions are not the main concern of this account (see Marquette, 2012, 2013).

implementation. The respondents were mostly urban, English-speaking and relatively well educated.

In both countries respondents agreed that religious teachings support moral behaviour and condemn corruption (see, for example, Aiyede et al., 2011; Simbine et al., 2011) but also observed that the nature, extent and mode of influence of religion varies between individuals and that religiosity (religious observance, performance of rituals) is not necessarily associated with moral uprightness. In addition, most regard religion as only part of a 'package' of moral upbringing that centrally involves the family, where children are socialised at an early age and where moral teaching is likely to be more effective than attempts to instil ethical values at a later age (for example, through official codes of conduct and workplace training). Indeed, in India, respondents generally considered family ideals and principles taught by parents to be more important than religion (Pavarala and Malik, 2012).

No respondents in either country expressed toleration of any form of corrupt behaviour, instead condemning all forms of corruption.[11] In India, this was expressed largely in secular and legal terms: corruption was considered to include any illegitimate behaviour related to getting things done (in administration or business), because it was considered to hinder implementation of policies, harm the governance system and undermine work culture, as well as damaging the society's core value system (Pavarala and Malik, 2012). Nigerian respondents, in contrast, explicitly linked religion and corruption, regarding the underlying moral question as one of good and evil. Thus they saw corruption as more than illegal behaviour, extending it to include all socially unacceptable behaviour, such as fraud, hoarding, cheating, nepotism and greed, and also 'corruption of the body' (e.g. adultery, homosexuality) (Simbine et al., 2011; see also Smith, 2007). Because of this wider conceptualisation and the explicit connection with questions of good and evil, support for resorting to the supernatural to tackle corruption was reportedly more widespread in Nigeria than in India (Aiyede et al., 2011).

In India, respondents' understandings of corruption can be broadly grouped into relatively narrow legalistic views and broader moralist views, although these are not distinct and some proposed that both conceptions are necessary to capture the full range of corrupt behaviour. Those advancing a legalistic view included activities involving monetary exchange under the heading of corruption (bribery, misappropriation of funds and

11 De Sardan (1999) notes that, despite the universal stigmatisation and denouncing of corruption in West Africa, there is widespread impunity, and condemnation rarely leads to legal proceedings or sanctions.

donations, the use of public office for private gain) but not gift giving, nepotism or adultery. They believe that social disapprobation of these types of corruption has declined, blamed by some on non-enforcement of the legal penalties. Those with a broader conceptualisation articulated "an abstract, amorphous morality, and list[ed] a large number of activities as corrupt" including gratification of one's needs (e.g. womanising) or obtaining a favour in return for giving something (e.g. a gift, tip or dowry) (Pavarala and Malik, 2012, p. 64).

Respondents in both countries see corruption as something 'other' people do while regarding what they themselves might do as making the best of a bad situation. They generally separate personal and public behaviour with respect to their own actions. Some had a sense that choosing to be incorrupt in a corrupt system would put their own families at a disadvantage. For example, a businessman in Amritsar (male, Sikh) admitted that "Although I am a God-fearing person conducting my business honestly, I gave five lakh rupees (approximately £6,700) for admission of my daughter [to an educational institution] because they were demanded of me. I had to arrange it by hook or crook" (Marquette, 2012, p. 22). Similarly, a respondent in Abuja (male, Christian) explained: "my religion influenced me not to be corrupt, but if you look at the future of yourself and your family, you just have to engage in it [corruption] to get what you want" (Marquette, 2012, p. 22). In both countries, respondents often used the word 'normal', as in 'Most people see corruption as normal', implying that 'systemic corruption' is widespread and deeply entrenched, so that a choice not to engage in it is regarded as illogical or even ridiculous. In addition, most believe that individual actions cannot make any difference. Respondents, who often described themselves as both religious and ethical, were therefore able to engage in a process of what Bandura calls 'selective moral disengagement' through 'diffusion of responsibility': "Where everyone is responsible, no one really feels responsible . . . and . . . harm can always be attributed largely to the behaviour of others" (Bandura, 2002, p. 107). This was true regardless of religion across both countries (Marquette, 2012, p. 13).

Moreover, almost all the respondents in both countries saw religious organisations as part of the problem, because they are themselves associated with corrupt behaviour, they contain wealth-seeking adherents and religious specialists who have engaged in corruption,[12] and they have received gifts alleged to be from the proceeds of corrupt actions (often given by

12 See also de Sardan's example of *marabouts* (Qur'anic scholars and religious teachers) in West Africa, who expect a preliminary gift in return for initiating therapeutic or magical action, such as tracing verses of the Qur'an for use in an amulet, and 'thanks' if the action turns out to be effective (de Sardan, 1999).

individuals to assuage their consciences). In India, many of the (largely middle-class) respondents perceived a moral vacuum at the heart of organised religion, noting that religious leaders rarely take a stand against corruption in public life. For example, "I know a large number of businesses who will donate five per cent to temple or maybe to some charity. This is how we try to overcome our guilty conscience" (director of an anti-corruption bureau in India). Similar allegations were made in Nigeria: "I have never heard of any church or mosque sending a corrupt person away from its premises; instead they are given [a seat in the] front row" (male, Muslim).

The research concluded that while, across the religions, adherents condemn corruption, their beliefs have little impact on corrupt behaviour, because the temptations provided by the ubiquity of opportunities for illicit gain outweigh religion's influence. Indeed, individuals often admit that they pray to God to make them richer and more successful and regard prosperity as a sign of God's favour. Thus in highly corrupt countries, people are likely to feel that their own participation in corrupt behaviour is inevitable (and even justified), given the systemic nature of corruption. Religion is, therefore, unlikely to be an effective entry point for tackling corruption which should, in Marquette's view, be seen as a collective action problem that cannot be effectively addressed by encouraging individuals to behave differently in everyday situations through legal changes or moral suasion or by enlisting the aid of religious organisations (Marquette, 2012).

Conclusion

A moral or ethical code implies a system of morality or an ethical framework that is coherent and consistent, often because it is linked to the teachings of a particular religion, which are often based on revelations and sacred texts and may or may not be reflected in law, for example the Qur'an and *sunnah* of Islam, the Ten Commandments of Christianity and the Five Precepts and Noble Eightfold path of Buddhism. The analysis in this chapter has focused on selected values that are central to both international conceptions of human rights and principles of religious teaching that are supposed to be realised in the behaviour of adherents and religious organisations: freedom, rights, tolerance, honesty and integrity. Although they are not identical, the values and moral norms associated with the traditions under review have considerable similarities, despite their different origins and the varying theological/philosophical underpinnings of their ethical teaching. At first sight, many seem to be broadly consistent with internationally recognised norms, although there are questions about how universalist ethical values are interpreted and transformed when applied in changing religious and socio-cultural contexts (Bochow et al., 2017). However, practical and methodological problems bedevil any attempt to assess whether and how values and norms are reflected in the day-to-day behaviour of individuals and groups, which is influenced by differences in interpretations and practices amongst adherents of a particular religion or denomination, the varying scope for active moral and ethical choices, and non-religious factors such as the wider and rapidly changing social and political context in which adherents live and their majority/minority status. Although the available evidence is sparse and patchy, it is clear that the links between discourses, religious institutions and social practices are complex and that the attitudes and behaviour of religious organisations, leaders and adherents are not always consistent with the values they espouse.

References

Abraham, A.J. (1998) Islam and human rights, *Journal of Third World Studies*, XV, 1.

ACN International (2016) *Religious Freedom in the World Report 2016, Executive Summary*, Sutton: Aid to the Church in Need (UK). www.religion-freedom-report-org-uk (downloaded 19/11/2017).

Aiyede, E.R., Simbine, A.T., Fagge, M.A. and Olaniyi, R. (2011) Religion, ethics and attitudes towards corruption in Nigeria: A historio-graphical review, *Research for Development*, 25, 1 & 2, 205–38.

Alhassan Alolo, N. and Connell, J.A. (2013) Indigenous religions and development: African traditional religion, in Clarke, M. (ed) *Handbook of Research on Development and Religion*, Cheltenham: Edward Elgar, 138–63.

An-Na'im, A.A. (1995) Toward an Islamic hermeneutics for human rights, in An-Nai'im, A.A., Gort, J.D., Jansen, H. and Vroom, H.M. (eds) *Human Rights and Religious Values: An Uneasy Relationship?* Amsterdam and Grand Rapids, MI: Editions Rodopi and William B Eerdmans Publishing Company, 229–42.

An-Na'im, A.A. (1996) Islamic foundations of religious human rights, in Witte, J. and van der Vyver, J.D. (eds) *Religious Human Rights in Global Perspective: Religious Perspectives*, Dordrecht: Martinus Nijhoff, 337–59.

Asad, T. and An-Na'im, A.A. (2009) *Religion, Law and the Politics of Human Rights*, Washington, DC: Georgetown University, Berkley Center for Religion, Peace, and World Affairs, Transcript of public conversation. https://berkleycenter.georgetown.edu/news/talal-asad-and-abdullah-an-na-im-discuss-human-rights accessed 16th October 2017.

Asia Foundation (2012) *National Values Survey, Sri Lanka 2011*, Colombo: The Asia Foundation.

Bandura, A. (2002) Selective moral disengagement in the exercise of moral agency, *Journal of Moral Education*, 31, 2, 101–19.

Beets, S.D. (2007) Global corruption and religion: An empirical examination, *Journal of Global Ethics*, 3, 1, 69–85.

Bochow, A., Kirsch, T. and van Dijk, R. (2017) Introduction: New ethical fields and the implicitness/explicitness of ethics in Africa, *Africa*, 87, 3, 447–61.

Bush, R. (2014) *A Snapshot of Muhammadiyah: Social Change and Shifting Markers of Identity and Values*, Singapore: Asia Research Institute WP 221.

Cantwell, C. and Kawanami, H. (2009) Buddhism, in Woodhead, L., Kawanami, H. and Partridge, C. (eds) *Religions in the Modern World: Traditions and Transformations*, London: Routledge, 67–102.

Chanock, M. (2002) Human rights and cultural branding: Who speaks and how, in An-Naim, A.A. (ed) *Cultural Transformation and Human Rights in Africa*, London: Zed Books, 38–67.

Clarke, M. (2011) *Development and Religion: Theology and Practice*, Cheltenham: Edward Elgar.

Curran, C.E. (2002) *Catholic Social Teaching, 1891-Present*, Washington, DC: Georgetown University Press.

Curran, C.E. (2005) Christian ethics, in Jones, L. (ed) *Encyclopedia of Religion*, Detroit: Macmillan, Vol. 3, 2nd ed, 1650–7.

Deneulin, S. and Bano, M. (2009) *Religion in Development: Rewriting the Secular Script*, London: Zed.

De Sardan, J.P.O. (1999) A moral economy of corruption in Africa, *Journal of Modern African Studies*, 37, 1, 25–52.

Devine, J. and Deneulin, S. (2011) Negotiating religion in everyday life: A critical exploration of the relationships between religion, choices and behaviour, *Culture and Religion*, 12, 1, 59–76.

Dulani, B., Sambo, G. and Dionne, K.Y. (2016) *Good Neighbours? Africans Express High Levels of Tolerance for Many, But Not for All*, Afrobarometer Round 16, Despatch No 74.

Duriez, B. and Soenens, B. (2006) Religiosity, moral attitudes and moral competence: A critical investigation of the religiosity-morality relation, *International Journal of Behavioral Development*, 30, 1, 76–83.

Ensor, J. (2005) Linking rights and culture – implications for rights-based approaches, in Gready, P. and Ensor, J. (eds) *Reinventing Development? Translating Rights-based Approaches from Theory into Practice*, London: Zed, 254–77.

Gatti, R., Paternostro, S. and Rigolini, J. (2003) *Individual Attitudes Towards Corruption: Do Social Effects Matter? XV Conferenza Societa Italiana di Economia Pubblica (SIEP)*, Pavia: Università di Pavia. http://siepweb.it/siep/oldDoc/wp/267.pdf (accessed 4/01/2019)

Hardy, P. (1987) Islam in South Asia, in Eliade, M. and Adams, C.J. (eds) *The Encyclopedia of religion*, New York: Macmillan, Vol. 7, 390–404.

Heywood, A. (2000) *Key Concepts in Politics*, Basingstoke: Macmillan.

Kim, K. (2007) *Concepts of Development in the Christian Traditions: A Religions and Development Background Paper*, Birmingham: University of Birmingham, Religions and Development WP 16.

Kroessin, M.R. (2008) *Concepts of Development in 'Islam': A Review of Contemporary Literature and Practice*, Birmingham: Religions and Development WP 20.

Loy, D.R. (2010) The karma of poverty: A Buddhist perspective, in Galston, W.A. and Hoffenberg, P.H. (eds) *Poverty and Morality: Religious and Secular Perspectives*, Cambridge: Cambridge University Press, 44–61.

Mageza, L. (1997) *African Religion: The Moral Traditions of Abundant Life*, Nairobi: Paulines Publications Africa.

Marquette, H. (2012) 'Finding God' or 'moral disengagement' in the fight against corruption in developing countries? Evidence from India and Nigeria, *Public Administration and Development*, 32, 1, 11–26.

Marquette, H. (2013) Corruption, religion and moral development, in Clarke, M. (ed) *Handbook of Research on Development and Religion*, Cheltenham: Edward Elgar, 220–37.

Maududi, A. (1976) *Human Rights in Islam*, Leicester: The Islamic Foundation.

Mayer, A.E. (1995) *Islam and Human Rights*, Boulder, CO: Westview.

Mensah, Y. (2014) An analysis of the effect of culture and religion on perceived corruption in a global context, *Journal of Business Ethics*, 121, 2, 255–82.

Mosse, D. (2018) Caste and development: Contemporary perspectives on a structure of discrimination and disadvantage, *World Development*, 1, 10, 422–36.

Nadkarni, M.V. (2008) *Hinduism: A Gandhian Perspective*, New Delhi: Ane Books, 2nd ed.
Norris, P. and Inglehart, R. (2004) *Sacred and Secular: Religion and Politics Worldwide*, Cambridge: Cambridge University Press.
Oliver, P. (2010) *World Faiths: An Introduction*, London: Hodder Education, 4th ed.
Pavarala, V. and Malik, K.K. (2012) Social constructions of religiosity and corruption, *Economic and Political Weekly*, XLVII, 1, 61–8.
Pelikan, J. (1987) Christianity: An overview, in Eliade, M. and Adams, C.J. (eds) *The Encyclopedia of Religion*, New York: Macmillan, Vol. 3, 348–62.
Pew Forum (2006) *Spirit and Power: A 10-Country Study of Pentecostals*, Washington, DC: Pew Forum, http://pewforum.org/uploadedfiles/Orphan_Migrated_Content/pentocostals-08.pdf
Pew Research Center (2013) *The World's Muslims: Religion, Politics and Society*, Washington, DC: Pew Research Center, http://assets.pewresearch.org/wp-content/uploads/sites/11/2013/04/worlds-muslims-religion-politics-society-full-report.pdf
Pew Research Center (2017a) *Global Restrictions on Religion Rise Modestly in 2015, Reversing Downward Trend*, Washington, DC: Pew Research Center, http://assets.pewresearch.org/wp-content/uploads/sites/11/2017/04/07161531/Pew-Research-Center-Religious-Restrictions-2017-FULL-REPORT.pdf
Pew Research Center (2017b) *Many Countries Favor Specific Religions, Officially or Unofficially*, Washington, DC: Pew Research Center, http://assets.pewresearch.org/wp-content/uploads/sites/11/2017/09/29162847/FULL-REPORT-FOR-WEB.pdf
Plant, S.J. and Weiss, D.H. (2013) Theology and development: Christian and Jewish approaches, in Tomalin, E. (ed) *The Routledge Handbook of Religions and Global Development*, London: Routledge, 53–67.
Rahman, F. (1987) Islam: An overview, in Eliade, M. and Adams, C.J. (eds) *The Encyclopedia of Religion*, New York: Macmillan, Vol. 7, 303–22.
Simbine, A.T., Aiyede, E.R., Olaniyi, R. and Fagge, M.A. (2011) What has religion got to do with it? Ethics and attitudes towards corruption in Nigeria, *Research for Development*, 25, 1 & 2, 239–76.
Smith, D.T. (2007) *A Culture of Corruption: Everyday Deception and Popular Discontent in Nigeria*, Princeton NJ: Princeton University Press.
Strong, J.S. (2005) Merit: Buddhist concepts, in Jones, L. (ed) *Encylopedia of Religion*, Detroit, MI: Macmillan, Vol. 3, 2nd ed, 5872–5.
Tibi, B. (2009) *Islam's Predicament with Modernity: Religious Reform and Cultural Change*, London and New York: Routledge.
Tomalin, E. (2009) *Hinduism and International Development: Religions and Development Background Paper*, Birmingham: Religions and Development WP 19.
Traer, R. (n.d.) *Muslims and Human Rights*, Berkeley, CA: Center for the Study of Ethics and Social Policy. www.religionhumanrights.com//Religion/Islamic/muslims.fhr.htm Accessed 30/10/17 (revised from Traer, R., 1991, Muslims, in *Faith and Human Rights: Support in Religious Traditions for a Global Struggle*, Washington, DC: Georgetown University Press).
Twining, W. (2009) Introduction to A. An-Nai'm, Readings, in Twining, W. (ed) *Human Rights, Southern Voices*, Cambridge: Cambridge University Press, 53–103.
Wilson, B. (2001) Salvation, secularization, and de-moralization, in Fenn, R.K. (ed) *The Blackwell Companion to the Sociology of Religion*, Oxford: Wiley-Blackwell, 39–51.
Wiredu, K. (2011) An African religions perspective, in Meister, C. (ed) *The Oxford Handbook of Religious Diversity*, Oxford: Oxford University Press, online ed.

Wolfe, R.W. and Gudorf, C.E. (1999) Introduction, in Wolfe, R.W. and Gudorf, C.E. (eds) *Ethics and World Religions: Cross-Cultural Studies*, Maryknoll, NY: Orbis Books, 1–20.

Wulff, D.M. (2001) Psychology of religion: An overview, in Jonte-Pace, D. and Parsons, W.B. (eds) *Religion and Psychology: Mapping the Terrain*, London: Routledge, 15–29.

Yamane, D. (1998) Charismatic movement (neo-Pentecostalism), in Swatos, W.H. Jnr (ed) *Encylopedia of Religion and Society*, London, Altamira: Sage, 80–2.

8 Well-being, poverty and inequality

Introduction

In Chapter 7, religious interpretations of the internationally accepted values of freedom and rights were explored in order to begin assessing their implications for everyday life. Realisation of rights, it was noted, requires tolerance, involves responsibilities and obligations and also implies justice, which applies to the distribution of 'goods' that may include freedom, rights, power and wealth. Justice is a different type of moral claim from, for example, charity or compassion, because it concerns entitlements – what can be demanded from others. Religious teachings generally pay attention to both, and both will be considered in this chapter, together with some examples of how they are understood by people of different religious traditions in SSA and southern Asia. Distinctions can be drawn between economic justice (the fair distribution of material burdens and benefits, especially income/wealth), environmental justice (the fair distribution of environmental burdens and benefits) and social justice (a morally justifiable distribution of rewards or punishments, which may require that people are rewarded according to how hard they have worked, i.e. how deserving they are) (Nussbaum, 2003).

A variety of approaches to achieving justice have been advocated and attempted. For example, procedural justice refers to independently adjudicated rules, generally embodied in law and implemented through a judicial system. The shortcomings of relying on the operation of market forces are the concern of distributive justice, which at a minimum requires that everyone's needs are met. One key issue concerns what should be (re-)distributed – welfare/well-being, resources or income/wealth (so people can choose how to use their share) or the capabilities that can enable people to live the lives they choose, as Nussbaum (2003) and Sen (2009) suggest, although Sen's approach has been criticised not only for providing few pointers to how individuals can achieve justice but also for failing to recognise that injustice is structural (Deneulin, 2011). A second issue is therefore how to ensure a fair distribution, including the role of different mechanisms and who should shoulder the responsibility. Only selected aspects of these issues can be discussed here.

In the previous chapter, the discussion of religious values was benchmarked against internationally accepted human rights. The human rights discourse is also relevant to considering religious conceptions of well-being, poverty and inequality. However, the discourse at governmental, societal and individual levels in Sub-Saharan Africa and southern Asia has also been strongly influenced by international and national debates on how to define and achieve socio-economic development goals, so these must be factored into the analysis.

Well-being

The concept of well-being in (development) theory typically lacks ethical and religious dimensions, perhaps because contemporary social theory is so closely associated with liberal ideas. In considering whether and for whom well-being is improving or worsening, much of the research and policy has focused on economic dimensions, namely household or individual income or consumption. Economists support such a conception, arguing that these measures are objective, the most satisfactory way of capturing the influence of many contributory factors, generate comparable data and enable progress to be monitored. Over the years, economic approaches and the data on which they draw have undoubtedly improved. Nevertheless, critics regard a monetary conception of poverty as reductionist, pointing out that people's own concepts of ill-being and well-being are multi-dimensional. Chambers (2006), for example, contrasts ill-being (characterised by the intersecting dimensions of powerlessness, insecurity, physical weakness/illness, material poverty and bad social relations) with well-being (characterised by freedom of choice and action, security, physical well-being, enough for a good life and good social relations).

Many of these dimensions/indicators potentially have links with religion, but health has received more attention than most. Biomedical approaches to health, based on scientific knowledge and expertise, often exclude the social and spiritual dimensions of illness and may regard religious practices as irrational, superstitious and ineffective. However, many people do not experience their physical and mental/emotional lives as separate from their spiritual lives, so often health is experienced as having physical, emotional and spiritual dimensions and ill-health attributed to a combination of causes. In addition, people do not see themselves as isolated individuals, so that illness of the body or mind is seen as reflecting social discord. Cochrane (2007) gives an example from southern Africa, where Sesotho does not contain words for 'religion' or 'health', instead providing *bophelo*, which signals comprehensive well-being and implies that many regard biomedical and other approaches to the treatment of illness, such as rituals and spiritual healing, as complementary.

A parallel debate concerns whether it is appropriate for well-being to be defined and assessed objectively by external analysts. Recognition that the dominant approaches neglect aspects of well-being that are not reflected in income or consumption has led to attempts to improve them, for example, the World Health Organisation's Quality of Life indicator. The WHO has also tried to incorporate spirituality, religion and personal beliefs into the indicator, although critics consider that the methodology has shortcomings, and it has not been widely applied in the global South (Moreira-Almeida and Koenig, 2006). Others suggest that subjective assessment is more likely to produce an accurate picture than objective approaches, although ideas about the most appropriate concepts, methods and indicators vary. One example is the growing body of research that uses the concept of happiness to capture subjective well-being (see for example Devine et al., 2017), although its conceptualisation of mental well-being is thought by critics to be inapplicable in many parts of the world and it also fails, for the most part, to take religious ideas into account (Joshanloo, 2013, 2014).

Some quantitative research, mainly in the North, finds that religious people report higher levels of well-being, suggesting possible

> mechanisms by which religious affiliation might contribute to greater mental and emotional well-being including: regulation of lifestyle and behaviour, such as restriction of intake of alcohol; provision of resources, such as social support and networks; promotion of positive self-esteem; acquisition of specific life-skills and coping resources such as a framework of understanding illness, stress or loss; generation of positive emotion, cultivation of disposition towards forgiveness, hope and transformation.
>
> (Swinton, 2007, in Graham, 2009, p. 12)

It is unclear whether such associations between religion and well-being are a result of religious belief and practice per se or of the social capacity derived from association with religious organisations (see, for example, Devine et al., 2017). In addition, there is a concern that concentrating on the potential contribution of religious adherence to subjective well-being may lead to a narrowly functionalist view of religion.

An alternative methodology for identifying conceptions of well-being uses an interpretive approach to reveal the influence of religious ideas and teaching, alongside other factors, on people's conceptions of ill-being and well-being. It was used in a study in Tanzania and Nigeria, some of the findings of which are summarised in Box 8.1.

Poverty

Both secular and religious traditions have wrestled with questions about poverty and the poor – what is poverty? Who are the poor and why are they poor? Is there a responsibility to reduce poverty? Whose responsibility is it? Who should receive assistance? Are there limits to individual and communal obligations (Galston and Hoffenberg, 2010)? The dilemma concerns the balance between pursuing one's self-interest and connections that bind members of society together – in ethical terms, between greed and virtue or between competition and cooperation. How this dilemma is resolved depends on the prevailing conceptions of a 'good life', self and society. The latter refers to the individuals and groups to whom moral obligations are owed, which inevitably involves tensions between the deserving and the non-deserving, or between all people (strangers) and those considered to belong to the group – insiders, who may be defined as family members, members of the same ethnic group and/or fellow religionists. In addition, views on the obligations owed to the poor depend on

- whether poverty is viewed as a moral and ethical issue
- perspectives on their rights, for example to exercise political voice and access essential resources and opportunities, and how these can be realised
- ideas about how poverty can and should be tackled
- perceptions of the ethics of wealth (especially if inherited)
- assumptions about whether poor people will contribute to society if their well-being is improved, and
- concerns about poor people's potential to threaten the socio-political order.

The relative importance of individual, cultural and structural factors in explaining poverty has been subject to much debate. By the 1960s, research seemed to have demonstrated that poverty can be better explained by structural factors (e.g. class) than the 'non-modern' values and cultural traits that had previously been thought to inhibit people's aspirations and achievements. However, in the 1980s, studies based on World Values Survey (WVS) data, which were first collected in 1981, demonstrated that values also have an influence and that these may be derived, inter alia, from religious beliefs (Wuthnow, 2008). For example, Inglehart and Baker (2000) used 1990s WVS data, first to compare the value systems of countries with a similar religious heritage and, second, to explain why national value sets, especially those derived primarily from religion, change or persist. They claim that the data show that the value systems of countries with a historical Catholic, Protestant, Confucian or Communist heritage differ more than those between religious groups within such societies, suggesting that national experience and culture trump religious difference in shaping values. Further, they suggest that value change is primarily a consequence of economic and technological change, which propels countries in a roughly predictable economic and social direction. As the level of economic development increases, they assert, national value sets change from an emphasis on achieving economic and physical security to prioritising subjective well-being and quality of life, attributing the persistence of religious values in less-developed countries to people's ongoing experience of existential insecurity and uncertainty (Norris and Inglehart, 2004). Haller (2002) provides a critique of these arguments.

Views about the best way of defining and measuring poverty differ. For example, Laderchi et al. (2007) review four approaches, the monetary, capability, social exclusion and participatory approaches, pointing out that each is a construction of reality, involving numerous judgements that are often not transparent. Different approaches are associated with different conceptions of poverty, identify different people as poor and have different implications for policy and action. As well as exploring the conceptions of well-being revealed by qualitative research in Tanzania and Nigeria, the discourse of local religious leaders and adherents on the nature of poverty is analysed in Box 8.1.

Inequality

For both practical and ethical reasons, addressing poverty implies that the distribution of well-being and wealth must be taken into account. First, unequal access to resources and opportunities are structural drivers of impoverishment and chronic poverty. In addition, "experience and research widely recognise that above certain income levels, greater prosperity is not matched by greater happiness, but is accompanied instead by greater social and individual distress, manifested for example in increasing crime and ill health, such as depression. Much evidence now also suggests that such trends are exacerbated by high levels of inequality in society" (Graham, 2009, pp. 5–6).

Four aspects of equality are important (Heywood, 2000):

- Foundational equality refers to the idea that humans are born equal and their lives are of equal moral value
- Formal equality refers to the status of individuals in society, in terms of their rights and entitlements, especially equality before the law and political equality (usually via universal suffrage)
- Equality of opportunity refers to equal life chances, implying that there is a distinction between inequalities that result from unequal social treatment (which is not regarded as legitimate), and those that result from the unequal distribution of merit, talent and willingness to work (which may be regarded as legitimate)
- Equality of outcome, implying the equal distribution of rewards, including income, wealth and other social goods.

As noted in Chapter 7, most religious traditions hold that individuals have equal spiritual and moral value, but the other dimensions identified above are likely to be reflected in different ways in religious teaching and the attitudes of religious people, not least because the dimensions are not always compatible. In addition, some, especially the last, may be controversial: equality of outcomes is thought by advocates to strengthen social cohesion, promote just treatment, and enlarge freedom, but opponents may object to it on the grounds that it removes incentives, caps aspirations and needs the state to achieve it, which may imply infringing individual liberty (Heywood, 2000). The available research gives some insight into the views of local religious leaders and adherents.

Well-being, poverty and inequality in the religious traditions

The religious traditions all have conceptions of well-being, wealth and poverty that emphasise spiritual and moral dimensions in addition to material welfare. It is not possible to discuss these conceptions exhaustively here, let alone to assess the various ways in which they may be measured or the characteristics and outcomes of alternative responses. Instead, an outline of the main ideas about well-being, wealth and poverty is provided for the various faith traditions. The findings of research that sought to explore the views and experiences of Muslims and Christians in Tanzania in East and Nigeria in West Africa are presented to illustrate whether and how local perspectives reflect the official or mainstream understandings. Perspectives from Hinduism, Buddhism and folk religious traditions will also be outlined and the latter illustrated with an example from rural Zambia.

Well-being, wealth and poverty in Islam

As discussed in earlier chapters, well-being in Islam is associated with submission to God and obedience to his law. To live a good life, one should have faith in God and put it into practice in daily living, tolerating hardship and illness with patience, as a test of faith. While worldly pleasures are legitimate, they are secondary. Instead, the ultimate state of individual well-being is contentment (Joshanloo, 2013). However, Muslim teachings also deal with the material aspects of well-being and poverty. In these respects, well-being is based on the fulfilment of necessities (*dharuruyyat*), needs or conveniences (*hajiat*) and comforts or refinements (*kamaliat*). Necessities refer to those aspects of life necessary for a person to adhere to the way of God and the five pillars of Islam (although the *hajj* is obligatory only if a person can afford it): protection of life (including access to healthcare); means for securing food, clothing, shelter and education; and the ability to earn a living and establish a family. An acceptable standard of living requires that all these are met at the barest minimum. Inability to meet even one means that the person is considered poor. Needs and comforts enable a person to avoid hardship and ensure a good life, and both have worldly and religious dimensions. Needs relate to religion (*deen*), life, the physical self (*nafs*), the mind (the intellect, knowledge, *aql*), progeny (*masl*) and wealth/property (*maal*) (Hashmi, 2010; Kroessin, 2008). Extremes of poverty and wealth are considered undesirable because they distract from God – conscious, moral life and the claims of the poor and obligations of the wealthy are referred to in hundreds of verses in the Qur'an (Hashmi, 2010).

Muslims are permitted to acquire property and wealth. However, the ways in which these are acquired are subject to religious injunctions. For example, speculation (including property dealing), gambling (including participation in capital markets), hoarding and destructive competition are banned (Kuran, 1986; Tripp, 2006).[1] Individual property rights are limited by various rules, relating for example to inheritance and commercial partnerships. Two of the five pillars of the faith (see Chapters 5 and 6) bear on wealth: first, the payment of alms to the poor (*zakat*, purification of wealth) is a religious duty and, second, fasting during Ramadan is intended in part to cultivate empathy with the poor, in line with an injunction to display solidarity with and mercy for fellow human beings (Kroessin, 2008, p. 44; Hashmi, 2010).

1 Only limited aspects of Islamic economic thinking and practice are explored here. These will be more extensively analysed in the companion volume.

Little attention is paid to the causes of poverty in the Qur'an (over and above greed and extravagance), but it does consider the characteristics of the poor. The most common terms used are *fuqara* and *masakin* (the poor and needy), both of which appear in the list of those entitled to receive *zakat* (see Chapter 6). Although the Qur'an does not provide criteria for distinguishing between the poor and needy, and some think the distinction is arbitrary, early interpreters agreed that the distinction is based on the severity of poverty or the capacity of poor people to ameliorate their situation, with the former stoically bearing their poverty and the latter forced to resort to begging. Begging, it is argued, is permissible only for the very poorest or the seriously indebted.

Qur'anic teaching and *shari'a* provide safeguards against poverty and ways of tackling it. Some *ahadith* prescribe work as a way of tackling poverty, with the authorities being held responsible for creating work opportunities if they are not available (Kroessin, 2008). The prohibition of *riba* is said to provide an additional safeguard against poverty. The ethical principle underlying the ban on *riba* (commonly translated as interest, although some think it should be confined to usury, or excessive interest) is that lending money with interest to a poor person violates the Qur'anic requirement that those with resources should provide charity to those in need. In addition, it is thought to encourage greed and inhumanity, as well as to impede the circulation of wealth in society. Alternative, supposedly less punitive, arrangements for borrowing money based on shared risk are advocated to prevent the spiralling of debt that tips many with a basic level of well-being into extreme poverty and destitution.

It is specified that a woman and her children are entitled to full maintenance from her husband while married. In addition, women are thought to be safeguarded by their property rights, especially as husbands are expected to provide not just necessities but sufficient to exempt women from needing to earn an income: a woman's property and the dowry she receives from her husband before marriage (*mahr/sadaq*) are her own, and women are entitled to inherit property, although a smaller share than male descendants because the latter are considered to have more extensive responsibilities to support their families (see also Chapter 9). While these provisions do mean that some women are adequately provided for and safeguarded from impoverishment, others are not protected. Injunctions governing inheritance (*wiratha*) also have a role to play – they provide for a wide and fair distribution of the estate of a deceased person (especially unearned wealth), not just to dependents and kin but also to the poor and needy, especially orphans (Hashmi, 2010).

States have a responsibility to tackle poverty and ensure well-being, although many Muslims think that only states based on *shari'a* are likely to fulfil this responsibility. However, above all, they are urged to tackle poverty through charity. This is regarded as an act of individual devotion and worship (Al-Qardawi, 1999; Benthall, 1999, 2003), which will "bring the giver closer to God and expiate sins" (Hashmi, 2010, p. 195). In addition to obligatory charitable giving, the Qur'an specifies a number of forms of voluntary giving (*sadaqa*) and provides guidance on who is entitled to both forms of charity. For an individual, the 'near poor' (one's kin other than dependents, who have a prior claim) take precedence over the 'far poor' – servants, neighbours, Muslims of one's community etc. (Hashmi, 2010). Qur'anic teaching on wealth and poverty is underlaid by the principle of distributive justice, which stresses that recipients, especially of *zakat*, are entitled to the proceeds of such redistribution. However, it is also argued by some that many of the mechanisms serve to temporarily alleviate extreme poverty and destitution rather than addressing the structural causes of poverty.

Views of wealth and poverty in Christianity

For Christians, ultimate fulfilment and salvation are given and received through the grace and love of God and Jesus Christ. Salvation of the cosmos is therefore regarded as a work of God in which humans are invited to share rather than a work of human hands. However, they believe that all human beings have been created in the image of God and are therefore obliged to manifest God's love in their behaviour and relations with others by demonstrating compassion for the disadvantaged and solidarity with the poor. In addition to witnessing to their faith, most believe that Christians have an obligation to work for social transformation, based on a belief in the desirability and achievability of human progress (Plant, 2009). Although they may disagree over the specific goals and how to achieve them, they are tasked with increasing human flourishing, through providing practical assistance to those in need and tackling social injustice (Deneulin and Bano, 2009).

Amongst Christians involuntary poverty may be seen as God's will (for testing or punishing individuals) or, more often, an evil (resulting from sin) to be eradicated, an inevitable characteristic of an unequal world or the outcome of political and economic injustice and oppression (Taylor, 2003; see also White and Tiongco, 1997). The churches generally regard rights to private property as valid and necessary and support the accumulation of wealth, providing this is done in ethical and legal ways, and it is used to ensure a basic minimum for everyone. Enabling people to become 'fully human' implies not only material sufficiency but also the ability to sustain a satisfactory level of well-being, based above all on education, especially literacy (Curran, 2002; Deneulin and Bano, 2009). The obligation rests with both individuals and states, which are regarded as being responsible for bringing about private prosperity and public well-being. Rather than challenging the capitalist system, therefore, Christian teaching focuses on the need for wealth to be acquired responsibly and used to help narrow the gap between the rich and the poor by paternalism or charity. States are expected to contribute by investing to achieve economic growth, which can trickle down to benefit the poor, and through taxation and state action (Kim, 2007; Taylor, 2000, 2003).

The relative emphasis placed on philanthropy, state action and structural change therefore varies. Traditional approaches have been paternalistic and philanthropic, giving rise to concerns about the potential contradictions between them and evangelism, as well as their effectiveness (Bradbury, 2013; Fountain, 2015). Increasingly, given the postcolonial experience of first Latin America and then other parts of the global South, the inadequacy of these approaches to tackling inequality and poverty was recognised by some Christians, especially within the mainstream denominations (Deneulin and Bano, 2009). Recognition that more is needed than charity is reflected in, for example, liberation theology and its critique of the prevalent model of capitalist development and church support for the politically powerful (especially authoritarian regimes). In response, redistribution between and within countries is often advocated, and the Old Testament principle of a jubilee every 50 years, when land reverts to its original owners and debt is cancelled, may be called upon to produce a more equitable distribution of resources (Kim, 2007).

In the late nineteenth and twentieth centuries Christian revivalism was in part a response to the strengthening of increasingly secular and liberal bases for social power. Revivalist movements of this era and their successors opposed social changes which appeared to be undermining religion and thought that the mainstream Christian denominations were focusing on social action at the cost of individual repentance, mission and evangelism. As discussed in Part II, such revivalists generally regard wealth as a blessing from God, even a reward for faith, in contrast to other Christians who may regard wealth as selfish, undeserved and possibly ill-gotten. Some renewalists openly preach a 'prosperity gospel', which holds that prosperity and success are signs of God's favour. They emphasise individual responsibility for family, neighbours and fellow church members rather than advocating structural economic and political change. All, including renewalist Christians, are expected to avoid greed and selfishness, instead stewarding their time, talents and material possessions as gifts of God and devoting a share of them to assisting the poor through both spiritual means (e.g. prayer, healing) and material giving (Curran, 2005; Kim, 2007). The Pew Forum's (2006) surveys of Christians in SSA revealed a widespread view that if enough people are brought to Christ, social ills will take care of themselves. Nevertheless, renewalist Christians almost all reported that they also believe that it is their duty "to work for justice for the poor" (p. 30).

Evidence on the views of ordinary Muslims and Christians in different parts of Sub-Saharan Africa is available from research undertaken in 2008/9 in Tanzania and Nigeria, which sought to reveal some of the beliefs and values held by religious leaders and ordinary adherents (Box 8.1).

Box 8.1 Well-being, wealth, and poverty in Tanzania and Nigeria

Semi-ethnographic research based on semi-structured interviews, focus group discussions, attendance at religious services and reading locally produced media and texts explored adherents' and local religious leaders' perceptions of values related to well-being, wealth, poverty and inequality in two districts in Tanzania and three cities and peri-urban areas in Nigeria (Dugbazah and Hassan, 2010; Mesaki et al., 2010). The main ethnic groups in rural, largely Christian Meru District and the mainly Muslim port city of Tanga are known in Tanzania as progressive and hardworking farmers, fishermen and workers. Meru District is a prosperous farming district experiencing increasing pressure on land, while Tanga has been hard hit by declining formal-sector industrial and service employment. The religious composition and prosperity of the three Nigerian study cities also varies: Jos, capital of Plateau State since 1976, is on the border between the largely Muslim north of Nigeria and the religiously and ethnically mixed Central Belt. The city has a largely Muslim population, with a sizeable Christian minority and some adherents of traditional religion. It was a relatively tolerant and peaceful city until 2011, since when there have been increasingly frequent conflicts over land rights, exacerbated by the spill-over effects of increased violent activity by the Islamist Boko Haram further north. Abuja, in the Federal Capital Territory, is located in the centre of Nigeria. Built from scratch in the 1980s, the city is religiously mixed. A large city and state capital, Ibadan in the south west is dominated by the Yoruba ethnic group and is largely Christian, although there are Muslim and ATR minorities in this prosperous, religiously moderate city, where Muslims follow a more liberal form of Islam than in the north.

In all the locations and amongst both Christians and Muslims, 'development' is seen as having both individual and social dimensions: for individuals it implies an enhanced ability to 'move forward in life' by increasing one's income and acquiring material possessions, while for localities it is conceived of as a process of economic, infrastructural and socio-cultural change that should benefit the whole community. For most, prosperity is a legitimate goal, although some add spiritual development as well. Thus in Tanzania, both individuals and local religious leaders stress that Christian teachings emphasise both spiritual and material well-being, the former to be achieved through belief in God and prayer, while Muslims see Islam as providing a foundation for the lives of both individuals and families, through the inculcation of good moral values. In Nigeria, many men focus on individual advancement, while women are more inclined to emphasise the well-being of the family unit – informants often refer to improvement compared to their parents' lives and tend to view modernisation/Westernisation in dress, speech etc. positively because of its association with advancement and social status.

The research revealed both similarities and differences between the locations and religious groups in attitudes towards well-being and prosperity and the strategies advocated to achieve these aims. Basic well-being is generally understood as having a good house, sufficient food and clothing and children who attend good schools. Over and above this, most believe that the teachings of their respective religions support wealth creation and accumulation, although their views then diverge somewhat. Amongst

Christians, Pentecostals are more likely to overtly encourage wealth creation, averring that believers can rely on God and prayer for material blessings. However, the view that well-being and prosperity are to be achieved through obtaining an education and hard work is more widely held. Christians in Tanzania, for example, often cited the parable of the talents, many quoted New Testament verses from 2 Thessalonians 3: 10–12 – *Asiya fanza kali na asile* (one who does not work should not eat) – and some pointed out that in the story of creation, God is portrayed as working for six days and only resting on the seventh (Mesaki et al., 2010, p. 8). In both countries, it is still widely believed that the most promising way of earning an adequate income is to secure formal wage employment, in part explaining the common emphasis on obtaining a good education. However, it is clear to all that the opportunities for doing so have shrunk, in both disadvantaged localities such as Tanga and Jos and in more economically buoyant contexts, necessitating alternative strategies such as setting up an informal sector enterprise or migration. The need for rural residents to migrate to urban areas in search of opportunities is widely thought to be putting families and communities under strain.

Despite the acceptability of wealth, concerns are expressed about its pursuit. Pentecostal Christians are more likely to regard this as religiously sanctioned and desirable, not least because of their obligation to tithe a proportion of their incomes to the church. In contrast, Lutherans and Catholics in Tanzania are more likely to regard wealth accumulation as sinful if it is sought for its own sake, gained by immoral or illegal means and not accompanied by charitable giving. Even some Pentecostal church leaders express concern at the potential adverse effects of covetousness and increased materialism, and a few respondents in both countries think that people should have few material possessions to enable them to focus on leading godly lives.

Men's inability to earn sufficient to support their families, especially in economically disadvantaged localities, has forced many women into the labour force. Sometimes this is well established and socially accepted, but in some contexts, it challenges prevalent religious and social norms. For example, in Tanga, Muslim women have started to work outside their homes, most commonly as informal-sector traders or food processors and sellers, although some are pushing the boundaries of acceptable behaviour by venturing into cross-border trade with Kenya or traditional male economic sectors such as fishing. In the past, disapproval of women working outside the home was based on the traditional (and Muslim) view of men as being responsible for providing for their families and women's role being to look after the home, as well as the Islamic practice of purdah. However, faced with dire economic need, both women and men have been forced to adapt (see also Chapter 9).

Informants characterised poverty as an inability to meet basic needs for food, shelter, clothing, education and employment. In both countries, not only Muslims (in line with Qur'anic teaching) but also Christians distinguished between the needy or destitute and the poor, using local adaptations of the Arabic terms, although the distinction is not clear cut in practice. Very few informants blamed poverty on the attitudes or behaviour of individuals, although one or two, especially in Jos, referred to parents' reluctance to send children to school or the practice of purdah, which restricts women's ability to engage in income-generating activities. Instead, it is attributed to the lack of opportunities for education or employment, environmental pressures on agriculture, competition for land and economic opportunities, the rising cost of food and, by some religious leaders in Nigeria, on endemic corruption, each of which is, in the view of residents in disadvantaged localities (especially the largely Muslim urban centres of Tanga and Jos), exacerbated by past and continued government discrimination and neglect. Some in Nigeria link poverty to traditional religious practices, although the direction of cause and effect was not specified, while others observe that educated and wealthy people also engage, often covertly, in such practices.

Some believe that poor people should accept poverty as their lot in life, and some, especially in the Evangelical Church of West Africa and Church of Christ in Nigeria, believe that 'spiritual emancipation', marked by more regular attendance at religious services and generous donations to the church, can reverse material poverty. Nevertheless, the vast majority of both leaders and adherents believe that it can only be overcome by hard work and perseverance, with the help of God. Hard work therefore needs to be accompanied by spiritual progress – developing a good relationship with God and improving one's religious knowledge will, it is thought, result in internal transformation, which will increase contentment, enable one to work hard and behave morally and (one day) yield material rewards.

Both leaders and adherents are acutely aware of inequality and stress the role of and obligation for individual charitable giving (see also below). They acknowledge that socio-economic inequality characterises not just the wider society but also religious organisations. For example, "In the church there is a big gap. The people who have and give [to] the church, they sit at the front. Those who do not have, they sit behind them at the back" (Interview with the Secretary of the Evangelical Lutheran Church of Tanzania, Meru Diocese). The view that wealth must not only be gained ethically but also shared is widely held: "it is un-Islamic for one to be gluttonous while his/her neighbour goes hungry" (an imam in Tanga). However, although there are always some wealthy adherents and ordinary people may be conscientious in their charitable giving, the amounts given are limited by widespread poverty, restricting the potential for alleviating poverty by this means.

Well-being, wealth and poverty in Hinduism

In Hinduism, life has four aims: *artha* (wealth, material prosperity), *dharma* (duty, role), *kama* (enjoyment) and *moksha* (spiritual liberation). These are linked in the *ashrama* system to the four stages of life for twice-born men, especially *artha* with the second, householder stage and *moksha* with the third and fourth stages (*vanaprastha*, living in the forest, and *samnayasin*, renouncer). These ideas reflect a view that there is a place for multiple aims in life, including material prosperity and enjoyment. The accumulation of wealth is therefore considered to be a legitimate goal, providing it is earned as a result of effort and not hoarded or used only for selfish interests (Nadkarni, 2008, p. 387). Indeed, the Hindu pantheon includes a goddess of prosperity, Lakshmi. Guidelines are provided for using wealth – one part, it is recommended, can be used for securing prosperity and fulfilling responsibilities (*dharma*), a second for *kama* (acceptable pleasures) and a third for accumulation. A distinction is made between involuntary and voluntary poverty. The first is considered unacceptable in society, to be addressed by 'the ruler', who should be responsible for providing assistance to the destitute (including widows and the disabled). However, there is also a strong tradition of choosing to live an austere life governed by higher standards of personal morality than the norm and renunciation for the sake of spiritual pursuits (*saanyasa*) (Sharma, 2010). Personal behaviour is regarded as important, not just because it has implications for *karma*, with increased prosperity in a future life believed to be related to actions in the present one, but also for others. This sense of moral responsibility for other people's well-being implies a shared obligation to relieve poverty, hunger, disease and unhappiness.

As philosophical thought and social and political organisation evolved, ideas about the aims and stages of life and their links with the class (*varna*) and caste system also changed. Vedanta philosophy holds that there is divinity in every creature, giving rise to ideas about equality and justice, while the pursuit of wealth is balanced by the obligation to fulfil duties (Nadkarni, 2008). Nevertheless there appear to be significant contradictions between some of the philosophical principles underlying Hinduism and the inequalities and injustice of the social system with which it is closely linked. For some, these contradictions can be addressed through philosophical thought, social reform and government action: "Reformers and critics of the oppressive hierarchies inherent within the Hindu tradition are attracted to the notion of *sadharana dharma* [everyday *dharma*] as the basis for a universal and egalitarian ethic with the capacity to bridge the gap (in culturally appropriate ways) between the stratified tradition of *svadharma* and modern secular human rights" (Tomalin, 2009, p. 23). However, for others, the contradictions are irreconcilable, leading to conversion to religions such as Buddhism and Christianity, which in the eyes of converts demonstrate more convincingly the equal dignity and worth of all human beings (see Chapter 6).

Wealth and poverty in Buddhism

From the outset, Buddhism considered that anyone is capable of reaching *nibbana/nirvana*, so it is essentially socially egalitarian. Not only do Buddhists believe that people can move beyond the social class into which they are born by means of their own talents and efforts, but also the monastic *sangha* is open to all regardless of their social standing (although not equally to both men and women).

In principle, the acquisition of property by laypeople is not despised, provided wealth is not acquired by immoral means, the owner of wealth is not unduly attached to it (leading to a critique of materialism and consumerism) and it is properly used to benefit people (Loy, 2010; Pryor, 1990). The Buddha spoke of four kinds of happiness attainable by householders: possessing enough material resources, enjoying those resources, sharing them with relatives and friends and not being in debt. However, material wealth is not equated with happiness, because a cause of *dukkha/duhkha* (ill-being, suffering) is *tanha/trsna* (craving). To cultivate non-attachment (possessing and using material things but not being possessed or used by them), it is thought that a person must be generous and give large parts of his/her wealth away, especially to monks. On the one hand, therefore, wealth is regarded as a means for gaining merit, with the expectation that a generous person will be reborn to a higher position of wealth and social status. However, on the other, renouncement of attachment to material things by joining the *sangha* is praised, and contentment is regarded as the greatest wealth (Loy, 2010).

Poverty may also be considered to be a result of deeds in a previous life, but it and indebtedness are also seen as undesirable because they cause *dukkha/duhkha* and impede a person's ability to follow the spiritual path (Loy, 2010; Tomalin, 2007).

> *Daliddiya* is [defined as] lacking the basic material requirements for leading a decent life free from hunger, exposure and disease. Buddhism recognises the importance of such minimum material needs even in the case of those who aspire to its spiritual goal, and in fact the basic needs of a monk or nun provide a useful benchmark for measuring that level of subsistence below which human beings should not be allowed to fall. The four requisites of a Buddhist renunciant are food sufficient to alleviate hunger and maintain one's health, clothing sufficient to be socially decent and to protect the body, shelter sufficient for serious engagement with cultivating the mind and healthcare sufficient to cure and prevent physical illness.
>
> (Loy, 2010, p. 47)

Poverty is presented as a potential cause of immoral behaviour (e.g. theft), and the poor are regarded as having the same responsibility as everyone else to transform their own motivations and deeds from greed, ill will and delusion to generosity, loving kindness and wisdom. It is acknowledged that debt is not necessarily incurred because of a greed for possessions and that it is often associated with poverty. The response in this instance should, it is thought, be compassion rather than either indifference or punishment. The solution to poverty is seen as being to ensure that people are able to meet their basic needs (and for the ruler to provide for the very poor and destitute) (Loy, 2010).

Socially engaged Buddhism has developed fresh perspectives on issues of wealth, poverty and inequality. The Vietnamese monk Thich Nhat Hanh, in particular, has revised and updated the traditional 5 precepts into 14, 5 related to refraining from accumulating wealth for wealth's sake and others calling for living simply and not profiting from the suffering of others (including non-human species) (Loy, 2010, p. 59; Tomalin, 2007). The ideas of socially engaged Buddhism have implications for both people's life choices and Buddhist organisations such as the Sarvódaya Shramadána Movement in Sri Lanka (Loy, 2010, p. 60; Daskon and Binns, 2012; Tyndale, 2006).

The realities of wealth and poverty in folk religious traditions

As discussed in earlier chapters, the cultural, social and religious principles and practices associated with folk religion are concerned to sustain the well-being of the social group and family unit, so individual action to accumulate wealth, while legitimate, is regarded as being justified by the role a prosperous individual can play in ensuring the prosperity of the wider society. The public display of wealth is generally permissible, even expected – it signifies superior social status, provided it is used and redistributed to ensure the well-being of the ethnic group, lineage or family. Some think that these mechanisms continue to operate, as seen, for example, through the role played by 'hometown associations' or diaspora networks in channelling resources from more prosperous rural–urban or international migrants to their families and areas of origin. However, large families and widespread poverty mean that such expectations may also impose what are experienced as excessive burdens on those perceived to be successful. Indeed, one of the appeals of Pentecostal Christianity is said to be its ability to enable church members to escape such obligations, on the grounds that they are associated with unacceptable religious beliefs and practices, enabling members to prosper (as well as supporting the church). Others perceive economic and political change as having undermined the traditional expectations, giving rise to social tensions, increased inequality and rising levels of corruption. The complex interactions between religio-cultural values and norms and day-to-day practices and between individual and communal interests are explored in various studies, for example in Zambia (Box 8.2).

Box 8.2 Navigating the accumulation of wealth and its obligations in rural Zambia

In a study of social and agricultural change in the 1980s among the Yombe in Isoka District in northern Zambia, Bond (1987) explores the complex relationships between existing forms of social organisation, cultural notions of the social person and changes to social relationships resulting from changes in productive activity and property relations. In this example, the main drivers of changes were the introduction of new crops and livestock, as well as policy and practical support for larger-scale commercial farming. At the time of Bond's research, these were leading to an increased demand for both male and female agricultural labour, on top of established patterns of largely male migration to areas such as the Copperbelt. Amongst the Yombe, the main unit of social organisation continued to be the agnatic kinship group (an exogamous descent group linking between four and six generations), which was the basis for gendered rights in property and social status. However, pressures on land and increased inequalities resulting from the emergence of commercial agriculture were giving rise to conflicts between different segments of the descent group, which were often expressed through the religio-cultural idiom of ancestor worship and witchcraft accusations. For example, junior male members of the group continued to be obliged to contribute labour to more senior members' agricultural activities. However, the proceeds of increased sales of surplus produce by the latter and the items purchased with this cash income were regarded by the successful individuals as their own rather than communal property. This was leading to increased consumption rather than redistribution and also to witchcraft accusations against those allegedly threatening a lineage by favouring personal advancement over collective interests. At the time of his research, Bond found that successful farmers were able to pursue commercial agriculture and other businesses based on individual private ownership while evading criticism and resentment by appearing to fulfil their obligations to the lineage group, for example by defending traditional customs and manipulating religious practices.

Altruism, philanthropy and service

Originally a Greek term, 'philanthropy' refers to the care people show for each other. It has strong implications of voluntarism – benevolence by the rich by choice, underpinned by the notion of private property rights, voluntarily given up (Retsikas, 2014). While altruism is a core value in most religious traditions, it is not always clear whether it is a religious value in itself or a means to secure a reward – religious merit. Its form and status vary, as do its theological underpinnings, purposes and practical arrangements. It includes both charitable giving and service, it may be mandatory or voluntary, and its outcomes vary. It has social implications, depending on the nature of the relationships involved (e.g. between donor and recipient), and also economic implications, depending on its scale, purpose and arrangements for the collection and disbursement of monetary donations or provision of services.

Views differ on the relative importance of ascertaining motivations and understanding the nature and effects of actions. While altruism and service are generally encouraged as aspects of the practice of goodness and compassion, they can also be motivated by a desire for social approval or feelings of guilt. If they are defined as a willingness to help others without expecting a return, they do not necessarily enhance self-esteem or alleviate feelings of guilt. They may also be related to the question of what adherents need to do to be 'saved', although as discussed in earlier chapters, the meaning of 'salvation' and ways of attaining it vary between religions (Inaba and Loewenthal, 2009). Altruism may be expressed by contributing in cash or kind to the needy, volunteering time or a person's choice of employment.

Research in the US in the 1960s and 1970s did not reveal a clear relationship between religiosity and altruistic behaviour, although since the 1980s, a positive relationship has been found more often. This is attributed to religious teachings and practices (usually those of Christianity) and the presence of role models in the relevant religious group (Inaba and Loewenthal, 2009). More recent findings from Gallup surveys of 240,000 adults (age 15+) in 145 countries found a positive relationship between generosity (respondents had 'donated money to a charity, volunteered time to an organisation or helped a stranger needing help') and religiosity (defined as attendance at a religious service in the previous week). This association was strongest with respect to donating money, perhaps because religious organisations provide regular opportunities to donate. It was detected across a number of religions, including Christianity, Islam, Judaism, Hinduism and Buddhism (Smith and Stark, 2009).

Philanthropic or altruistic practices involve discerning needs, choosing between competing priorities, and identifying appropriate activities and recipients (known/unknown, near/far, own family/others, the worthy/the needy, organisations that are religious/non-religious). They may be personalised, involving interaction between the donor and recipient, or, increasingly, institutionalised, although practising altruism through organisations may conflict with some religious values and become bureaucratised (Vacek, 2010). Here, the discussion will primarily focus on the motivations of those involved, with some attention to the relationship between religious teachings and the type of activities which organisations consider appropriate, leaving more detailed consideration of the choice of actions and their outcomes to the companion volume.

Charitable giving in Islam

As noted above, the right of the poor to receive support and the duty of the wealthy to provide it is central to Islam. Islamic teaching specifies acceptable motivations for giving (in humility, with no expectation of benefit, except for God's favour), how it ought to be disbursed (in secret) and what proportion of wealth should be given (a set proportion of wealth annually for *zakat* and unsparing generosity with respect to voluntary giving or *sadaqa*, once an individual's obligations to dependents and kin have been fulfilled). Some of the activities for which charitable funding is considered appropriate and the motives of some of those involved in such organisations are illustrated in Box 8.3.

An important form of *sadaqa* is *waqf* (pl *awqaf*, endowment/s). *Waqf* appears to have emerged quite early on as a means of fulfilling charitable requirements. A *waqf* (pious foundation) is endowed by designating a source of revenue to be used for some public good in perpetuity (effectively a trust). It is often landed property, the revenue from which provides for the building and upkeep of a charitable institution such as a public water fountain, religious school, mosque, clinic, orphanage, refugee camp or shelter. As well as being a way of fulfilling the charitable requirements of Islam, *waqf* can provide a means of accruing prestige, circumventing inheritance laws, and providing an income to one's descendants as guardians of a mosque or shrine or teachers in a madrasa (Benthall, 2003; Hashmi, 2010).

Giving at particular times of year is encouraged and may be thought to yield greater spiritual rewards. For example, Muslims are expected to feed the destitute while fasting themselves during Ramadan, including giving *zakat al-fitr* (sufficient to feed one person in need or its monetary equivalent) to mark the Id el-Fitr feast at the end of Ramadan. In addition, during Id el-Adha, which commemorates the occasion when Allah is believed to have substituted a ram for sacrifice instead of the son of the Old Testament prophet Abraham, Muslims are required to sacrifice a domestic animal (*qurbani*), the meat from which is given to the poor (Kroessin, 2008). Proceeds from the sale of the hides may also be donated.

Charitable giving and service are thus religious imperatives in Islam, and there are thousands of examples throughout the Muslim world. They may be channelled through organisations with no overt political affiliation or motive (Box 8.3) or those whose core ideology includes a commitment to establishing an Islamic state. The welfare wings of religious political parties ostensibly have different motives from other philanthropic organisations – it might be anticipated that their primary rationale is to recruit members/voters for the political party concerned. However, research on the Jama'at-i-Islami demonstrates greater complexity in practice (Box 8.4).

Box 8.3 Philanthropy and service in Pakistani voluntary organisations

In Karachi, Pakistan, case studies of three voluntary organisations, the Edhi Foundation, the Saylani Welfare Trust and the Behbud Association, show how Islam influences the activities undertaken and the motives of founders, staff and volunteers (Kirmani and Zaidi, 2010; Kirmani, 2012). The founders of the Edhi Foundation, which provides a range of services, including an ambulance service, homes for the destitute, feeding programmes and healthcare, identified their motivation as rooted in submission to God, which arouses a sense of compassion, expressed through humanitarian work that does not discriminate between needy people on the basis of their religion.

The Saylani Welfare Trust, which is affiliated to the Barelvi school of thought, provides free meals and other material assistance to the poor, runs health services and madrasas, has an employment programme for men and provides free *istikhara* services (a form of religious guidance for solving personal problems). Operational staff in its main office and donation centre reported that it organises religious education sessions for its staff, who prefer a religious working environment and express a sense of performing a religious duty through their work – doing *nek kaam* (pious work). Similarly, for a woman employee in the vocational training centre of the Behbud Association, which provides education, health and vocational training services, working for the organisation is, for her, a way of making God happy and gaining spiritual peace. These organisations, like others, provide middle- and upper- class people with a way of fulfilling their religious duties to the less privileged in a manner that is perceived to be 'modern' and efficient.

Box 8.4 Welfare and politics in a religious political party: Jama'at-i-Islami in Pakistan and Bangladesh

Jama'at-i-Islami, the largest religious political party in Pakistan and Bangladesh, was established in 1941 (Bano, 2009, 2012). Its main objective is to work towards the establishment of a state based on the injunctions of *shari'a*. In Pakistan the party formed a welfare wing in the early 1950s. This is registered separately from the political party and today maintains a large and complex network of social and emergency relief services, including education, health, water and sanitation and poverty relief. Its welfare activities are undertaken by various organisations, of which the Al-Kidmat Foundation is the flagship, while others run more specialist programmes. J-i-I in Bangladesh is also involved in and supervises a range of Islamic institutions involved in welfare work.

Following field research in 2008/9, Bano concluded that J-i-I's primary motive is to win the trust of party members by demonstrating its commitment to implementing religious precepts. The party's ultimate aim is to capture state power in order to establish a just society subject to God's rule. This is believed to require both reform of the state and individuals taking responsibility for the welfare of their fellow human beings. The organisations attached to the party enable them to do this. In Pakistan, the welfare organisations are established by or affiliated with the party and registered as voluntary organisations. In Bangladesh, the party's political status is fragile, and so it has few affiliated organisations (except some engaged in emergency relief). However, members take on duties in many organisations that do not have formal links with the party, acting as advisers, board members and directors and providing technical assistance. In addition to making financial contributions, members of J-i-I (and its non-political affiliates, which include students', workers' and peasants' organisations) volunteer to manage and work for the welfare organisations, which also employ paid staff. Those involved as donors, volunteers and staff reported that they are driven by their belief that welfare work is a means to fulfil God's instructions, a search for other-worldly rewards and commitment to the goal of bringing about a just social order: "In totality the motive of members to get involved is one: to seek the pleasure of God" (senior members of Al-Kidmat Foundation, quoted in Bano, 2009, p. 22). Members report that their engagement in welfare activities gives them a sense of achievement and personal religious gain.

Altruism in Christianity

Christians are expected to engage in altruism and service to tackle problems of social injustice, inequality and need in order to ensure that God's purpose in the world can be achieved. For many, as discussed in Chapter 6, the purpose of religious giving is to enable churches to cover their operating costs and expand by planting new congregations nearby or further afield, including international mission. As discussed above, some Christians emphasise the need for state action and structural change to reduce poverty and redistribute wealth. However, charitable giving and voluntary service remain an important expectation for all.

Early Catholic teaching emphasised human beings' dependence on God's provision, but since the 1980s the Catholic Church has recognised humans' responsibility as co-creators with God. While states are expected to ensure justice and well-being, the Church has always played important roles in the provision of services (especially education) and the alleviation of poverty. Educational activities are motivated in part by a desire to meet the needs of both the Church and government but are also considered important means of communicating religious teachings, instilling values and attitudes (especially related to gender and the family), training church workers, influencing potential social leaders and enabling the poor to improve their well-being. Altruism is based on the theological concept of *agape*, love for the sake of the beloved,[2] who may not be known to the giver. This involves emotion (caring about those helped), participation (interaction between those in need and those with resources, time and talents) and realisation of the good (flourishing of the beloved, without creating dependence or undermining his/her dignity). It is distinguished from giving as a

2 In contrast to *eros*, love for an individual's own sake, including love of God, or *philia*, love for fellow members of a community.

self-interested bargain in which people act for their own sake and derive satisfaction and other benefits from giving (Vacek, 2010). The theological imperative to give is expressed through the activities of a range of international and local monastic orders and lay organisations, such as Caritas, which is, through its local branches, involved in disaster relief, welfare and community development programmes. A study of the Pakistani branch found that employees saw their service as a religious imperative (Kirmani and Zaidi, 2010, pp. 51–3; Kirmani, 2012).

Altruism, philanthropy and service are also important in the Protestant tradition. In Europe, almsgiving and philanthropy were initially stressed as responses to the social ills of industrial society but later extended to fund overseas mission, which from the outset included educational, health and other social activities. For these churches, good deeds are regarded as themselves a communication of the Christian message, so how they are carried out is considered to be at least as important as their outcomes. Difficult questions include whether charitable activities should be aimed at needy church members or the wider population and whether they are purely altruistic demonstrations of witness or a means of proselytising. Later approaches, both single and multi-denominational, have incorporated participation and partnership to try and avoid paternalism and dependency, although many altruistic initiatives continue to have these characteristics.

Research in Nigeria found that many churches promote the value of altruism, are engaged in service provision and have welfare wings (e.g. the Evangelical Church of West Africa, the Christian Association of Nigeria, New Life for All, the Fellowship of the Churches of Christ in Nigeria). In addition, members of fellowship groups in churches, especially women's groups, referring to the biblical example of Dorcas, reported that they are motivated by Biblical teachings on the stewardship of wealth, sharing and economic justice. These lead them to voluntarily donate cash and their time. However, in both Christianity and Islam, as discussed by Bradbury (2013) and Fountain (2015), there are potential contradictions between altruism and proselytisation.

Altruism in Hinduism and Buddhism

The primary emphasis of the closely related religious traditions of Hinduism and Buddhism is on personal spiritual enrichment and inner peace, pursued primarily through self-reflection, *bhakti* (devotion) and meditation. However, spiritual attainment is believed to be empty without love and compassion, which may be expressed through engagement in service and welfare activities. *Dana* (charity or giving) to both monks and nuns (and, in the case of Hinduism, priests or mendicants) and, directly or indirectly, for welfare activities, is central to both traditions. It refers to actions that generate good *karma* and is an obligation for all.

Like work, which can be done for the good of all, for the good only of oneself or for malicious or harmful reasons, in Hinduism the motive for giving is crucial: *satvik* means giving without expecting anything in return, *rajasik* giving and expecting something in return and *tamasik* giving with contempt or with the intention of harming. Clearly, the first is preferable, although in practice, people's motives for *dana* may be mixed (Nadkarni, 2008, p. 182). The concept of *seva* has its origins among worshippers of Vishnu: it referred both to helping in his temples and the worship of God, expressed through rituals, but also to serving others. In earlier phases of Hinduism, charity and hospitality were regarded as individual duties, but more recently, these have been increasingly emphasised and organised on a larger scale (Nadkarni, 2008). Today in Hinduism, *seva* refers to both devotion to God and service to the less fortunate, with the latter being regarded as a *sadhana*, a religious path based on the realisation that all humans are equal (Sanford, 2013; Tomalin, 2009, p. 44). Bradley's study of the discourses, ideologies and action programmes of three Hindu organisations reveals that, while their claimed motivations are the religious values of love, compassion and service (to God and the poor), their ideologies, welfare activities and girls' education programmes serve less to reduce poverty or challenge inequality than to reinforce the gender and caste status quo, with the partial exception of the Guru Mata Amritanandamayi Mission (Bradley, 2011).

In Buddhism, the central act of giving is the offering of alms to the monastic *sangha* for food, robes and shelter, which is thought to create merit for lay Buddhists (Goonasekera, 2013, p. 122; Tomalin and Starkey, 2013). In practice, in some contexts, more merit is derived from giving to male than female mendicants. For example, in Thailand, the white-robed nuns (*mae chis*) typically receive less alms than the orange-robed monks (*bhikkhus*). Monastics in turn engage in *dana* by teaching, giving advice and providing the example of their own lives. In addition, many monasteries provide shelter, food, healthcare and education, so in giving to monks, lay Buddhists are contributing to these welfare activities, even if their primary motivation may be the religious merit accrued rather than a desire to help the poor. Whether giving alms beyond those to the *sangha* is mandated is unclear. However, the Buddhist values of *metta* (loving kindness and generosity) and *karuna* (compassion), which encourage adherents to overcome the distinction between themselves and others, seem to support a concern for the needy, leading many in practice to give directly to welfare organisations. Some analysts portray traditional Buddhist thought and practice as socially passive, whereas others claim that this is a Western orientalist construction and that in fact socially active Buddhism predated the colonial era, so that for Buddhists, social engagement is nothing new, although it has become more prominent in recent decades. This is a response to both Western influences and contemporary social problems. Given the underlying value of compassion, the motive of social engagement is to help human beings to liberate themselves from the suffering they experience in their lives, social relationships and social systems. Examples include the provision of humanitarian assistance and delivery of social services (Tomalin, 2007; Loy, 2010). However, the relative importance of engagement in activities to improve welfare and work for social justice varies between Buddhist traditions, as revealed by Bhatewara and Bradley's (2012) comparison of Tibetan Buddhism in Dharamsala and Ambedkarite Navayana Buddhism in Pune, India.

Conclusion

Discourse, attitudes and actions related to well-being, the accumulation of wealth and the nature and experience of poverty are influenced by a variety of cultural, religious, economic, political and social dimensions of people's lives and the contexts in which they live. Conceptions of well-being are embodied in the basic teachings and discourses of all the religions, along with values related to the acceptability of wealth and inequality. In the African and Asian contexts considered here, discourses and attitudes have also been strongly influenced by the policy discourses of 'development' adopted by international agencies, newly independent governments and emerging civil society organisations, which are imbricated in the day-to-day understandings of local religious leaders and adherents, even though those associated with the religious traditions also identify contradictions between the visions and values they promote and those advocated by development actors. Informants in the research reported on in Box 8.1 reveal a general understanding and conventional interpretation of the teachings of Christianity and Islam that is reflected in their ideas, attitudes and values towards well-being, wealth and poverty, although it is not possible to assess from the research how these influence the behaviour of respondents differently positioned within their societies. Similarities between the African countries under study, the research locations within each country and the religious traditions examined outweigh the differences between them, although more detailed case studies of teachings, attitudes and behaviour might reveal more differences.

The role played by prosperity and wealth accumulation in improving well-being and the associated contradictions (poverty and inequality) have implications for both individual and collective behaviour and state policies. It is methodologically difficult to assess whether and how the behaviour and actions of people and groups complies with their averred values, and much of the research reported on here does not attempt to do so, concentrating instead on analysing people's own perspectives and, in some cases, advancing psychological, social, economic or political explanations for those ideas and attitudes. The discussion in this chapter does not unpick in any depth the extent to which the views of local religious leaders and adherents comply with official/international doctrines. Nor does it analyse the strategies they advocate for tackling poverty and inequality and ensuring widespread well-being, which for most are issues beyond the scope of religious teaching and practice. While respondents did not explicitly advance ideas about justice, compassion and altruism emerged as core religious values and important sources of ideas about how to alleviate poverty. Some of the ways in which altruism is expressed through charitable giving and service are documented in research studies. These reveal that adherents' participation in such activities appears to be influenced as much by the religious merit yielded by their engagement as by the likely efficacy of the altruistic behaviour in addressing the structural drivers of injustice, inequality and poverty.

References

Al-Qardawi, Y. (1999) *Fiqh az-Zakat: A Comparative Study*, London: Dar-al-Taqwa.

Bano, M. (2009) *Marker of Identity: Religious Political Parties and Welfare Work – The Case of Jama'at-i-Islami in Pakistan and Bangladesh*, Birmingham: University of Birmingham, Religions and Development WP 34.

Bano, M. (2012) Welfare work and politics of Jama'at-i-Islami in Pakistan and Bangladesh, *Economic and Political Weekly*, XLVII, 1, 86–93.

Benthall, J. (1999) Financial worship: The Qur'anic injunction to almsgiving, *Journal of the Royal Anthropological Institute*, 27, 1.

Benthall, J. and Bellion-Jourdan, J. (2003) *The Charitable Crescent: Politics of Aid in the Muslim World*, London: IB Tauris.

Bhatewara, Z. and Bradley, T. (2012) Buddhist engagements with social justice: A comparison between Tibetan exiled Buddhists in Dharamsala and Dalit Buddhists of Pune, *Economic and Political Weekly*, XLVII, 1, 69–76.

Bond, G.C. (1987) Religion, ideology and property in northern Zambia, in Markovitz, I.L. (ed) *Studies in Power and Class in Africa*, New York and Oxford: Oxford University Press, 170–88.

Bradbury, S. (2013) Mission, missionaries and development, in Clarke, M. (ed) *Handbook of Research on Religion and Development*, Cheltenham: Edward Elgar, 413–29.

Bradley, T. (2011) The politicisation of mothering in Hindu missions, *Politics, Religion and Ideology*, 12, 2, 161–77.

Chambers, R. (2006) What is poverty? Who asks? Who answers? In Ehrenpreis, D. (ed) *Poverty in Focus*, Brasilia: UNDP International Poverty Centre, 3–4.

Cochrane, J.R. (2007) Seeing healthworlds differently, *Religion and Theology*, 14, 6–27.

Curran, C.E. (2002) *Catholic Social Teaching, 1891-Present*, Washington, DC: Georgetown University Press.

Curran, C.E. (2005) Christian ethics, in Jones, L. (ed) *Encyclopedia of Religion*, Detroit: Palgrave Macmillan, Vol. 3, 2nd ed, 1650–7.

Daskon, C. and Binns, T. (2012) Practising Buddhism in a development context: Sri Lanka's *Sarvodáya* movement, *Development in Practice*, 22, 5–6, 867–74.

Deneulin, S. (2011) Development and the limits of Amartya Sen's *The Idea of Justice*, *Third World Quarterly*, 32, 4, 787–97.

Deneulin, S. and Bano, M. (2009) *Religion in Development: Rewriting the Secular Script*, London: Zed.

Devine, J., Hinks, T. and Naveed, A. (2017) Happiness in Bangladesh: The role of religion and connectedness, *Journal of Happiness Studies*. https://doi.org/10/1007/s10902-017-9939-x

Dugbazah, J. and Hassan, I. (2010) *The Relationship between Religious Teachings and Development in Nigeria: An Exploratory Study of Poverty, Debt and Credit*, Birmingham: University of Birmingham, Religions and Development Research Programme, unpublished draft paper.

Fountain, P. (2015) Proselytizing development, in Tomalin, E. (ed) *The Routledge Handbook of Religions and Global Development*, London: Routledge, 80–98.

Galston, W.A. and Hoffenberg, P.H. (2010) Introduction, in Galston, W.A. and Hoffenberg, P.H. (eds) *Poverty and Morality: Religious and Secular Perspectives*, Cambridge: Cambridge University Press, 1–14.

Goonasekera, S. (2013) Buddhism in South Asia: Practicing tradition today, in Pechilis, K. and Raj, S.J. (eds) *South Asian Religions: Tradition and Today*, London and New York: Routledge, 101–42.

Graham, E. (2009) Health, wealth or wisdom? Religion and the paradox of prosperity, *International Journal of Public Theology*, 3, 5–23.

Haller, M. (2002) Theory and method in the comparative study of values: Critique and alternative to Inglehart, *European Sociological Review*, 18, 139–58.

Hashmi, S.H. (2010) The problem of poverty in Islamic ethics, in Galston, W.A. and Hoffenberg, P.H. (eds) *Poverty and Morality: Religious and Secular Perspectives*, Cambridge: Cambridge University Press, 180–203.

Heywood, A. (2000) *Key Concepts in Politics*, Basingstoke: Palgrave Macmillan.

Inaba, K. and Loewenthal, K. (2009) Religion and altruism, in Clarke, P.B. (ed) *The Oxford Handbook of the Sociology of Religion*, Oxford: Wiley-Blackwell, online ed.

Inglehart, R. and Baker, W.E. (2000) Modernization, cultural change, and the persistence of traditional values, *American Sociological Review*, 65, 1, 19–51.

Joshanloo, M. (2013) A comparison of Western and Islamic conceptions of happiness, *Journal of Happiness Studies*, 14, 6, 1857–74.

Joshanloo, M. (2014) Eastern conceptualizations of happiness: Fundamental differences with western views, *Journal of Happiness Studies*, 15, 2, 475–93.

Kim, K. (2007) *Concepts of Development in the Christian Traditions: A Religions and Development Background Paper*, Birmingham: University of Birmingham, Religions and Development WP 16.

Kirmani, N. (2012) The role of religious values and beliefs in charitable and development organisations in Karachi and Sindh, Pakistan, *Development in Practice*, 22, 5–6, 735–48.

Kirmani, N. and Zaidi, S. (2010) *The Role of Faith in the Charity and Development Sector in Karachi and Sindh, Pakistan*, Birmingham: University of Birmingham, Religions and Development WP 50.

Kroessin, M.R. (2008) *Concepts of Development in 'Islam': A Review of Contemporary Literature and Practice*, Birmingham: Religions and Development WP 20.

Kuran, T. (1986) The economic system in contemporary Islamic thought: interpretation and assessment, *International Journal of Middle East Studies*, 18, 2, 135–64.

Laderchi, C.R., Saith, R. and Stewart, F. (2007) Does it matter that we do not agree on the definition of poverty? A comparison of four approaches, *Oxford Development Studies*, 31, 3, 243–74.

Loy, D.R. (2010) The karma of poverty: A Buddhist perspective, in Galston, W.A. and Hoffenberg, P.H. (eds) *Poverty and Morality: Religious and Secular Perspectives*, Cambridge: Cambridge University Press, 44–61.

Mesaki, S., Kipacha, A. and Dugbazah, J. (2010) *Religious teachings and development concepts in Tanzania: Insights from Tanga and Arumeru*, Birmingham: University of Birmingham, Religions and Development Research Programme, unpublished draft paper.

Moreira-Almeida, A. and Koenig, H.G. (2006) Retaining the meaning of the words religiousness and spirituality: A commentary on the WHOQOL SRPB group's 'A cross-cultural study of spirituality, religion, and personal beliefs as components of quality of life' (62:6, 2005, 1486–1497), *Social Science and Medicine*, 63, 843–45.

Nadkarni, M.V. (2008) *Hinduism: A Gandhian Perspective*, New Delhi: Ane Books, 2nd ed.

Norris, P. and Inglehart (2004) *Sacred and Secular: Religion and Politics Worldwide*, Cambridge: Cambridge University Press.

Nussbaum, M. (2003) Gender and governance: An introduction, in UNDP (ed) *Essays on Gender and Governance*, New Delhi: United Nations Development Programme, Human Resource Development Centre, 1–19.

Pew Forum (2006) *Spirit and Power: A 10-Country Study of Pentecostals*, Washington, DC: Pew Forum, http://pewforum.org/uploadedfiles/Orphan_Migrated_Content/pentocostals-08.pdf

Plant, S.J. (2009) International development and belief in progress, *Journal of International Development*, 21, 844–55.

Pryor, F.L. (1990) A Buddhist economic system – in principle, *American Journal of Economics and Sociology*, 59, 3, 339–49.

Retsikas, K. (2014) Reconceptualising zakat in Indonesia: Worship, philanthropy and rights, *Indonesia and the Malay World*, 42, 124, 337–57.

Sanford, A.W. (2013) Hinduism and development, in Clarke, M. (ed) *Handbook of Research on Development and Religion*, Cheltenham: Edward Elgar, 79–96.

Sen, A. (2009) *The Idea of Justice*, London: Allen Lane.

Sharma, A. (2010) Hinduism and poverty, in Galston, W.A. and Hoffenberg, P.H. (eds) *Poverty and Morality: Religious and Secular Perspectives*, Cambridge: Cambridge University Press, 160–79.

Smith, B.G. and Stark, R. (2009) *Religious Attendance Relates to Generosity Worldwide*, www.gallup.com/poll/122807/Religious-Attendance-Relates-Generosity-Worldwide,aspx?g_source=position4&g_medium=related&g_campaign=tiles

Swinton, J. (2007) Researching spirituality and mental health, in Coyte, M.E., Gilbert, P. and Nicholls, V. (eds) *Spirituality, Values and Mental Health*, London: Jessica Kingsley Publishers, 292–305.

Taylor, M. (2000) *Poverty and Christianity: Reflections at the Interface between Faith and Experience*, London: SCM Press.

Taylor, M. (2003) *Christianity, Poverty and Wealth: The Findings of 'Project 21'*, Geneva and London: World Council of Churches and Society for Promoting Christian Knowledge.

Tomalin, E. (2007) *Buddhism and Development: A Background Paper*, Birmingham: University of Birmingham, Religions and Development WP 18.

Tomalin, E. (2009) *Hinduism and International Development: Religions and Development Background Paper*, Birmingham: Religions and Development WP 19.

Tomalin, E. and Starkey, C. (2013) Buddhism and development, in Clarke, M. (ed) *Handbook of Research on Religion and Development*, Cheltenham: Edward Elgar, 31–50.

Tripp, C. (2006) *Islam and the Moral Economy: The Challenge of Capitalism*, Cambridge: Cambridge University Press.

Tyndale, W. (2006) *Visions of Development*, Aldershot: Ashgate.

Vacek, E. (2010) A Catholic theology of philanthropy, in Smith, D.H. (ed) *Religious Giving: For the Love of God*, Bloomington, IN: Indiana University Press, 99–124.

White, S. and Tiongco, R. (1997) *Doing Theology and Development: Meeting the Challenge of Poverty*, Edinburgh: Saint Andrew Press.

Wuthnow, R. (2008) The sociological study of values, *Sociological Forum*, 23, 1, 333–43.

9 Gender equality, sexuality and the family

Religions are intertwined in the construction of gender and sexuality, through their teachings, practices and institutions, in ways that have implications for individual behaviour, including sexual relations and fertility behaviour, and the functioning of social units, especially families. However, as in social analysis more generally, in the past much discussion of the links between religion and society was gender blind. Thus, while attention has been drawn to gender aspects of religious beliefs and practices in Chapters 3–8, the discussion was less systematic than desirable because much of the writing is gender blind, the available evidence not disaggregated by sex and the analysis not gendered.

In recent years, more attention has been paid to women's roles and perspectives. Their disadvantage and subordination have been recognised not just in the economic, political and wider social spheres but also within the religious field, and attempts have been made to redress gender inequality. Avishai et al. (2015), however describe how these analyses, while drawing attention to women's perspectives and situations, often tend to adopt feminist stances that regard all religious traditions (especially the 'fundamentalist' strands within them) as oppressive and assume that all women share similar experiences of marginalisation. Not only do they tend to treat 'religion' and 'women' as (often undifferentiated) categories, they tend to see the world in binaries (women/men, feminine/masculine, oppressed/empowered). Gradually, it has been recognised that women's narratives reveal diverse social realities, expectations and ability to exercise agency with respect to their everyday lives, interactions with social and political institutions and participation in religion.

Gender differences and relationships have both biogenetic and socio-cultural aspects, but the roles and relative importance of biology and culture continue to be contested (Bruce and Yearley, 2006; Moore, 2002; Nussbaum, 2003). Because gendered bodies and relationships are socially defined, they vary between social and cultural contexts, intersecting with other axes of social difference, such as class and age. They are also contested and changeable. Religious communities play a role in producing and reproducing gendered attitudes, identities and institutions, which men and women negotiate in different ways. In particular, the widespread construction of women as the bearers of culture, religion and tradition means that their sexuality is laden with symbolic meaning, with implications for the norms defining appropriate behaviour and gendered social roles (Imam, 2005; Pereira and Ibrahim, 2010). Increasing recognition that gender is constructed and relational has been accompanied by increased attention to men and the construction of masculinity (White and Tiongco, 1997; Van Klinken, 2011a, 2011b, 2012). The gendered nature of relations between individuals is important in the 'private' sphere of the family and also in the 'public' sphere, where it can play an important role in the construction of communal identities (Desai and Temsah, 2014).

Sex and sexuality are important in people's lives – they refer to the most intimate expression of embodied selves; they are the foundation of social reproduction and kinship structures; and they are potential economic resources (Cornwall and Jolly, 2006). Sexuality is essentially private. However, at the same time, as the primary means of socio-biological reproduction, it is a public affair that no society believes it can leave unregulated, and religious institutions play important roles in enforcing the rules and norms governing sexual activity (Avishai et al., 2015; Casanova, 2009; Cornwall and Jolly, 2006; Sakai and Yasmeen, 2016). Views about whether sexual activity can be a source of joy and pleasure or should only be concerned with reproduction differ. Often, in practice, considerations of sexuality are limited to the physical aspects of sexual activity and their consequences for reproduction and health.

Gender equality refers to women and men having the same opportunities in life, including the right and ability to participate in the public sphere and religious life. At the international level, this has been addressed through incorporating gender into the human rights discourse, through international agreements such as the UNDHR and the 1979 Convention on the Elimination of All Forms of Discrimination Against Women (CEDAW), to which most countries in SSA and southern Asia have signed up, albeit sometimes with significant reservations and caveats. Increasingly, it is agreed internationally that an approach to sexual rights is needed which takes account of the diversity of sexual identity and expression, not just protection of sexual minorities, although that is important. Such an approach, it is argued, should provide people with rights over their bodies, including a negative right not to be discriminated against and positive rights to enjoyment and respect (Cornwall and Jolly, 2006). Since the elaboration of the Yogyakarta Principles for the Application of Human Rights in relation to Sexual Orientation and Gender Identity by an independent international group in 2006, the issue has been increasingly reflected in the deliberations of UN bodies (Corrëa, 2015; O'Flaherty and Fisher, 2008). International agreements on human rights can play an important role in specifying the entitlements and protections due to all human beings, because of the roles they can play in national debates, the formulation and implementation of state law and holding political and social organisations to account. Like the internationally recognised rights to freedom, justice and well-being discussed in Chapters 7 and 8, gender equality and sexual rights are principles against which to benchmark the following discussions of religious teachings and the social practices with which they are associated.

Gender and sexuality interact with religious teachings and institutions in complex ways, and there has been an explosion of research and evidence on some aspects of these interactions in recent years. It is not possible in the space available to attempt a comprehensive review of the evidence from SSA and southern Asia. Instead, this chapter will start with a brief consideration of how the principles of gender equality and sexual rights are reflected in religious teachings. It will then examine, first, individual attitudes and behaviour with respect to gender and sexuality, with a focus on bodies and sexual behaviour, reproductive health behaviour, HIV/AIDS and homosexuality and, second, the links between religion, marriage and the family. In each instance, the ways in which the four domains of religion and the personal, social and organisational dimensions of the links between religion and society interact at the individual, small-scale and wider societal levels will be considered.

Gender equality: the religious starting points

The monotheistic religions (Christianity and Islam) share the central principle that men and women are equal before God, that they are spiritually and morally equal. In Islam, the often-quoted Qur'anic verse 4:1 states that "Allah created you [men] from a single soul, and from the same soul created his mate". Women and men are supposed to participate equally in the religious life of the *ummah* and are expected to fulfil the same religious duties (worship, prayer, fasting, almsgiving and pilgrimage) (Qazi, 2015). In Christianity, women and men are both held to be made in the image of God and so to be equal in their personhood (van Klinken, 2011b). However, interpretations of the Genesis creation story vary. It may be interpreted literally – Eve was created from Adam's rib and is thus "a flawed female helpmate extracted from him as an afterthought or utility" (Qazi, 2015). Alternatively, it may be regarded as an analogy or disregarded in favour of a belief that men and women were created from the same essence (see Qibtiyah, 2012).

In each tradition, the central figures were surrounded by women as well as men and are portrayed in the texts as treating both with care and respect (Casanova, 2009). My choice of the word 'portrayed' here is deliberate, since the key texts were recorded, written and interpreted by men, whose writings reflected their historical and cultural contexts and the notions and practices of their time with respect to women's and men's biological and psychological differences, roles and deportment. As noted in Chapter 3, these were reflected in their choice of material (in the *ahadith* and *sunnah* and the Bible), the symbolism they used and the norms and rules they set out to govern individual and social behaviour (Jenkins, 2006; Qazi, 2015). In addition, the texts have often been read selectively to reflect and reinforce the cultural perceptions and practices associated with the wider gender hierarchical socio-cultural order, not least because religious specialists in both traditions have been overwhelmingly male.

In the polytheistic religions (Hinduism, folk religions), the oral and written sources generally portray the deities as male, but there are also female deities in the pantheon. In Hinduism, male deities are more numerous and often more powerful, but this is not always the case, and the female deities, as portrayed in the myths and epics, demonstrate a variety of characteristics and provide potential role models for both women and men. However, the Brahmanic priestly tradition is male. Because folk religions have an oral tradition, which was, for the most part, recorded during the colonial period, the available accounts reflect the gender assumptions of the writers, and so it is unlikely that we have a full understanding of the numbers, roles and characteristics of female deities and spirits on which contemporary adherents of these traditions can draw (Amadiume, 2005; Njoh and Akiwumi, 2012; Para-Mallam, 2010). Missionary Christianity, Jenkins (2006) observes, looked askance at the spiritual roles women in folk religious traditions played as prophets, priests, mediums, diviners and healers.

In the nontheistic religion considered here, Buddhism, "the central teachings . . . can be considered gender neutral (since they concern the ways in which the individual – regardless of sex – may overcome *dukkha* (suffering)" (Tomalin and Starkey, 2013, p. 41). In addition, in principle, both women and men may choose to become monastics, although, as noted in earlier chapters, interpretations vary between different strands of Buddhism, and practices often fall short of the principle.

A central value espoused by many of the religions considered here is therefore gender equality, although in practice, their teachings, attitudes and practices are ambivalent and contradictory, as will be explored below.

Individual attitudes and behaviour with respect to gender and sexuality

Despite their ostensible adherence to the idea that women and men are spiritual equals, religious teachings, practices and organisations assume and support a gender hierarchy that is remarkably similar across traditions and contexts, reflecting a predominantly patriarchal social order, which is often assumed to be and legitimised as 'natural' and 'God-given' (Casanova, 2009; Jenkins, 2006; Kirmani and Phillips, 2011; Ladbury and Khan, 2008; Mir-Hosseini, 2011; Para-Mallam, 2010; White and Tiongco, 1997).[1] This does not, of course, imply that there are no differences in attitudes within religious traditions (Para-Mallam, 2010; Qibtiyah, 2012).

In this social order, biological differences and the roles related to them are reflected in constructions of gender, masculinity and femininity: men are regarded

1 Seguino (2011) uses World Values Survey data to demonstrate that all the religions have gender-inequitable beliefs/attitudes, which have gender-inequitable outcomes. The data show that men, less well educated and poorer people hold more gender-inequitable views, regardless of religion. However, religiosity (intensity of belief and frequency of religious participation) is consistently more negatively correlated with gender-inequitable beliefs and outcomes than religion per se. This analysis suffers from the usual limitations: the WVS indicators of religiosity are Christian oriented, its indicators of gender equality have a Western bias, and Seguino acknowledges that the available indicators of gender well-being are inadequate.

as being naturally suited to providing for and protecting women through their leadership roles in families and social groups. Because childbearing is crucial to the continuation of families, religions and societies, maintaining control over it is a concern for men, through marital arrangements that provide for control over women's bodies and sexual behaviour and constructions of femininity that label tasks associated with childrearing and home-making as essentially female. Religious teachings play a central role in socialising people into complying with these social expectations and men, including fathers, husbands and religious leaders, seek to enforce them through moral teachings and various disciplinary mechanisms.

Bodies and sexuality: attitudes and regulation

Because of their roles in social (and religious) reproduction, bodies and sexuality are important concerns for religious groups, which attempt to regulate them. The differences between men and women, sexuality and appropriate behaviour are all dealt with extensively in religious texts. The discourse in both Islam and Christianity treats 'sex drives' as 'natural', providing they are channelled toward marriage and reproduction. Thus men's pursuit of women is regarded as a natural instinct, while women's sexuality is widely portrayed as dangerous, capable of leading men astray. The role of families in the process of social reproduction is reinforced and regulated by a variety of different social, religious and legal provisions in every society, some of which will emerge in the course of this chapter. Those relevant to this subsection include regulations governing acceptable sexual partners, moral teachings about sexual behaviour and social sanctions designed to enforce appropriate behaviour and sexual relations.

For all the religions, the primary purpose of sexual relations is reproduction, and they are prohibited outside marriage. Nevertheless, Islam has a relatively positive attitude to sexuality, which is regarded as a God-given means of creating a loving relationship, with wives being as entitled to sexual gratification as their husbands, a position closer to that of Protestant than traditional Catholic views in Christianity (Halim, 1999; Mantell et al., 2011). Both mainstream and revivalist teachings regard women as posing the main threat to this view that sexual relations are legitimate only within marriage (Pereira and Ibrahim, 2010). Teachings and practices designed, inter alia, to deter extra-marital sexual relations include initiation rituals, circumcision, guidance and sanctions related to pre- and extra-marital sex, the expectation that all will marry and restrictions on women's presence in the public sphere via seclusion practices and dress codes.

Initiation rituals intended to assist in the transition from youth to adulthood and contribute to conceptions of masculinity/femininity and sexual morality are common in many societies and religious traditions (Jenkins, 2006). During traditional rituals, for example in many African societies, boys and girls are separated and, amongst other things, given advice on their future roles as adults, spouses and parents. Girls were/are given advice, mostly by mothers, aunts, traditional birth attendants and herbalists, on how to ensure male sexual pleasure, although they are not taught that they are entitled to pleasure themselves and may be stigmatised if they show it (Aken'Ova, 2008).

Circumcision is often an important part of initiation, for both boys and girls. In Islam, for example, according to one *hadith*, it is obligatory for Muslim boys and favoured for girls. While it symbolises the transition to adulthood, it is also a means of enforcing female chastity, albeit with a weak basis in religious teachings. In Africa, it is most common amongst Muslims in countries with a Muslim majority (although not in Senegal) or a large Muslim population. However, in Ethiopia, Nigeria, Tanzania and Kenya, roughly equal proportions of Muslim and Christian parents reported that they had had a daughter circumcised, indicating that in the African context, it is a socio-cultural practice that predates widespread conversion to Islam or Christianity (Pew Forum, 2010, p. 54). It is particularly controversial because, although there is some recent evidence that male circumcision can help prevent HIV transmission, evidence of its adverse health impacts on girls and women is extensive. In largely Muslim Indonesia, for example, a relatively non-invasive and ritualistic form is almost universal, even though it is considered to be dangerous to the health of nearly three-quarters of the girls concerned. A 2003 study, in which mothers were asked why they arranged for their daughters to be circumcised, found that over half said 'religion', even though they did not know the specific *hadith* (White and Anshor, 2007). In recent years, not only has female genital mutilation become an international human rights issue, awareness of the absence (in Christianity) or weakness (in Islam) of a religious justification for the practice has enabled campaigners to draw on their governments' endorsement of international human rights agreements and increased religious literacy to build opposition to it as a cultural practice that infringes on the rights of girls and women.

Puberty and initiation subject young people's religious ideas about and their negotiation of sexual relationships to new social pressures. Studies reveal that they often have romantic ideals, hoping for sexually exclusive relationships based on emotional attachment that lead to companionate marriage, for example in the low income areas of Cape Town, Kampala and Dakar (Burchardt, 2011; Sadgrove, 2009; Van Eerdewijk, 2012). In practice, however, they have difficulty reconciling these hopes with their desire for increased material consumption and the real world of sexual relationships. In all the case study cities, despite the attempts of religious bodies to preach pre-marital sexual abstinence and male responsibility (as husbands and household heads), it is widely regarded as necessary/acceptable for boys to demonstrate their potency and virility by engaging in pre-marital sex, often with multiple partners, while girls are expected to remain virgins in order to fulfil their anticipated/desired roles as child bearers and homemakers. In practice many girls engage in sex before marriage, despite the risk of pregnancy, HIV infection and damage to their reputation, to win the affection of a prospective marriage partner or alleviate their financial insecurity. Everywhere, therefore, intimate relationships are transactional – they are characterised by exchanges of gifts for sexual pleasure, men and women both distinguish between casual liaisons and prospective marriage partners, and young women are particularly vulnerable to violence, infidelity and unsafe sex. Young men's inability to meet girls' expectations may lead the latter to seek relationships with older, better-off, often married men (Sadgrove, 2009), in turn undermining young men's view of their own masculinity (Van Eerdewijk, 2012). Religious bodies, the studies reveal, may or may not provide young men and women with help to negotiate this tricky terrain. For example, in Kampala, Sadgrove (2009) documents how young Pentecostal Christians use the theology of sin and repentance to deal with their sexual behaviour and their desire for social acceptance in a church community to help them fulfil the moral expectation of celibacy, although neither of these strategies are unproblematic.

Measures to deter pre- or extra-marital sex include moral condemnation, early marriage and seclusion. A large majority of respondents in the Pew Forum's international surveys, most of them Christians or Muslims, consider sex between unmarried people to be morally wrong (Pew Forum, 2010). In Islam, like other religions, extra-marital sexual relations (*zina*) are prohibited. While some Muslims hold that men and women should bear equal responsibility and be equally punished, many interpretations criminalise the behaviour and lay the blame for it on women (Halim, 1999). For example, in all 12 of the Northern Nigerian states which extended the reach of *shari'a* following democratisation, *zina* was criminalised. In the early years after the laws were introduced, both men and women were found guilty of fornication and whipped, but only women were found guilty of adultery, because a higher standard of evidence applies to them. Not only do women have to prove their innocence, the Maliki school of law, which prevails in this region, stipulates that pregnancy outside marriage is itself evidence, rendering women more likely to be accused and opening the door to treating rape as a form of *zina* (Imam, 2005; Pereira and Ibrahim, 2010). In addition, a significant minority of Muslims in a number of largely Muslim countries think that so-called honour killing, while not condoned in the Qur'an or *ahadith*, can sometimes be justified (46% in Pakistan, 41% in Malaysia, 36% in Bangladesh and 31% in southern Thailand). Indonesia, where only 18% think this, is an outlier (Pew Forum, 2010).

The expectation that women will marry, preferably at an early age, reinforces moral teachings and other sanctions against pre- and extra-marital sex. Those who choose to remain single or leave unsatisfactory marital partnerships are often socially stigmatised, even labelled as immoral and, in some contexts, pressured to (re-)marry, although sometimes they may be able to use religious and social norms to resist such expectations. Davies (2018) describes how debates on gender, morality and sexuality are played out in the recruitment and regulation of policewomen in largely Muslim Indonesia.

Social and religious norms that seek to regulate women's public presence include seclusion (in the home or single-sex education or work environments) and dress codes, both predominantly linked to Islam, although not confined to

Muslims.[2] Thus in India, both Muslims and Hindus practise seclusion, which is rooted in socio-cultural norms related to family honour and respectability, although the former justify it in religious terms. Seclusion practices vary with social status and caste, and female household members may be able to play an active role in upholding or resisting its imposition (Desai and Temsah, 2014; Kantor, 2002; Vera-Sanso, 1995). Perhaps because dress is such a visible symbol of gender relations and women's status, it has received more research (and media) attention than most other aspects of female sexuality. Research in Nigeria, for example, discussed how, in the 1980s, the idea that 'indecent dressing' was a major social problem became widespread. This 'moral panic', fuelled by the media, was rooted in the view that exposure of women's bodies is dangerous and offensive and that men cannot control themselves, ideas that came to be accepted among Christians as well as Muslims (Mahdi, 2007; Pereira and Ibrahim, 2010). The issues are discussed in multiple studies in, for example, Indonesia, Malaysia, Sri Lanka and India.[3]

The Qur'an instructs women to dress modestly but does not refer to veiling, although the latter is referred to in several *ahadith*. In practice, a wide range of practices which women embrace with varying degrees of freedom are included in the collective term 'veiling'. Critics consider it to be a symbol of female oppression, a male tool to control women (and their sexuality). They fear that it may be used to justify discriminatory laws and practices, erode women's freedom of choice and increase their vulnerability to arbitrary enforcement (e.g. through vigilante attacks). However, controversy rages amongst Muslims over whether it should be regarded as a religious or cultural practice and whether women should be free to choose whether or not to veil. In addition, the reasons women themselves advance for veiling and their experience of adopting modest dress vary.

In 10 of the 15 African countries with substantial Muslim populations surveyed by the Pew Forum, fewer than half the respondents held the view that women should have the right to decide whether or not to veil for themselves, ranging from 29% in the DRC to 43% in Liberia, although in Senegal two-thirds thought that they should have this right. The proportion of Muslim men holding this view was smaller than the proportion of women, except in Senegal, where 58% of men thought that women should be able to decide for themselves (Pew Forum, 2010, p. 54). Thus in two-thirds of the countries included, at least half

2 For example, Desai and Temsah's (2014) comparison of Muslims and Hindus in India found marked differences in public symbolic performance, including veiling and participation in public life, including wage work. However, at the household level, Hindu–Muslim differences with respect to gender segregation and decision-making power were absent or insignificant. Instead these vary by class, with educated women in both groups more likely to play a role in decision making but also, amongst Muslims, more likely to wear the veil.

3 Indonesia (Adeney-Risakotta, 2016; Beta, 2014; Brenner, 1996; Bucar, 2017; Jones, 2007; Nisa, 2012; Rinaldo, 2008; Robinson, 2012), Malaysia (Mouser, 2007; Olivier, 2016; Ong, 1995; Tong and Turner, 2010), Sri Lanka (Asia Foundation, 2012; Haniffa, 2008), India (Anand, 2014; Desai and Temsah, 2014; Kirmani, 2008a, 2008b, 2009).

Table 9.1 The views of southern Asian Muslims on women's right to choose whether to veil

	All respondents	*Women*	*Men*
	%	%	%
Indonesia	79	81	77
Southern Thailand	79	84	69
Malaysia	77	81	74
Pakistan	70	81	60
Bangladesh	56	64	49

Source: Pew Research Center, 2013, p. 97.

the respondents said that women should be governed by the social group as a whole when deciding whether to veil. In southern Asia, much larger proportions of survey respondents in most of the countries for which data are available thought that women should have the right to decide for themselves, ranging from almost two-thirds of women in Bangladesh to about 80% in the other countries (see Table 9.1).

The significance of modest dress varies. For revivalist Muslims, it is a sign of piety and an expression of identity and commitment to membership of a national and/or global community. It may even be a political statement signifying rejection of colonial and Western values (Anand, 2014; Kirmani, 2008a, 2008b, 2009; Ladbury and Khan, 2008). In addition, for many Muslims, it has practical value, enabling them to enter the sexually mixed public spheres of education, work, politics, public transport etc. (see, for example, Anand, 2014; Kirmani, 2008a, 2008b, 2009, on Delhi; Haniffa, 2008, on Sri Lanka; Mahdi, 2007, on Nigeria; Mouser, 2007, on Malaysia; and Rinaldo, 2008, on Indonesia). In many places, modest dress has been more widely adopted in recent years, fuelled by Middle Eastern/Wahhabi influence; expanding educational, work and political opportunities; increasingly accessible international communications; and fashion (see, for example, Bucar, 2017, on Indonesia). Commonly, the trend has been driven by educated middle- or upper-class urban women, as the options open to poorer (and rural) women are more limited. It can become a source of conflict. For example, in Nigeria, many Christians resent government policy on girls' school uniforms, which they regard as promoting Muslim practices (Mahdi, 2007; Pereira and Ibrahim, 2010).

Fertility, contraception and abortion

Two categories of religious values are relevant to the issue of fertility: specific norms that seek to regulate behaviour directly connected to the proximate determinants of fertility and broader values whose effect is indirect (McQuillan, 2004). The first are usually specified in teachings, especially those related to the use of contraception and abortion, the most significant proximate determinants in many societies, although they may also include norms governing entry into sexual unions (such as the age of entry or number of spouses). The broader messages can be grouped into two: the high value placed on large families and general views on the nature of family relationships, especially women's rights and roles. To determine whether and how socio-religious factors influence attitudes and practices related to fertility, it is necessary, if difficult, to specify the pathways by which they do so.

Attempts to examine the links between religion and fertility tend to adopt either small-scale qualitative approaches, especially semi-structured interviews with a limited number of adherents, religious leaders or health workers, or the traditional quantitative methods used by demographers. For a variety of reasons, many quantitative studies fail to find clear links between religion, sexual behaviour and reproductive health (McQuillan, 2004), but their inconclusive results do not imply that no such links exist (Olivier and Wodon, 2015).

The religious traditions are generally pronatalist: they regard children as a sign of God's blessing and as conferring socio-religious status on their parents. This view is associated with positive encouragement of high fertility, although the benefits of child spacing may be recognised (Gyimah et al., 2008). High fertility is encouraged by early marriage, religious groups' need to reproduce themselves and often, a desire to increase the number of adherents. It may also reflect a need for the labour that children can provide, especially in rural societies. Further, it is sometimes argued that minority groups are pronatalist in order to protect and improve their socio-political position. High fertility may also be underpinned by a desire for sons because of traditional succession and inheritance practices that favour men and, on the part of women, a need for support in widowhood and old age (Farrell et al., 2014). It may be positively encouraged through the practice of polygyny. Polygynous marriages, whether associated with Islam or traditional religion (especially in SSA), seem to be associated with higher fertility, generally attributed to competition between wives, which encourages childbearing to sustain a husband's favour and discourage him from taking an additional wife or contemplating divorce.

In line with their pronatalist stance, religious groups often attempt to restrict the use of practices intended to reduce fertility, at least at the level of official doctrine, although teachings and levels of fertility vary between and within religious traditions and between different contexts. Within Christianity, the strongest opposition is expressed by the Roman Catholic Church, which forbids most forms of contraception (other than the rhythm method) and any recourse to abortion. Other Christian denominations have gradually moved towards more relaxed views on contraception, but some strongly oppose abortion, partly because of a desire for high fertility but also because of teachings about the stage of development at which a foetus is said to become a person or have a soul. The Qur'an contains no explicit texts regarding contraception, although some of the *ahadith* do, so that while some traditions in Islam forbid it, most permit various forms. Attitudes to abortion vary between different jurisprudential schools, with some leaders opposing it. In Hinduism and Buddhism, while concern about some practices, especially abortion, has been expressed by religious leaders, there are no formal rules that forbid recourse to abortion or any sense that those who seek or perform abortions should face punishment (McQuillan, 2004). Various studies, some of which are summarised below, have attempted to identify links between religious affiliation, religiosity, contraceptive use and fertility.

In 11 largely Muslim Francophone West African countries, fertility remains very high, and modern contraceptive methods are little used, leading to a perception that Islam and reproductive behaviour are linked. It is argued that Islamic beliefs and values support large families, in which women seek to please their husbands by bearing children, have low status and little power and cannot use contraception without their husbands' permission. Both men and women believe that the number of children produced by a marital union reflects God's will and should not be interfered with. Nevertheless, an attempt to test these potential explanations of contraceptive use was inconclusive, because the available data do not take into account differences in religiosity or women's influence on household decision making (Farrell et al., 2014). Similarly, in southern Asian countries surveyed, the majority of Muslim respondents think family planning is unacceptable for married couples (53% in Bangladesh, 57% in Malaysia, 73% in Pakistan and 76% in southern Thailand) (Pew Research Center, 2013, p. 86). Indonesia, where 61% of Muslims think birth control is acceptable and a large number of women seek abortions (which are increasingly carried out in maternity clinics and hospitals), is an exception (Blackburn, 2008; White and Anshor, 2007).

Various studies have reported differences in fertility and contraceptive use between religions, for example Christians and Muslims in Nigeria, Tanzania and Ghana and Buddhists and Muslims in Thailand (Agadjanian, 2013). One study in Ghana, for example, shows significant religious differences, with the fertility of traditionalist women exceeding that of both Christians and Muslims, although the differences diminish after controlling for socio-economic variables, especially educational levels and urban/rural residence (which influence the use of contraception) and residence in the north or the south of the country (which reflects the relative poverty of northern regions and differential access to health services) (Gyimah et al., 2008).

In India, in addition to fertility differences between the Hindu majority and Muslim minority (total fertility rates 2.8 and 3.6 respectively in 1998/9), the country's sex ratio is remarkably low: 933 in 2001 as against 1022 in SSA and 1050 in Europe and North America (Iyer, 2003; Borooah and Iyer, 2005). The 'reversed sex ratio' is generally explained by women's disadvantaged access to healthcare, female infanticide and, more recently, sex-selective abortion of female foetuses (Tomalin, 2013). In her most recent analysis, Iyer argues that the relative costs and benefits of having sons and daughters are affected by Hindu and Muslim teachings and practices, with son preference and daughter aversion more marked and desired family size smaller amongst the former. Although there are differences in fertility between poorer (Scheduled Caste and Scheduled Tribe) and richer (higher-caste) Hindus, the difference between them and Muslims is still significant (Iyer and Joshi, 2013).

Various studies demonstrate that fertility behaviour also varies between Christian denominations, for example, in rural Mozambique and Zimbabwe. In predominantly Christian Gaza Province, Mozambique, where the Catholic Church was large and influential in the past, African Initiated Churches (Zionist, Apostolic, Pentecostal) have proliferated since the 1970s, and the Catholic and mainstream Protestant churches have become minority denominations. Membership of the Catholic or an African Initiated Church is associated with high fertility. Agadjanian and Yabiku (2014) attribute this to the churches' ability to reinforce their pronatalist theological teaching by a hierarchical organisational identity and strong institutional culture. In this province, 22% of women of childbearing age use modern family-planning methods, ranging from almost a third of both Roman Catholics and traditional Protestants to about a fifth of renewalist Christians. Differing levels of contraceptive use are not, Agadjanian suggests, a result of official religious teaching, since, although church leaders generally praise parenting, especially mothering, they rarely mention contraception, which is seen as a matter outside their agenda. The reasons for the higher rates of contraceptive use are, instead, the higher levels of education amongst Catholics and traditional Protestants, their ability to access formal healthcare and the presence of well-developed and influential women's groups, through which members are exposed to new ideas (Agadjanian, 2013). In Zimbabwe also, fertility is higher amongst members of AICs than traditional Protestant denominations. This is attributed to a preference for large families, early marriage (typically of girls married to older men in polygynous marriages) and less use of contraception (which is seen as interfering with God's purpose) (Gregson et al., 1999; Dodzo et al., 2016). However, Agadjanian (2013) warns, differences in the take-up of contraceptives do not necessarily align in the same way with religious traditions or denominations in every country or region, because use also depends on differences within traditions, the teachings of particular religious leaders, differences in religiosity and socio-economic determinants such as education, age and rural or urban location.

McQuillan (2004) concludes that "religious values, while important, are likely to play a critical role in shaping demographic behaviour only when religious authorities have at their disposal a menu of rewards and sanctions that will encourage the faithful to conform" (p. 47). This is most likely when religious traditions or denominations have organisational arrangements that command loyalty and intensive participation, are able to ensure adherence to their teachings, play a formative role in the lives of adherents and also exercise influence over other socio-political institutions (Agadjanian and Yabiku, 2014).

In the intense debates concerning reproductive behaviour, conservative religious morality is often pitted against human and women's rights and gender equality discourses, raising questions about who should have moral authority in politics and whether states should use religion to enforce particular moral ideas. Use of the law to curtail reproductive rights (a woman's right to determine when she will have children and how many) is not confined to any one country or religious tradition. In the Philippines, for example, the dominant Catholic Church, backed by the Vatican and alongside the much smaller Orthodox and Evangelical churches, blocked the introduction of reproductive health legislation that aimed to guarantee access to modern contraception, sexual education and maternal healthcare for 15 years (Imam et al., 2017; Parmanand, 2014). Further, a recent analysis of the characteristics of 190+ countries with strong anti-abortion policies and laws found that those with a large share of Muslims, Catholics and Buddhists in their population are more likely to have restrictive abortion laws, regardless of the extent of political rights for women or the degree of religious freedom. However, only religiosity is *significantly* correlated with anti-abortion laws. For example, Muslim countries with low religiosity have permissive abortion laws (Forman-Rabinovici and Sommer, 2018).

Religious teachings and state laws that prohibit birth control and criminalise abortion clearly have adverse effects on women's rights and health, making both those who seek abortions and those who perform them vulnerable to unsafe practices and prosecution. However, in some traditions and contexts, opposition to religious teachings that are inconsistent with internationally agreed reproductive health rights is growing. For example, in Indonesia, Fatayat NU, the young women's wing of the mass Muslim organisation Nahdlatul Ulama, alongside women's rights groups, has pushed to improve women's reproductive rights (White and Anshor, 2007), and many Roman Catholic parish priests and adherents, especially in areas of high fertility, high infant and maternal mortality and rising HIV/AIDS prevalence, have departed from the centrally mandated teaching and practices, forcing the Church as a whole to modify its teaching.

Gender, sexuality and attitudes to HIV/AIDS

Attitudes towards gender and sexuality are also central to the understanding and treatment of sexually transmitted infections, especially HIV/AIDS. There has been much social research into the drivers of the epidemic, differences in prevalence, gendered vulnerabilities to infection and approaches to prevention, treatment and care, especially in Africa. However, less attention has been paid to the links between these characteristics and dynamics and the religious beliefs and attitudes of individuals than to the influence of the religious right in the US, which has skewed much of the governmental and non-governmental funding for HIV/AIDS-related initiatives and influenced the role played by local and foreign religious organisations in prevention, care and treatment (Becker and Geissler, 2009).

Ten years ago, AIDS was regarded as punishment for immoral sexual behaviour by a large minority of Christians (about four in every ten, with no major differences between Pentecostal and other denominations). In Kenya, a majority held this view (62% of Pentecostal Christians, 69% of Charismatics and 53% of other Christians) (Pew Forum, 2006, p. 37). However, today locals generally attribute increasing HIV prevalence to the frequency of extra-marital sexual relations, poverty, the breakdown of family values and families and violence against women. Economic decline and impoverishment have interacted with social change, affecting household strategies and gender relations. Not only do women lack economic autonomy and power with respect to sexual relations, rendering them more vulnerable to infection, economic change has reduced men's ability to fulfil traditional masculine responsibilities, undermining their commitment to their families, exacerbating the oppression of women within marriage and fuelling men's irresponsible sexual behaviour (Burchardt, 2010; Van Klinken, 2011a). Men's limited sense of responsibility towards their families, especially in urban areas, is linked by some religious groups, such as the Roman Catholic Church in Zambia, to their lack of involvement in religious life (van Klinken, 2011a). Some also believe that stigma is a main driver of the epidemic, leading, for example, to an unwillingness to be tested (Campbell et al., 2010). However, in 2014/15 surveys in 33 countries, an average of 68% of respondents said that they would like or not care if they had a person living with HIV and AIDS (PLWHA) as a neighbour. This percentage is higher in those countries with higher prevalence and amongst younger, more educated, urban and Christian respondents, demonstrating that people have become more tolerant as familiarity and understanding has grown (Dulani et al., 2016).

Much AIDS education focuses on and problematises women. Like family planning, they are generally held to be responsible for preventing HIV infection, despite their relatively powerless position and desire to comply with cultural and religious expectations (Van Eerdewijk, 2012). Risky behaviour, Becker and Geissler note, "arises from lack of choice and imbalanced power relations, but also. . . [the risks may be] less important than ritual obligations to engage in bodily intercourse. [In practice] People may give religiously sanctioned relations precedence over concerns with the individual body and life-expectancy, and prioritise the wish to have children" (2009, p. 5). Religious teachings generally focus on restrictions as a means of preventing transmission, though whether these can reduce the risk of acquiring HIV/AIDS is, in Becker and Geissler's view, an open question, with many studies showing that "the famous ABC (Abstinence, Be faithful, use a Condom) formula for HIV prevention, with its emphasis on 'choices', barely scrapes the surface of the way people experience, practise and think about sex" (2009, p. 5).

Attitudes vary between and within religious traditions and between conservative and liberal wings, leaders and ordinary adherents, women and men and PLWHA and their families and others. For example, Agadjanian (2005) shows how in Mozambique, perceptions of risks, exposure to prevention messages and prevention practices vary between both Christian denominations and men and women. As prevalence has increased, some religious leaders have recognised that the use of condoms is more effective than abstinence and fidelity and/or desirable in marriages in which one partner is HIV positive (Mantell et al., 2011). However, many religious groups refuse to promote the use of condoms, not only because they prioritise childbearing but also because condoms are thought to encourage promiscuity, even though a message that promotes fidelity for a woman on religious grounds, regardless of a man's unfaithfulness, may effectively be a death sentence if her partner refuses to use a condom (Burchardt, 2010; Mantell et al., 2011; Pearson and Tomalin, 2008). Nevertheless, the links between HIV/AIDS and the growth of revivalist movements in Christianity and Islam, with their conservative, restrictive attitudes towards gender and sexuality, are complex (Becker, 2009; Behrend, 2009; Svensson, 2009; van Dijk, 2009).[4] For example, Pentecostal churches condemn extra-marital sexual relations and reject condoms, but there is contradictory evidence about whether born-again Christianity disempowers women because it subjects them to strong patriarchal control and exposes them to unprotected sex or empowers them by providing a community that helps them to protect themselves against unwanted sex and cope with the illness (Becker and Geissler, 2009; Prince, 2009; Pugh, 2010). In addition, widespread homophobia and the criminalisation of same-sex relationships (see below) deter gay people from being tested for HIV (Jenkins, 2006; Mantell et al., 2011).

HIV/AIDS is regarded by many Africans not so much as a new and distinctive illness as one of a long succession of epidemics and general misfortunes, for which religion has always offered explanations and solutions (Becker and Geissler, 2009). Amongst the cures, faith healing is rooted in the relations between humans and spiritual beings associated with folk religion but today is promoted not only by adherents of traditional religion but also by evangelical Christians, often as an alternative when biomedical cures fail (Ashforth, 2011). To many (e.g. Horn, 2012), belief in the efficacy of 'miracle' cures undermines the health and survival of PLWHA and hinders the roll-out of effective treatment. However, faith healing may also be combined with biomedical understanding of the disease and its treatment, especially since ARVs have become more widely (if unevenly) available, contributing to changes in attitudes (Dilger, 2009).

Despite the increasing availability of effective medical treatment, many PLWHA face long-term ill health. Some ways of coping rely on religious beliefs and practices. Although the moralistic attitudes and gender ideology associated with many religious groups perpetuate stigma, they also contribute to the care and support of PLWHA and create social spaces for challenging stigma (Campbell et al., 2010). For example, the Ghanaian women interviewed by Perry et al. (2010) referred to their belief in the power of God to protect, support and deliver them, the power of prayer and the benefits of church attendance. In South Africa, like elsewhere, Burchardt (2010) points out that, while caring for the sick is still universally perceived to be a female duty, women are able to draw on gendered forms of sociality to create and mobilise networks of solidarity. In addition,

4 The longstanding SSA cultural practice of widow inheritance (levirate marriage), which is accompanied by various rituals, is designed to ensure continuation of the family line. However, it is opposed by many widows and women's rights advocates on a number of grounds, including its potential for increasing HIV transmission, leading some religious groups to participate in attempts to modify the practice (Becker and Geissler, 2009; Prince, 2009).

research has started to show the multiple ways in which PLWHA and those who care for them are developing new attitudinal and practical approaches to living positively, even within judgemental religious communities. For example, Beckman (2009) shows that rather than seeing the infection as divine punishment for individual misbehaviour, HIV+ Muslims in Zanzibar speak of it as a trial they are undergoing on behalf of a whole community that has fallen short of God's ideals, allowing them to claim a place for themselves *within* the Muslim community.

Attitudes to homosexuality

Heteronormative sexuality is the ideal for all the faith traditions, based on the view that the purpose of sexual activity is procreation rather than pleasure. Homosexuality is widely regarded as sinful and morally wrong, paving the way for widespread homophobic attitudes, discrimination against people who express alternative sexualities and resistance to the introduction or enforcement of laws that make homosexual acts between consenting adults legal.[5]

Respondents in surveys of renewalist Christians in SSA nearly all believe that homosexuality is unacceptable, except in South Africa, where attitudes are slightly more liberal (Pew Forum, 2006). Pentecostal Christians view the world as a site of war between God and the devil. They believe that 'God's order of creation' is heterosexual, consider homosexuality to be an abomination in the eyes of God and call for gay rights to be the work of the devil. These extreme views are reflected in the debates in Zambia, the only self-declared 'Christian nation' in SSA, where influential Pentecostal and African Initiated Churches portray homosexuality as a threat to the purity of the nation and a litmus test of its Christian character and use homophobia as a political weapon, even though the views of these churches are not shared by the mainstream Protestant and Catholic denominations (van Klinken, 2013, 2014). That such views are not confined to renewalist Christianity or Zambia is demonstrated by findings from the World Values Surveys 2010–14, which show that in many African and southern Asian countries a higher proportion of highly religious people than the population as a whole believe that homosexuality is never justified.[6] For example, 90% or more Muslims in most countries and 70% or more in the rest share the view that homosexuality is unacceptable (Pew Research Center, 2013, p. 81).

Some believe that there was greater tolerance of alternative sexualities in the past, including in earlier traditions of Hinduism and some of those associated with folk religions. For example, although evidence is scarce, Epprecht (2012)

5 In 2008, of 242 countries worldwide, 78 had laws criminalising male same-sex relationships, and in 45 female same-sex relationships were illegal, according to data from the International Lesbian, Gay, Bisexual, Trans and Intersex Association, quoted in Mantell et al., 2011.

6 The majority of highly religious people in many countries believe that homosexuality is never justified (Ghana 85%, Zimbabwe 73%, India 73%, Pakistan 73%, Nigeria 64%, Rwanda 63%, Malaysia 61%), although some countries are more liberal (e.g. Thailand, Philippines, South Africa) (www.worldvaluessurveyorg/WVSOnline.jsp, accessed 10/8/2017).

suggests that contemporary homophobia ignores the implicit toleration of same-sex relationships that existed in various precolonial African societies (see also Mantell et al., 2011, p. S201).

That some of the main attitudinal differences are found within religious traditions can be illustrated by the contested debate within Hinduism. In India, which legalised homosexual relations in 2009, rival factions reference Hinduism to support their points of view. The diversity and ambiguity of Hindu texts, myths and cultural artefacts (such as paintings and sculpture) with respect to non-heterosexuality, as well as the variety of people who express alternative sexualities, enables both supporters of liberalisation and social conservatives to use religious references. Conservative Hindu cadres and rightist political parties, already alarmed by liberalising tendencies, appeal not just to the secular ethic of citizenship but also to Hindu religio-cultural motifs to mobilise support for their opposition to legal reform (Hunt, 2012).

In societies in which most people are religious, most homosexuals are adherents despite the homophobic attitudes they experience (van Klinken and Chitando, 2015). Little is known about how they navigate these difficult waters, although van Klinken's studies of small groups of gay men in Lusaka have begun to throw some light on this. Resisting discourses that demonise them, his research subjects humanise themselves by referencing the universal category of love – their belief that they have been created in the image of God who is love and their own desire for loving relationships. In the Zambian context, this enables them to claim full citizenship of the 'Christian nation' and to portray homophobia, indeed any form of discrimination, as distorting the image of God (Van Klinken, 2015; van Klinken and Phiri, 2015).

In many countries, discussions of homosexuality are heated and resistance to improving sexual rights widespread. However, it cannot be assumed the issue is equally prevalent everywhere or that the homophobic attitudes that fuel campaigns against legal reform are universal and unchanging. In Uganda, for example, homosexuality and sexual rights have been subject to debate at the national level for the last 20 years. Vociferous opposition to legal reform has been expressed, not least by the powerful Catholic Church. However, in the Acholi Region of Northern Uganda, there has been almost complete silence on this issue. Alava (2017) attributes this lack of public debate to the desire of the Church to maintain its popularity in a region where persistent cultural norms undermine its teachings on marriage and the family and to promote a discourse of peace and unity in this war-torn environment. Intolerance of homosexuality does indeed remain widespread, as shown by surveys in 33 SSA countries, where on average 78% of respondents said that they would somewhat or strongly dislike having a homosexual as a neighbour. However, majorities in four countries (South Africa, Mozambique, Namibia, Cape Verde) said that they would not mind: in South Africa, homosexuality was decriminalised in the post-apartheid constitution and in Mozambique just before the survey. The finding that, overall, younger, more educated, urban respondents are more tolerant implies that tolerance is likely to increase over time (Dulani et al., 2016).

Religion, marriage and the family

Attitudes and behaviour are never solely individual – always relational, one of the main spheres in which they are played out is the family. Gender relationships within families are closely linked to ideas about sexuality and reproductive rights and practices, which have been discussed above. Here, the links between ideas of the family and marital relations are discussed further and extended to briefly consider the available evidence on religious influences on aspects of work, inheritance and the socialisation of children, focusing on attitudes to education.

Marital relations and the family

Gender and other relations within the family are influenced by social custom, state law and religious law/guidance to varying extents between contexts and over time, affecting norms and practices related to marriage and marital breakdown. In precolonial rural societies, for example in SSA, productive and reproductive arrangements were typically reflected in widely accepted kinship and marriage patterns, generally expressed and reinforced by rituals. These assigned authority and power to men and valued motherhood, although in practice, conventions governing marital arrangements, including the allocation of labour, access to property and inheritance practices varied, and in some contexts, women were permitted wider roles in agriculture, trade and even politics (Amadiume, 2005).

In Islam, the family is regarded as a natural unit, provided by God and the basis of a good society. The ideal relationship between men and women is seen as marriage, based on complementarity rather than equality, with each spouse having rights over the other. Backed by specific texts, men are regarded as breadwinners and maintainers of the family unit, in return for which they can expect obedience from their wives. The Qur'anic injunction that makes men financially and morally responsible for women is intended to free the latter from wage labour so they can concentrate on the task of raising the next generation. The proscription of women being forced into marriage against their will and the possibility of a marital contract were improvements on the previous situation, in which women were regarded as chattels (Qazi, 2015).

Today, evidence from a variety of studies shows that many Muslims respect these values and social structures. For example, a study of everyday middle-class family life in Muslim areas in Mumbai reveals that, despite the increased status and authority that education has conveyed on women, they support the idea of complementary roles for men and women within the family and the traditional religiously sanctioned division of labour (Anand, 2014). Contemporary Muslims almost all believe that wives must always obey their husbands: around 90% in the southern Asian countries with large Muslim populations (Pew Research Center, 2013, p. 93). Male headship and female obedience are accompanied by a belief that it is acceptable for husbands to discipline their wives and that wives must not reject their husbands' sexual advances. However, views vary, with literalist interpretations of Islamic teaching being particularly strongly supported among socially conservative revivalist groups, such as the middle class women who are members of the religious political party, Partai Keadilan Sejahtera, in Indonesia (Rinaldo, 2008).

Nevertheless, there are many variations in local practice, partly because customary family arrangements persist and have influenced secular state law and/or *shari'a* (see, for example, Robinson, 2012, on Indonesia). In addition, in many countries, women's education and entry into the paid labour force is associated with greater economic autonomy, an increased average age of marriage and declining birth rates, so that many women are less preoccupied with home and family than in the past (Blackburn, 2008; see also Adeney Risakotta, 2016). In southern Asian countries, changing social constructions of the family have been negotiated between religious leaders, employers, politicians and adherents, often as part of the state family values projects (see, for example, Brenner, 2011; Parmanand, 2014; Platt et al., 2018; Stivens, 2006). Changing perceptions may be reflected in legal change. For example, in Indonesia, a 2004 Act outlawed domestic violence and included marital rape in the definition despite earlier religious opposition (Blackburn, 2008).

The idea that male headship is fundamental to the order of creation is also central in Christianity. Men, it is believed, have been given primary responsibility as providers for their families. This is typically interpreted as implying male dominance and wifely submission, even though Jesus's life demonstrated the notions of servant leadership and respect for women (van Klinken, 2011b). In practice, national sample surveys show that the proportion of Christians who believe that wives must always obey their husbands varies, from almost all in some countries (e.g. Nigeria and Kenya, followed by India) to a slightly lower proportion in others (e.g. South Africa, the Philippines), with renewalist Christians being slightly more likely than others to think this (Pew Forum, 2006, p. 42).

In India, traditional Hindu law as codified by the British colonial government specifies that one of the central prerogatives, and indeed duties, of the householder is strict control over the women of the house (Nussbaum, 2003). Such ideas have often been invoked to back local resistance to social reform, for example increasing the minimum age of marriage or reducing the prerogatives of husbands (including violence towards their spouses).

Polygynous marriage is permitted under Islamic law and is often part of customary social practice. Although an arrangement by which men are permitted multiple partners but women are not is seen by many as discriminatory, the practice and support for it is widespread in SSA, where at least six in ten people think that it is acceptable. It has wide Muslim support, especially amongst men, with no consistent differences between younger (< 35) and older (> 35) men in the countries surveyed (Pew Research Center, 2012). In some contexts, the practice has declined, because socio-economic change has reduced the labour-related benefits and made it more difficult for Muslim men to adhere to the requirement that they can afford to take more than one wife and can treat all their wives equally. For example, a survey in rural areas in Indonesia showed that only a minority of respondents support polygyny (Mujani and Liddle, 2004, p. 114). In contrast, revivalist Muslims may support it more strongly. For example, in Malaysia, recent Islamisation has been associated with an increase in the practice amongst middle-class Malaysians (Olivier, 2016). Christian attitudes vary, depending on how strongly particular churches are influenced by customary marriage practices. Most Christians, especially members of Pentecostal and charismatic churches, disapprove of polygyny, although some African Initiated Churches favour it (Pearce, 2012).

Divorce is widely seen as morally wrong or unacceptable, especially amongst Christians (Pew Forum, 2010). Muslims' views vary between countries, including amongst those who want *shari'a* to be enshrined as state law. In the southern Asian countries for which data are available, the proportion of Muslims who regard divorce as morally wrong varies from a minority in some (ranging from 13% in Thailand to 42% in Indonesia) to a majority in others, especially Pakistan (71%), with no consistent differences by age or gender. The proportion in SSA countries who believe divorce is morally wrong also varies widely, from 19% in Niger to 72% in Liberia (Pew Research Center, 2012, p. 82). Access to means of dissolving a marriage through divorce are often gender unequal. In Islam, for example, according to *shari'a* men may divorce their wives by repudiation (*talaq*), but women do not have such an entitlement and instead have to seek a decree of *khula* (Sait and Lim, 2006). Muslims disagree over whether a woman should be able to initiate a divorce: in most of the small number of countries for which relevant data are available, only a minority support this. Bangladesh, where 62% of Muslims do so, is an exception. Surprisingly, women are often no more supportive of the idea that they should be able to initiate divorce than men (Pew Research Center, 2013, p. 94). However, national averages may conceal attitudinal differences, as shown by a survey of Muhammadiyah and NU members in Indonesia, where state law permits both men and women to initiate divorce proceedings. Forty-five percent of the more urban, educated and middle class Muhammadiyah members thought that women should have this right, compared to only 33% of more rural, traditionalist NU members (Bush, 2014, p. 21). That issues like this can be used in politics is illustrated by the case of India, where Muslims are governed by *shari'a* personal law. By demanding unified civil law, the Hindu right has been able to portray itself as a protector of Muslim women while ignoring the ways in which civil law perpetuates gender injustices (Desai and Temsah, 2014). In practice, divorced and separated women are often stigmatised, so that generally, women choose to stay in difficult marriages if they can.

Widowed women may also be stigmatised, although this varies between contexts and depends on age. A particularly contentious issue in many African contexts, especially those with high HIV/AIDS prevalence, is levirate marriage. Because most African kinship systems are patrilineal and patrilocal, and a wife's family has generally paid bridewealth, children are regarded as belonging to the husband's family, which is generally held to be responsible for maintaining the widow until the children are of age. The institution of levirate marriage (in which a male relative, usually the brother of a woman's late husband, becomes responsible for her, her children and her property, with or without a sexual relationship) evolved in response. Often resisted by widows, it is also under strain because of economic change. Typically the churches disapprove of widow inheritance. In the largely Christian district of Busia in eastern Uganda, for example, the Pentecostal churches reject it outright, while the Roman Catholic Church will accept a levirate marriage that does not involve a sexual relationship and also helps widows resist re-marriage on the grounds that they are religiously committed and/or wish to remain celibate. For example, widows may refer to themselves as being 'married to Christ', evoking the Catholic tradition of female orders of (celibate) nuns (Christiansen, 2009).

Influenced by international human rights discourse and social change, especially literacy and access to education, industrialisation and urbanisation, inherited social institutions and laws have, increasingly, been challenged not only by equal-rights advocates but also by ordinary women (Phillips, 2009). For example, informants in a 2002–4 study of grassroots Christian and Muslim women in and around the city of Jos in Plateau State, Nigeria, saw unequal gender relations stemming from the religious and customary laws which govern their daily lives, including the practice of seclusion and male domination of household decision making, as exacerbating their poverty. The women studied regard domestic violence as normal, justified by religious teaching and exacerbated by virilocality, economic dependence and the stigma attached to separation and divorce (Para-Mallam, 2006). Their view that customary law and practice discriminate against women is echoed by a large majority of respondents in a 2012 survey, who thought that women should have equal rights rather than being subject to traditional law: 68% in West Africa, 74% in southern Africa and 82% in East Africa, although the proportion varied from 87% in Tanzania to a minority in Mali and Niger (Chingwete et al., 2014). For Para-Mallam's informants, statutory law was less problematic, although they regarded some provisions as undesirable. For example, the provision that all married women aged under 18 are considered adults (a concession to northern Muslim political interests) provides implicit support to early marriage (Para-Mallam, 2006).

Work, property and inheritance

Family units and marriages are important sites for the organisation of productive and reproductive activities, allocation of the proceeds of labour and income, and entitlements to property during and after the end of a marital union. The religious traditions generally support a division of labour that allocates responsibility for reproduction to women and production to men, even though women often play significant roles in the latter, especially in rural economies. Most of the contradictions faced by women seem to be associated with this customary conception of the household division of labour and with Islam. However, economic changes have influenced the opportunities for earning incomes and accumulating property, in turn influencing religious perceptions of appropriate roles for women and men within marriage.

Women tend to be more disadvantaged in patrilineal kinship systems with patrilocal residence patterns, which characterise most African ethnic groups. In Southeast Asia, in contrast, gender equality and women's autonomy tend to be greater because many societies are characterised by a bilateral descent system and matrilocality. Vietnam, which is mostly patrilineal and patrilocal, is an exception (Falk, 2015). In this region, women are likely to be engaged in economic activities (especially as more men than women become monastics in Buddhist societies) and so have more control over money than elsewhere, are less likely to be considered a financial burden on a family and are more likely to inherit on equal terms with men.

Women's labour force participation is also influenced by class, rural/urban residence and religion. Para-Mallam (2006, 2010), for example, demonstrates how the opportunities open to and restrictions placed upon ordinary and elite women in Nigeria vary. Typically, seclusion practices mean that Muslim women are less likely to participate in the labour force, although this is modified by their access to education, whether they are members of farming families and the availability of home-based work. In India, as noted above, both Muslims and Hindus practise seclusion, affecting women's labour force participation and mobility, although in ways that are differentiated by social status, household composition, poverty and age (Kantor, 2002; Vera-Sanso, 1995). Generally, Muslim women are less likely than women in other religious groups to participate in the labour force, especially formal-sector wage work, because of seclusion practices, their lower education levels and their role as symbols of Muslim community identity in a Hindu nationalist climate (Desai and Temsah, 2014). Ghatak's (2006) study of women and girl *beedi* rollers in villages in West Bengal illustrates some of these dynamics. Poverty and landlessness mean that women have little choice but to participate in this form of piecework. However, they also see it as providing an opportunity to assemble dowries for their daughters' marriages and train them in useful skills.

Paid work, especially wage labour, can give rise to tensions between religious and social expectations. Women workers have developed a variety of strategies to negotiate such tensions, often by positioning themselves as not only good workers but also good wives/daughters and good Muslims. These often include physical elements (e.g. working in sex-segregated spaces, living in special accommodations, adopting modest behaviour). Even educated/elite women have to navigate their needs and opportunities in relation to Islamic injunctions with care, as shown in a study of middle-class married Muslim businesswomen in Indonesia (Sakai and Fauzia, 2016). Employers of young Muslim female factory workers in Tangerang, west of Jakarta, provide dormitory accommodations and support Qur'anic recitation groups to help employees navigate work and religious requirements and the expectations of their rural parents (Warouw, 2016). In Malaysia, growing demand for factory labour has provided opportunities for young women to enter the wage labour force, encouraged by the government. However, increased female autonomy and assertiveness have been accompanied by calls for them to adhere to stricter behavioural norms (for example with respect to Islamic dress)

(Ong, 1995). In Pakistan, educated female social organisers working away from their homes in the Hazara region of the Khyber Pakhtunhwa Province (formerly the Northwest Frontier Province) travel with male colleagues, work in segregated offices, limit their field visits to daylight hours, live in women's hostels and are careful to adopt correct behaviour while at the same time portraying themselves as modern, open-minded, and capable (Grünenfelder, 2013). Similarly, in Darfur in Sudan, young women teachers seek to maintain their position as 'good Muslim women' by living in a hostel, wearing modest dress and mentoring pupils in an adjacent boarding house, claiming that they had more time to do so because they were not married. Attributing their unmarried state to a lack of suitable educated local men, they legitimise it by a 'moral motherhood' (Willemse, 2007).

New work demands and opportunities can therefore lead to attitudinal change. For example, a recent survey of Muhammadiyah members in Indonesia found that today, 82% of men and 77% of women think that wives have a right to earn an income and to be consulted on economic issues (Bush, 2014). However, where men see women as competitors (e.g. in Nigeria, Mahdi, 2007) there can also be a backlash, in which gender norms are reinforced and new restrictions introduced (e.g. against teachers recruited from the first cohort of women in low-income communities to receive education in Pakistan, Khurshid, 2015).

The inter-gender and inter-generational transfer of wealth and property is generally governed by social custom and statutory law. However, amongst Muslims, it also involves Islamic law. Customary and state law may or may not protect married women's property rights. Muslims argue that *shari'a* benefits women in this respect, in that *mahr*, a payment of cash and/or property that a husband is obliged to make to his wife on marriage, is hers to use and dispose of, even if her husband divorces her (Sait and Lim, 2006). However, in other respects, *shari'a* law discriminates, and this may be reflected in state law. For example, the Qur'an stipulates that a son should receive two shares of an inheritance for every one share inherited by a daughter. Contemporary Muslims (including women) disagree about whether this is appropriate. In the southern Asian countries for which data are available, the figures show that the proportion of Muslims believing that men and women should have equal inheritance rights varies: a minority in Malaysia (36%) and Bangladesh (46%), a small majority in Pakistan (53%), and larger majorities in multi-religious Thailand (61%) and Indonesia (76%). The proportion of women who think they should have inheritance rights equal to those of men invariably (and often significantly) exceeds the proportion of men (Pew Research Center, 2013, p. 99).

Compliance with *shari'a* is always influenced by social relationships, so it operates less straightforwardly than the religious injunctions suggest. For example, in Nigeria and elsewhere, discriminatory practices associated with patrilineal inheritance systems undermine the Muslim provisions (Para-Mallam, 2006). In India, Kabeer (1999) describes how Muslim women generally prefer or are encouraged to waive their rights to parental property in favour of their brothers, strengthening their future claims on those brothers following widowhood or marriage breakdown (see also Sait and Lim, 2006). Muslim women's ability to claim their inheritance rights, therefore, invariably involves an interplay of "custom, gender, kinship and property" (Sait and Lim, 2006, p. 125).

The education of children

Education is a key determinant not just of people's access to knowledge and life opportunities but also of their ability to understand the world in which they live and challenge discriminatory social norms and institutions. Yet it is not equally available for all, despite the evidence that education for girls contributes to economic growth, improved health, reduced fertility, increased farm productivity and female participation in household and political decision making (UNICEF, 2005). Persistent gender inequality in access and educational achievement is often attributed, amongst other factors, to religious beliefs and values, which are said to influence the attitudes of parents, social groups, religious organisations and governments to the provision, purpose and content of education (both religious and secular). In this subsection, the analysis seeks to disentangle the influence of religious and other factors on attitudes and practices with respect to education for girls, especially amongst Muslims, whose supposedly lower demand for education, especially for girls, is alleged to have held back the development of Muslim countries and population groups.

Christian missionaries established educational facilities long before colonial governments did so and continued to operate these alongside pre- and post-independence government systems until, in many places, governments decided to nationalise church educational establishments. Initially, the primary purpose of their educational provision was to enable existing and prospective Christians to read the Bible in the vernacular, and this type of religious education is still important (see Chapter 6). However, from early on, schools sought to provide a grounding in the subjects required by governments and other potential employers, initially by providing basic and technical education and then, as pressure for independence mounted, secondary/higher education for a small elite. Education was valued for both girls and boys, although the perceived destiny of girls as wives and mothers often restricted both the years of schooling provided and the subjects taught in girls' schools.

In Islam, Qur'anic education has long been provided for both girls and boys, although advanced religious learning was in the past almost always confined to men. Basic religious education (learning to recite the Qur'an) is often the only type available at the local level. Further basic and advanced religious education is offered in madrasas, which provided the staff for government bureaucracies in Muslim states, in India and elsewhere. However, the graduates of madrasas did not meet the requirements of colonial governments, and so, during the nineteenth century, the education they provided became less useful to and less valued by prospective students (Qazi, 2015). More recently, some madrasa education systems have sought to address contemporary needs by offering non-religious subjects in addition to Islamic subjects and increasing the numbers of female madrasas.

Since colonial times, there has been a tension between the provision of modern secular education for girls and their perceived destiny as wives and mothers. This tension is exacerbated within religious communities by their belief that women are responsible for the inter-generational transmission of religious beliefs and values within families. The perception that education, especially for girls and women, is valued less by Muslims than by other religious communities is examined here through summaries of the findings of two very different studies.

McClendon et al. (2018) seek to analyse this perception at a global level, using a quantitative approach. To assess whether lower levels of enrolment and attainment in Muslim countries, especially amongst girls and women, indicate lower demand for education amongst Muslims, they examine female enrolment in 'formal' schools (i.e. those that offer secular subjects instead of or in addition to religious ones) and female educational attainment. The data show that globally, the enrolment and attainment of Muslim women have both increased. However, some regions, including SSA and South Asia, lag behind others.[7] Rather than discrimination in family law, whether or not a country has a democratic government and whether Muslims are a majority or minority, the main factor explaining the lag in Muslim women's enrolment and attainment in these regions is their low level of economic development. This is borne out by the greater increase in both enrolment and attainment amongst Muslim women in wealthier Indonesia and Malaysia, where the gender gap in education has been reversed, than in Pakistan or India, where educational attainment is low overall, reflecting these countries' less well-developed education systems. The study notes that, in India, although they have made gains, levels of enrolment and attainment are also low among Hindu women, attributing this, as with Muslims, to the low overall level of educational development. The data also show that enrolment and attainment are higher amongst younger women. McClendon et al. conclude, first, that the lag in female enrolment and attainment is probably unrelated to religion and, second, that it may well not be permanent.

This analysis can be complemented by a qualitative study in Pakistan, Nigeria and Tanzania, which sought to provide a more nuanced understanding of attitudes towards education for girls and women than that provided by quantitative survey data, again with a focus on Muslim attitudes. The research was based on interviews and focus-group discussions with a range of key informants, including male and female religious leaders, teachers, ordinary people, government officials, community leaders and some senior students in 2009. It explored how religious values shape respondents' views on the educational entitlement of girls and boys and the characteristics of suitable educational provisions. In largely

7 Other data for SSA bear this out and show the gender differences. In large parts of the subcontinent, especially in West Africa, a significant proportion of men (and an even larger proportion of women, especially in West and East Africa) have not received any formal education, and only a small proportion of both men and women (about 10%) have received any post-secondary education (Chingwete et al., 2014).

Muslim Pakistan, all except a few of the informants were practising Muslims. The research explored commonalities and differences in views within the religious tradition, by including both Sunni and Shi'ite Muslims and working in Lahore and Peshawar (the latter supposedly more conservative in its religious and social values and norms than the former) (Bradley and Saigol, 2012). In Nigeria, the research was carried out in largely Muslim Jos in northern and religiously mixed Ibadan in southern Nigeria, and in Tanzania in the Muslim port of Tanga and the mainly Christian agricultural district of Meru, both in the north of the country (Dugbazah and Labeodan, 2010; Dugbazah et al., 2010; Hassan, 2010; Kipacha and Dugbazah, 2010).

Across the religions and locations, the study showed that most people today believe that girls are as entitled to both secular and religious education as boys, with Muslims everywhere noting the importance attached in the Qur'an to acquiring knowledge and, like Christians, emphasising the role of education in transmitting both academic, practical and religious knowledge and moral values. In Pakistan, the responses were similar amongst Sunni and Shi'ite Muslims and in Lahore and Peshawar, despite the initial assumption that people in the latter would be less amenable to the idea of education for women. In both Nigeria and Tanzania, respondents universally expressed support for education for both girls and boys, with the exception of Muslims influenced by fundamentalist groups such as Boko Haram in northern Nigeria, who see education for girls as unimportant, even un-Islamic.

However, views about the purpose of education, its content and appropriate modes of provision differ. These differences are often gendered, reflect respondents' position on the socio-religious conservative–liberal spectrum, and are influenced by the history of religious and government involvement in education provision and changing socio-economic opportunities and pressures.

In all three countries, views of the purpose of education for boys and girls differ, reflecting the supposed 'natural' and 'social' differences between men and women, with the primary purpose of education for the latter seen by most religious leaders, education providers and ordinary adherents, both Muslim and Christian, as being to equip them for their roles as wives and mothers, with content focused on religious education, home economics and the domestic sciences. However, differences of view were also expressed. For example, in Pakistan, male respondents often favour medicine and teaching as suitable professions for women, because, unlike occupations such as driving, engineering or industrial employment, these are considered respectable and well suited to women's nature, and also because female doctors and teachers are needed to attend to female patients and instruct girls. In contrast, female respondents from across the religiously conservative–liberal spectrum believe that women should be free to choose the subjects they study and the occupation they choose, invoking Muslim history to show that women have, in the past, fulfilled political, judicial and other roles. Home economics, they suggest, is better taught in the home than the school (Bradley and Saigol, 2012). In Nigeria and Tanzania, some (mostly male) respondents fear that education may lead women to assert themselves, encouraging non-compliance with religious practices (modest dress, seclusion etc.) and immoral behaviour, undermining the norm of female submissiveness and harming women's prospects of marriage. In response, some men refuse permission for their daughters and wives to attend educational institutions, and some parents withdraw post-pubertal girls from education and insist on early marriage. That views within religious traditions vary considerably is borne out by a study of three Hindu movements engaged in providing education for girls in Pune, India. The comparison of their attitudes towards girls' education attests to the need for a nuanced understanding of the contents and interpretations of religious teaching (Bradley, 2011).

While opinions differ on the purpose and content of education, concerns about co-education are fairly universal, especially at the post-primary level. In Pakistan, for example, most religious respondents (leaders and parents) prefer segregated education to avoid perceived threats to girls' morals and chastity, although secular and female respondents, especially Shi'ite women, have a more relaxed attitude (Bradley and Saigol, 2012). The preference for sex-segregated schooling has fuelled an increase in the number of madrasas for girls and young women in recent years, although critics accuse female madrasas of reinforcing patriarchal beliefs and values by their stress on religious, household and mothering education (Bano, 2010).

Girls' access to education, the characteristics of educational institutions and attitudes to co-education are also affected by government policies. In Pakistan, Islam is the state religion, so differences in the education provided in religious and state schools is less marked than in countries with secular governance (Ali, 2008; Saigol, 1999, 2003). In addition, government schools use the same curriculum for boys and girls, so even though the state educational system reinforces religious values and norms that subordinate women, it is considered by many, especially secular informants, to be preferable to madrasa education (Bano, 2010). This is also true in Tanzania, where the government has prioritised investment in education since independence, and madrasa education, although free, is generally regarded as having little practical value in today's labour market (Mesaki and Kipacha, 2008). In Nigeria, despite the expansion of state education since independence, economic liberalisation, declining government funding and Muslim–Christian political competition have adversely affected the potential for the public education system to reduce the north–south, religious and gender differences in educational levels (Hassan, 2010).

Attitudes to education for girls have also changed in response to changes in economic opportunities. In some contexts, boys are still given preference and girls' and women's participation in education constrained by religious norms and practices, perceived social and moral dangers, poverty, the demand for family labour and the educational opportunities available. However, elsewhere, the increasing availability of income-earning opportunities for women has helped to change attitudes. Thus the evidence shows that, while some Muslims would prefer to limit education for girls to religious education, increasingly parents want a wider education for their daughters, not just to fulfil their roles as family educators and moral exemplars but also to train them for occupations outside the home

(Qazi, 2015). Attitudinal changes on the part of girls, young women and parents are reflected in the emergence of new ways of enabling them to pursue their education, ranging from informal early-morning and evening schools for working children from poor families and girls who marry young in Tanzania and the women's study groups discussed in Chapter 6 to changes in the content of religious education in the extensive network of *pesantren* (Islamic boarding schools) in Indonesia (Qazi, 2015).

Conclusion

The analysis in this chapter has covered selected aspects of gender, sexuality, family relationships, religious traditions and contexts, reflecting the scope and complexity of the issues and the patchy availability of research studies. It reveals commonalities and differences between and within the religious traditions with respect to discourses and practices related to gender equality and sexual rights. In principle, all the religions recognise the spiritual equality and worth of women and men. However, patriarchal interpretations of religious teachings dominate, and all the traditions reflect and reinforce a system of gender relations that institutionalises male physical, social and economic power over women and considers sex and sexuality that does not have a reproductive purpose to be immoral and illegitimate. This religio-social view of gender roles and responsibilities as complementary rather than equal influences women's expectations, their social relationships and their ability to exercise agency, adversely affecting their lives and opportunities. Critics of fundamentalist/conservative trends, in particular, tend to regard all religions as irredeemably oppressive (e.g. AWID et al., 2015; Balchin, 2009; Horn, 2010; Vaggione, 2008).

However, the evidence demonstrates that there are also differences between and within the religious traditions, especially along a religiously and socially conservative/liberal spectrum. As suggested in the analytical framework, the nature of religious organisation is important to understanding these differences, because it helps to explain where authority lies and how power is exercised, for example with respect to how gender institutions (norms and rules) are established, interpreted and enforced, or how alternative discourses are legitimised. Channels of communication and the exercise of religious and social authority and power have local, national/regional and international dimensions. Particularly relevant in the context of gender equality and sexual rights are interactions between religious organisations and local societies, the influence exerted by religious organisations on law and policy, and the roles played by the UN, the Organisation of Islamic Conference and other international religious bodies. Differences between and within religious traditions intersect with class and other differences, influencing attitudes and practices with respect to gender equality and sexual rights. These have implications for individual attitudes and behaviour, regulatory practices, marriage and the family. They may reinforce moral and social norms and enforce behaviours but also influence how societies adapt to change. Socio-economic and political changes at the local, national and international levels have influenced the opportunities available to women (for both good and ill), leading to changes in attitudes and practices. While religious organisations may react by resisting and opposing change, they may also alter their own attitudes and adopt innovative practices that enable adherents to cope with new challenges, such as HIV/AIDS.

Challenges to prevailing gender ideologies and inequalities have increased in recent decades, both within the religious traditions and in wider societies (see, for example, Kirmani and Phillips, 2011; Mir-Hosseini, 2011; Pearson and Tomalin, 2008). These employ different conceptualisations of empowerment and agency (Avishai, 2008; Burke, 2012; Kabeer, 1999) and include attempts to

- redress the gender imbalance in religious scholarship and leadership, to permit reinterpretations of religious teachings, promote the emergence of alternative discourses and influence the exercise of pastoral care, authority and power within religious traditions and organisations
- reconceptualise femininity and masculinity within religious traditions and in wider society
- expand the roles and realise the potential of women's organisations, including religious orders, piety movements, study groups, local fellowships and national organisations
- build on the experience and achievements of movements campaigning for gender and sexual rights, both within the religious traditions and through participation in wider social and political movements.

Such advocacy and effort are often opposed by conservative religious movements with traditional gender ideologies, the scale and influence of which have grown in many places. As a result, the outcomes of contextual and religious change vary, and their implications for women's empowerment, gender equality and sexual rights are complex and contradictory. Nevertheless, some broad attitudinal shifts, both within the religious traditions and in the wider societies in which they are embedded, seem unlikely to be reversed.

References

Adeney-Risakotta, B. (2016) Traditional, Islamic and national law in the experience of Indonesian Muslim women, *Islam and Christian-Muslim Religions*, 27, 3, 303–18.

Agadjanian, V. (2005) Gender, religious involvement, and HIV/AIDS prevention in Mozambique, *Social Science and Medicine*, 61, 1529–39.

Agadjanian, V. (2013) Religious denomination, religious involvement, and modern contraceptive use in Southern Mozambique, *Studies in Family Planning*, 44, 3, 259–74.

Agadjanian, V. and Yabiku, S.T. (2014) Religious affiliation and fertility in a Sub-Saharan context: Dynamic and lifetime perspectives, *Population Research Policy Review*, 33, 673–91.

Aken'Ova, D. (2008) Sexual pleasure and safer sex: The under-explored synergy, *ID21 Insights* 75, p. 4 (see International Centre for Sexual Reproductive Rights (INCRESE) www.increse.org

Alava, H. (2017) Homosexuality, the holy family and a failed mass wedding in Catholic Northern Uganda, *Critical African Studies*, 9, 1, 32–51.

Ali, N. (2008) Outrageous state, sectarianized citizens: Deconstructing the textbook controversy in the Northern Areas, Pakistan, *South Asia Multidisciplinary Academic Journal*, 2, http://samaj.revues.org/index1172.html (accessed 2/3/12).

Amadiume, I. (2005) Family and culture in Africa, in Essed, P., Goldberg, D.T. and Kobayashi, A. (eds) *A Companion to Gender Studies*, Malden, MA: Wiley-Blackwell, 357–69.

Anand, A.S. (2014) Ethical selfhood and the status of the secular: Muslim identity in Mumbai, *Culture and Religion*, 15, 4, 377–98.

Ashforth, A. (2011) AIDS, religious enthusiasm and spiritual insecurity in Africa, *Global Public Health*, 6, Supp 2, S132–47.

Asia Foundation (2012) *National Values Survey, Sri Lanka 2011*, Colombo: The Asia Foundation.

Avishai, O. (2008) 'Doing religion' in a secular world: Women in conservative religions and the question of agency, *Gender and Society*, 22, 4, 409–33.

Avishai, O., Jafar, A. and Rinaldo, R. (2015) A gender lens on religion, *Gender and Society*, 29, 1, 5–25.

AWID, ARROW, SRI AND WCC (2015) *The Impact of Religious Fundamentalisms and Extreme Interpretations of Religion on Women's Human Rights*, Association for Women's Rights in Development, Asia-Pacific Research and Resource Centre for Women, Sexual Rights Initiative and World Council of Churches, www.awid.org/sites/default/files/atoms/fles/rfs_cedaw_briefing _paper_no15.pdf

Balchin, C. (2009) *Religious Fundamentalisms on the Rise: A Case for Action*, Action for Women in Development (AWID) https://www.awid.org/sites/default/files/atoms/files/religious_fundamentalisms_on_the_rise_a_case_for_action.pdf accessed 07/01/2019

Bano, M. (2010) *Madrasas in Pakistan: A Response to Modernity*, Birmingham: University of Birmingham, Religions and Development WP 45.

Becker, F. (2009) Competing explanations and treatment choices: Muslims, AIDS and ARVs in Tanzania, in Becker, F. and Geissler, P.W. (eds) *AIDS and Religious Practice in Africa*, Leiden: Brill, Studies of Religion in Africa, Vol. 36, 155–88.

Becker, F. and Geissler, P.W. (2009) Introduction: Searching for pathways in a landscape of death: Religion and AIDS in Africa, in Becker, F. and Geissler, P.W. (eds) *AIDS and Religious Practice in Africa*, Leiden: Brill, Studies of Religion in Africa, Vol. 36, 1–25.

Beckman, N. (2009) AIDS and the power of God: Narratives of decline and coping strategies in Zanzibar, in Becker, F. and Geissler, P.W. (eds) *AIDS and Religious Practice in Africa*, Leiden: Brill, Studies of Religion in Africa, Vol. 36, 119–54.

Behrend, H. (2009) The rise of occult powers, AIDS and the Roman Catholic Church in Western Uganda, in Becker, F. and Geissler, P.W. (eds) *AIDS and Religious Practice in Africa*, Leiden: Brill, Studies of Religion in Africa, Vol. 36, 29–48.

Beta, A.R. (2014) Hijabers – how young urban Muslim women redefine themselves in Indonesia, *The International Communication Gazette*, 76, 4–5, 377–89.

Blackburn, S. (2008) Indonesian women and political Islam, *Journal of Southeast Asian Studies*, 39, 1, 83–105.

Borooah, V.K. and Iyer, S. (2005) Religion, literacy, and the female-to-male ratio, *Economic and Political Weekly*, 40, 5, 419–27.

Bradley, T. (2011) The politicisation of mothering in Hindu missions, *Politics, Religion and Ideology*, 12, 2, 161–77.

Bradley, T. and Saigol, R. (2012) Religious values and beliefs and education for women in Pakistan, *Development in Practice*, 22, 5–6, 675–88.

Brenner, S. (1996) Reconstructing self and society: Javanese women and 'the veil', *American Ethnologist*, 23, 24, 673–97.

Brenner, S. (2011) Private moralities in the public sphere: democratization, Islam, and gender in Indonesia, *American Anthropologist*, 118, 3, 478–90.

Bruce, S. and Yearley, S. (2006) The SAGE Dictionary of Sociology, London: Sage (electronic book).

Bucar, E. (2017) *Pious Fashion: How Muslim Women Dress*, Cambridge, MA: Harvard University Press.

Burchardt, M. (2010) Ironies of subordination: Ambivalences of gender and religion. AIDS interventions in South Africa, *Oxford Development Studies*, 38, 1, 63–82.

Burchardt, M. (2011) Challenging Pentecostal moralism: Erotic geographies, religion and sexual.

Burke, K.C. (2012) Women's agency in gender-traditional religions: A review of four approaches, *Sociology Compass*, 6, 2, 122–33.

Bush, R. (2014) *A Snapshot of Muhammadiyah: Social Change and Shifting Markers of Identity and Values*, Singapore: Asia Research Institute WP 221.

Campbell, C., Skovdal, M. and Gibbs, A. (2010) Creating social spaces to tackle AIDS-related stigma: Reviewing the role of church groups in Sub-Saharan Africa, *AIDS and Behaviour*, 15, 1204–19.

Casanova, J. (2009) Religion, politics and gender equality: Public religions revisited, in Razavi, S. (ed) *A Debate on the Public Role of Religion and Its Social and Gender Implications*, Geneva: UNRISD, Gender and Development Programme Paper Number 5, 1–33.

Chingwete, A., Richmond, S. and Alpin, C. (2014) *Support for African Women's Equality Rises*, Afrobarometer, Policy Paper #8, www.afrobarometer.org

Christiansen, C. (2009) The new wives of Christ: Paradoxes and potentials in the remaking of widow lives in Uganda, in Becker, F. and Geissler, P.W. (eds) *AIDS and Religious Practice in Africa*, Leiden: Brill, Studies of Religion in Africa, Vol. 36, 85–117.

Cornwall, A. and Jolly, S. (2006) Introduction: Sexuality matters, *IDS Bulletin*, 37, 5, 1–11.

Corrëa, S. (2015) Reproductive and sexual rights: Transnational trends from a global South perspective, Wright, J.D. (ed) *International Encyclopedia of the Social & Behavioral Sciences*, 2nd ed, Elsevier, 457–67.

Davies, S.G. (2018) Skins of morality: Bio-borders, ephemeral citizenship and policing women in Indonesia, *Asian Studies Review*, 42, 1, 69–88.

Desai, S. and Temsah, G. (2014) Muslim and Hindu women's public and private behaviors: Gender, family, and communalized politics in India, *Demography*, 51, 6, 2307–32.

Dilger, H. (2009) Healing the wounds of modernity: Salvation, community and care in a neo-Pentecostal church in Dar es Salaam, Tanzania, in Becker, F. and Geissler, P.W. (eds) *AIDS and Religious Practice in Africa*, Leiden: Brill, Studies of Religion in Africa, Vol. 36, 255–82.

Dodzo, M.K., Mholyi, M., Moyo, S. and Dozo-Masawi, M. (2016) Paying until death: Apostolicism, delays and maternal mortality in Zimbabwe, *PLOS One*, 11, 8, e0160170. Doi: 10.1371/journal.pone.0160170 Open access online journal.

Dugbazah, J. and Labeodan, K. (2010) A promise or threat: The interface between religion, gender relations and women's education in Nigeria, *Research for Development*, 24, 1 & 2, 133–56.

Dugbazah, J., Labeodan, K. and Dadughun, S. (2010) *Envisioning Knowledge: Christian and Muslim Perspectives and Women's Education in Nigeria*, Birmingham: University of Birmingham, Religions and Development Research Programme, unpublished draft paper.

Dulani, B., Sambo, G. and Dionne, K.Y. (2016) *Good Neighbours? Africans Express High Levels of Tolerance for Many, But Not for All*, Afrobarometer Round 16, Despatch No 74.

Epprecht, M. (2012) Religion and same-sex relationships in Africa, in Bongmba, E.K. (ed) *The Wiley-Blackwell Companion to African Religions*, Malden and Oxford: Wiley-Blackwell, 515–28.

Falk, M.L. (2015) Religion and gender in developing South-East Asia, in Tomalin, E. (ed) *The Routledge Handbook of Religions and Global Development*, London: Routledge, 266–79.

Farrell, M., Masquelier, A., Tissot, E. and Bertran, J. (2014) Islam, polygyny and modern contraceptive use in Francophone sub-Saharan African countries, *African Population Studies*, 28, 3, 1389–98.

Forman-Rabinovici, A. and Sommer, U. (2018) An impediment to gender equality? Religion's influence on development and reproductive policy, *World Development*, 105, 48–58.

Ghatak, A. (2006) Faith, work, and women in a changing world: The influence of religion in the lives of *beedi* rollers in West Bengal, *Gender and Development*, 14, 3, 375–83.

Gregson, S., Zhuwau, T., Anderson, R.M. and Chandiwana, S.K. (1999) Apostles and Zionists: The influence of religion on demographic change in rural Zimbabwe, *Population Studies*, 53, 2, 179–93.

Grünenfelder, J. (2013) Negotiating gender relations: Muslim women and formal sector employment in Pakistan's rural development sector, *Gender, Work and Organization*, 20, 6, 599–615.

Gyimah, S.O., Takyi, B. and Tenkorang, E.Y. (2008) Denominational affiliation and fertility behaviour in an African context: An examination of couple data from Ghana, *Journal of Biosocial Science*, 40, 3, 445–58.

Halim, A.M.A. (1999) Reconciling the opposites: Equal but subordinate, in Howland, C.S. (ed) *Religious Fundamentalisms and the Human Rights of Women*, Basingstoke: Palgrave Macmillan, 203–13.

Haniffa, F. (2008) Piety as politics amongst Muslim women in contemporary Sri Lanka, *Modern Asian Studies*, 42, 2–3, 347–75.

Hassan, I. (2010) *Muslim Women in Development: Perceptions from Muslims in Jos, Central Nigeria*, Birmingham: University of Birmingham, Religions and Development Research Programme, unpublished draft paper.

Horn, J. (2010) *Christian Fundamentalisms and Women's Rights in the African Context: Mapping the Terrain*, Action for Women's Rights in Development (AWID) https://www.awid.org/sites/default/files/atoms/files/feminists_on_the_frontline_-_christian_fundamentalisms_and_womens_rights_in_the_african_context.pdf accessed 7/01/2019

Horn, J. (2012) *Not as Simple as ABC: Christian Fundamentalisms and HIV and AIDS Responses in Africa*, Toronto and Mexico City: AWID (Association for Women's Rights in Development). www.awid.org/sites/default/files/atoms/files/not_as_simple_abc.pdf

Hunt, S.J. (2012) The queer case of Hinduism: Religious discourse and the legitimacy of non-heterosexuality, in Hunt, S.J. and Yip, A.K.T. (eds) *The Ashgate Companion to Contemporary Religion and Sexuality*, Aldershot: Ashgate, 223–39.

Imam, A. (2005) Women's reproductive and sexual rights and the offence of Zina in Muslim laws in Nigeria, in Chavkin, W. and Chesler, E. (eds) *Where Human Rights Begin – Health, Sexuality and Women in the New Millennium*, New Brunswick: Rutgers University Press.

Imam, A., Gokal, S. and Marler, I. (2017) The devil is in the details: A feminist perspective on development, women's rights and fundamentalisms, *Gender and Development*, 25, 1, 15–36.

Iyer, S. (2003) Religion, reproduction and development in contemporary India, *Development*, 46, 4, 50–6.

Iyer, S. and Joshi, S. (2013) Missing women and India's religious demography, *Journal of South Asian Development*, 8, 3, 301–31.

Jenkins, P. (2006) *The New Faces of Christianity: Believing the Bible in the Global South*, New York: Oxford University Press.

Jones, C. (2007) Fashion and faith in urban Indonesia, *Theory: The Journal of Dress, Body and Culture*, 11, 21, 211–32.

Kabeer, N. (1999) Resources, agency, achievements: Reflections on the measurement of women's empowerment, *Development and Change*, 30, 435–64.

Kantor, P. (2002) Female mobility in India: The influence of seclusion norms on economic outcomes, *International Development Planning Review*, 24, 2, 145–59.

Khurshid, A. (2015) Islamic traditions of modernity: Gender, class, and Islam in a transnational women's education project, *Gender and Society*, 29, 1, 98–121.

Kipacha, A. and Dugbazah, J. (2010) *Discourses on Muslim Women's Education and Development in Tanzania: Insights from Tanga*, Birmingham: University of Birmingham, Religions and Development unpublished paper.

Kirmani, N. (2008a) Competing constructions of 'Muslim-ness- in the South Delhi neighbourhood of Zakir Nagar, *Journal of Muslim Minority Affairs*, 28, 3, 355–70.

Kirmani, N. (2008b) Constructing 'the other': Narrating religious boundaries in Zakir Nagar, *Contemporary South Asia*, 16, 4, 397–412.

Kirmani, N. (2009) Deconstructing and reconstructing 'Muslim women' through women's narratives, *Journal of Gender Studies*, 18, 1, 47–62.

Kirmani, N. and Phillips, I. (2011) Engaging with Islam to promote women's rights: Exploring opportunities and challenging assumptions, *Progress in Development Studies*, 11, 2, 87–99.

Ladbury, S. and Khan, S. (2008) *Increased Religiosity among Women in Muslim Majority Countries*, Birmingham: Governance and Social Development Resource Centre, issues paper.

Mahdi, H. (2007) The *hijab* in Nigeria, the woman's body and the feminist private/public discourse, paper for conference on Islam and the Public Sphere in Africa, Institute for the Study of Islamic Thought in Africa, Northwestern University, 17–19 May 2007.

Mantell, J.E., Correale, J., Adams-Skinner, J. and Stein, Z.A. (2011) Conflicts between conservative Christian institutions and secular groups in sub-Saharan Africa: Ideological discourses on sexualities, reproduction and HIV/AIDS, *Global Public Health*, 6, sup2, S192–S209.

McClendon, D., Hackett, C., Potanĉoková, Stonawski, M. and Skirbekk, V. (2018) Women's education in the Muslim world, *Population and Development Review*, 44, 2, 311–42.

McQuillan, K. (2004) When does religion influence fertility? *Population and Development Review*, 30, 1, 25–56.

Mesaki, S. and Kipacha, A. (2008) *The Relationship between Religious Teachings, Values and Beliefs and Concepts of Development in Tanzania*, Birmingham: Religions and Development Research Programme, unpublished research notes.

Mir-Hosseini, Z. (2011) Beyond 'Islam' vs 'Feminism', *IDS Bulletin*, 42, 1, 67–77.

Moore, H.L. (2002) Understanding sex and gender, in Ingold, T. (ed) *Companion Encyclopedia of Anthropology*, London and New York: Routledge, 2nd ed., 813–30.

Mouser, A.E. (2007) Defining 'modern' Malay womanhood and the coexistent messages of the veil, *Religion*, 37, 2, 164–74.

Mujani, S. and Liddle, W. (2004) Politics, Islam and public opinion, *Journal of Democracy*, 15, 1, 109–23.

Nisa, E.F. (2012) Embodied faith: Agency and obedience among face-veiled university students in Indonesia, *The Asia-Pacific Journal of Anthropology*, 13, 4, 366–81.

Njoh, A.J. and Akiwumi, F.A. (2012) The impact of religion on women empowerment as a Millennium Development Goal in Africa, *Social Indicators Research*, 107, 1–18.

Nussbaum, M. (2003) Gender and governance: An introduction, in UNDP (ed) *Essays on Gender and Governance*, New Delhi: United Nations Development Programme, Human Resource Development Centre, 1–19.

O'Flaherty, M. and Fisher, J. (2008) Identity and international human rights law: Contextualising the Yogyakarta Principles, *Human Rights Law Review*, 8, 2, 207–48.

Olivier, B. (2016) The Malaysian Islamization phenomenon: The underlying dynamics and their impact on Muslim women, *Islam and Christian-Muslim Relations*, 27, 3, 267–82.

Olivier, J. and Wodon, Q. (2015) Religion, reproductive health and sexual behaviour in Ghana: Why statistics from large surveys don't tell the whole story, *The Review of Faith and International Affairs*, 13, 2, 64–73.

Ong, A. (1995) State vs Islam – Malay families, women's bodies and the body politic in Malaysia, in Ong, A. and Peletz, M. (eds) *Bewitching Women, Pious Men: Gender and Body Politics in Southeast Asia*, Berkeley: University of California Press.

Para-Mallam, F.J. (2006) Faith, gender and development agendas in Nigeria: Conflicts, challenges, and opportunities, *Gender and Development*, 14, 3, 409–13.

Para-Mallam, F.J. (2010) Promoting gender equality in the context of Nigerian cultural and religious expression: Beyond increasing female access to education, *Compare: A Journal of Comparative and International Education*, 40, 4, 459–77.

Parmanand, S. (2014) Mapping the path to Philippines reproductive rights legislation: Signs of progress amidst obstacles, *Social Transformations*, 2, 1, 61–80.

Pearce, T.O. (2012) Reconstructing sexuality in the shadow of neoliberal globalization: Investigating the approach of Charismatic churches in southwestern Nigeria, *Journal of Religion in Africa*, 42, 4, 345–68.

Pearson, R. and Tomalin, E. (2008) Intelligent design? A gender-sensitive interrogation of religion and development, in Clarke, G. and Jennings, M. (eds) *Development, Civil Society and Faith-Based Organisations: Bridging the Sacred and the Secular*, Basingstoke: Palgrave Macmillan, 46–71.

Pereira, C. and Ibrahim, J. (2010) On the bodies of women: The common ground between Islam and Christianity in Nigeria, *Third World Quarterly*, 31, 6, 921–37.

Perry, T.E., Davis-Maye, D. and Onolemhemhen, D.N. (2010) Faith spirituality, fatalism and hope: Ghanaian women coping in the face of HIV/AIDS, *Journal of HIV/AIDS and Social Services*, 6, 4, 37–58.

Pew Forum (2006) *Spirit and Power: A 10-Country Study of Pentecostals*, Washington, DC: Pew Forum, http://pewforum.org/uploadedfiles/Orphan_Migrated_Content/pentocostals-08.pdf

Pew Forum (2010) *Tolerance and Tension: Islam and Christianity in Sub-Saharan Africa*, Washington, DC: Pew Forum, http://pewforum.org/uploadedFiles/Topics/Belief_and_Practices/sub-saharan-africa-full-report.pdf

Pew Research Center (2012) *The World's Muslims: Unity and Diversity*, Washington, DC: Pew Research Center, http://assets.pewresearch.org/wp-content/uploads/sites/11/2012/08/the-worlds-muslims-full-report.pdf

Pew Research Center (2013) *The World's Muslims: Religion, Politics and Society*, Washington, DC: Pew Research Center, http://assets.pewresearch.org/wp-content/uploads/sites/11/2013/04/worlds-muslims-religion-politics-society-full-report.pdf

Phillips, A. (2009) Religion: Ally, threat or just religion? in Razavi, S. (ed) *A Debate on the Public Role of Religion and its Social and Gender Implications*, Geneva: UNRISD, Gender and Development Programme Paper Number 5, 34–58.

Platt, M., Davies, S.G. and Bennett, L.R. (2018) Contestations of gender, sexuality and morality in contemporary Indonesia, *Asian Studies Review*, 42, 1, 1–15.

Prince, R. (2009) Christian salvation and Luo tradition: Arguments of faith in a time of death in Western Kenya, in Becker, F. and Geissler, P.W. (eds) *AIDS and Religious Practice in Africa*, Leiden: Brill, Studies of Religion in Africa, Vol. 36, 49–83.

Pugh, S.A. (2010) Examining the interface between HIV/AIDS, religion and gender in Sub-Saharan Africa, *Canadian Journal of Development Studies*, 44, 3, 624–43.

Qazi, M. (2015) Islam and women: The new paradigm, *New Age Islam*, 20 Sept. www.newageislam.com/islam,-women-and-feminism/moin-qazi,-new-age-islam/islam-and-women – the-new-paradigm/d/104651

Qibtiyah, A. (2012) The conceptualisation of gender issues among gender activists and scholars in Indonesian universities, *Intersections: Gender and Sexuality in Asia and the Pacific*, 30.

Rinaldo, R. (2008) Muslim women, middle class habitus and modernity in Indonesia, *Contemporary Islam*, 2, 23–39.

Robinson, K. (2012) Islam and gender relations in Indonesia, with a special focus on Eastern Indonesia, *Intersections: Gender and Sexuality in Asia and the Pacific*, 30.

Sadgrove, J. (2009) 'Keeping up appearances': Sex and religion among university students in Uganda, in Becker, F. and Geissler, P.W. (eds) *AIDS and Religious Practice in Africa*, Leiden: Brill, Studies of Religion in Africa, Vol. 36, 223–54 (earlier published in the *Journal of Religion in Africa*, 37, 1, 116–44, 2007).

Saigol, R. (1999) Homemakers and homebreakers: The binary construction of women in Muslim nationalism, in Thiruchandran, S. (ed) *Women, Narration and Nation: Collective Images and Multiple Identities*, New Delhi: Vikas.

Saigol, R. (2003) His rights, her duties: Citizen and mother in the civics discourse, *Indian Journal of Gender Studies*, 10, 3, 379–404.

Sait, S. and Lim, H. (2006) *Land, Law and Islam: Property and Human Rights in the Muslim World*, London: Zed.

Sakai, M. and Fauzia, A. (2016) Performing Muslim womanhood: Muslim businesswomen moderating Islamic practices in contemporary Indonesia, *Islam and Islam-Christian Relations*, 27, 3, 1–21.

Sakai, M. and Yasmeen, S. (2016) Narratives of Muslim womanhood and women's agency, *Islam and Christian-Muslim Relations*, 27, 4, 371–7.

Seguino, S. (2011) Help or hindrance? Religion's impact on gender inequality attitudes and outcomes, *World Development*, 39, 3, 1308–21.

Stivens, M. (2006) 'Family values' and Islamic revival: Gender, rights and state moral projects in Malaysia, *Women's Studies International Forum*, 29, 354–67.

Svensson, J. (2009) 'Muslims have instructions': HIV/AIDS, modernity and Islamic religious education in Kisumu, Kenya, in Becker, F. and Geissler, P.W. (eds) *AIDS and Religious Practice in Africa*, Leiden: Brill, Studies of Religion in Africa, Vol. 36, 189–221.

Tomalin, E. (2013) Gender, religion and development, in Clarke, M. (ed) *Handbook of Research on Religion and Development*, Cheltenham: Edward Elgar, 183–200.

Tomalin, E. and Starkey, C. (2013) Buddhism and development, in Clarke, M. (ed) *Handbook of Research on Religion and Development*, Cheltenham: Edward Elgar, 31–50.

Tong, J. K.C. and Turner, B.S. (2010) Women, piety and practice: A study of women's religious practice in Malaysia, *Contemporary Islam*, 2, 1, 41–59.
UNICEF (2005) *Childhood Under Threat: The State of the World's Children 2005*, New York: UNICEF.
Vaggione, J.M. (2008) *Shared Insights: Women's Rights Activists Define Religious Fundamentalisms*, AWID (Association for Women's Rights in Development) www.awid.org
Van Dijk, R. (2009) Gloves in times of AIDS: Pentecostalism, hair and social distancing in Botswana, in Becker, F. and Geissler, P.W. (eds) *AIDS and Religious Practice in Africa*, Leiden: Brill, Studies of Religion in Africa, Vol. 36, 283–307.
Van Eerdewijk, A. (2012) Virginity in a multidimensional perspective: Negotiating premarital sex in Dakar, in Hunt, S.J. and Yip, A.K.T. (eds) *The Ashgate Companion to Contemporary Religion and Sexuality*, Aldershot: Ashgate, 143–56.
Van Klinken, A.S. (2011a) St. Joachim as a model of Catholic manhood in times of AIDS: A case study on masculinity in an African Christian context, *Crosscurrents*, December, 467–79.
Van Klinken, A.S. (2011b) Male headship as male agency: An alternative understanding of a 'patriarchal' African Pentecostal discourse on masculinity, *Religion and Gender*, 1, 1, 104–24.
Van Klinken, A.S. (2012) Men in the remaking: Conversion narratives and born-again masculinity in Zambia, *Journal of Religion in Africa*, 42, 215–39.
Van Klinken, A.S. (2013) Gay rights, the devil and the end times: Public religion and the enchantment of the homosexuality debate in Zambia, *Religion*, 43, 4, 519–40.
Van Klinken, A.S. (2014) Homosexuality, politics and Pentecostal nationalism in Zambia, *Studies in World Christianity*, 20, 3, 259–81.
Van Klinken, A. (2015) Queer love in a 'Christian nation': Zambian gay men negotiating sexual and religious identities, *Journal of the American Academy of Religion*, 83, 4, 947–64.
Van Klinken, A. and Chitando, E. (2015) Masculinities, HIV and religion in Africa, in Tomalin, E. (ed) *The Routledge Handbook of Religions and Global Development*, London: Routledge, 127–37.
Van Klinken, A. and Phiri, L. (2015) 'In the image of God': Reconstructing and developing a grassroots African queer theology from urban Zambia, *Theology and Sexuality*, 21, 1, 36–52.
Vera-Sanso, P. (1995) Community, seclusion and female labour force participation in Madras, India, *Third World Planning Review*, 17, 2, 155–67.
Warouw, N. (2016) Negotiating modernity: Women workers, Islam and urban trajectory in Indonesia, *Islam and Christian-Muslim Relations*, 27, 3, 283–302.
White, S. and Anshor, M.U. (2007) Islam and gender in contemporary Indonesia: Public discourses on duties, rights and morality, in Fealy, G. and White, S. (eds) *Expressing Islam: Religious Life and Politics in Indonesia*, Singapore: Institute of Southeast Asian Studies, 137–58.
White, S. and Tiongco, R. (1997) *Doing Theology and Development: Meeting the Challenge of Poverty*, Edinburgh: Saint Andrew Press.
Willemse, W. (2007) 'In my father's house: Gender, Islam and the construction of a gendered public sphere in Darfur, Sudan, *Journal for Islamic Studies*, 27, 73–115.

10 Conclusion

Religion is entangled in most people's lives in Sub-Saharan African and southern Asian countries, but our knowledge of the complex interactions between it, everyday living and social relationships in these diverse contexts is uneven. This book is not concerned with advocacy: it does not argue for or against a positive role for religion in society. Rather, it seeks to develop a better understanding of the similarities and differences between and within religious traditions and how these influence the thinking, religious practices and lifeworlds of people who live in contexts in which religious affiliation is widespread, religious beliefs and practices are persistent (but not static) and the implications of religion in rapidly changing social circumstances are tricky to negotiate.

To improve understanding of the ways in which the teachings of various religious traditions are understood, embodied in practices and pervade social relationships in these religiously diverse regions of the world, the first requirement is basic literacy in the discourses, practices and organisational arrangements of the main religious traditions. Second, a review of available research can reveal the areas in which our knowledge is extensive and it is possible to move towards general explanations and conclusions, and those where the evidence is patchy or absent. This book provides an introduction to the beliefs, practices and values associated with the main religious traditions for social scientists unfamiliar with the religious field (or with traditions other than their own). Its main contribution, however, is to review the available evidence on how these are interpreted, reflected in religious and social practices, and interact with social and cultural characteristics and dynamics in a variety of Sub-Saharan African and southern Asian contexts.

To achieve these aims, it provides an analytical framework that specifies the domains and dimensions of the religious and social fields and their intersections at various levels, from the individual to the international. One axis of the analytical matrix distinguishes between three dimensions: a personal dimension that seeks to understand what religion means to individuals and how it informs their views, motivations and actions; a social dimension that reflects the ways in which religion interacts with social processes through the operation of institutions that regulate social relationships; and an organisational dimension that recognises the ways religious traditions seek to sustain themselves (and sometimes expand their reach), as well as influencing the practices of adherents, relations between religious groups and interactions between such groups and other social institutions and structures. The analysis concentrates on the personal dimension, examining how people engage with religious discourses and practices as individuals and how this engagement is reflected in their social interactions in basic social units, especially families, and at the local level, while acknowledging that these cannot be understood without acknowledging wider religious, social, cultural and political dynamics and influences – the second axis of the framework.

Intersecting with these dimensions and levels is a third axis, which draws on Lincoln's conceptualisation of social fields as having four domains, with the religious field distinguished from others by its "desire to speak of things eternal and transcendent with an authority that is equally transcendental and eternal" (Lincoln, 1996, p. 225). The first of these domains is an idiom or discourse that makes it possible for people to describe an experienced reality in which material and spiritual dimensions are intertwined, often captured in a set of beliefs which provide a meaning system and ways of dealing with some of the questions that preoccupy human beings (such as death, evil and the relative significance of fate/agency). The second religious domain, Lincoln suggests, is a set of practices that embody the religious discourses. In this book, an analytical distinction has been made between religious and social practices, although adherents may not perceive them as separate and distinct. For many, the former, which include a range of ritual practices, play important roles in the construction of their identity and personhood, as well as in their daily lives, although of course the nature, frequency and regularity of participation and the significance of religious practices in people's conceptions of themselves varies between religions and individuals. It is generally assumed that those who share common beliefs, values and norms associate themselves with a particular religious tradition, sect, denomination or organisation and that their beliefs, norms and behaviour can be read off from that association (Lincoln's third domain). In practice, association with a religion can take a variety of forms, ranging from loose affiliation to formal membership. As has become clear, the nature of such association influences attitudes to official teaching and the intensity and frequency of engagement in religious practices, as well as the relative importance of religion and other factors in influencing attitudes, behaviour and social relationships, often varying as much within religions as between them. Finally, Lincoln suggests that, in order to understand the ways in which religion is imbricated in everyday life, it is necessary to examine the institutions that reproduce, regulate and/or modify a religious discourse, set of practices and community. These institutions provide guidance on 'correct' interpretations of the relevant texts, doctrines and myths and on specified religious and social practices, as well as regulating compliance. Religious organisations seek to set out and implement the rules through specifying values and moral norms that are supposed to govern individual behaviour and day-to-day social relationships, as well as developing blueprints for their desired relationships with socio-political regimes.

As a historian of religion, Lincoln has used this conceptualisation in his own work on the ways in which religion, culture and politics interact, focusing on the ways in which religious ideology, expressed through various forms of discourse, influences the construction, deconstruction and reconstruction of social groups and collective identities and provides a launching pad for and ingredient in political conflict (Lincoln, 2006, 2014).[1] From the outset, while acknowledging that the social, cultural and political roles of religion can only be understood by examining both the characteristics of each domain and interactions between them, Lincoln himself seems to blur the conceptual distinctions, for example portraying discourse as a separate domain from ritual practice in his earlier work (Lincoln, 2006) but later presenting ritual as a form of discourse (Lincoln, 2014). While much of his empirical work is historical, he has also applied this conceptualisation to analyses of contemporary socio-political phenomena, especially American evangelical Christianity and political Islam, which he regards as maximalist strands which hold that religious should infuse all of life, including politics (Lincoln, 2006). Here, my focus on individuals and social groups means that, although these are related to larger-scale social phenomena (religious traditions, societies, politics) and trends (globalisation, persistent religiosity and revivalist movements within the religious traditions), some elements of Lincoln's concerns with culture, societies and politics are less relevant than his basic conceptualisation. In the analysis in this volume, the conceptualisation of religion as having four interrelated domains has demonstrated its value, although, as argued in my initial review and subsequent application, practices can usefully be disaggregated into religious and social practices (while acknowledging that they overlap and interact).

Religions are thought to provide a framework of values, ethics and moral norms which guide social practices, the lives of individual adherents and the teachings and behaviour of religious organisations, especially in societies in which most people associate themselves with a religious tradition, sect or denomination. However, the theoretical and practical links between religious affiliation, the values espoused, the translation of values into ethical and moral ideas and prescriptions and the ways in which these may (or may not) be reflected in day-to-day actions and social relationships cannot be assumed. The links between values, morals and behaviour are extremely complex. Not only are they influenced by many factors other than religious teachings, it is methodologically difficult to identify them with any degree of confidence, and few studies attempt to do so except by relying on the researcher's own observations and interpretations. In addition, as noted above, when considering how religious groups are governed and reproduced, a distinction must be made between institutions (formal and informal social rules) and organisations, in line with recent social science theorising.

1 Some discussions of Lincoln's work are specific to the internal dynamics of and debates within religious studies and are neither drawn upon nor critiqued in this book.

The introduction to the beliefs, practices and values associated with each of the religious traditions presented in this volume draws on a variety of secondary sources. Guided by the analytical framework, which informs the thematic structure of the book, material is presented on teachings and beliefs, religious practices, values and guides to behaviour that will be new to some readers and well known to others. For each tradition, the evidence about how teachings are interpreted and reflected in social relationships at the local level has been sourced from available research using both quantitative and qualitative approaches.

Recent international datasets yield evidence on the prevalence of certain beliefs, practices and attitudes in a varying range of African and southern Asian countries. However, they rarely provide in-depth comparisons, analyses or explanations. In contrast, individual research studies generate illustrative case studies of particular phenomena and religious groups. Many provide thick description, useful insights and persuasive explanations, but few use a comparative methodology, and their coverage is often context and religion specific. They are studies of particular towns, neighbourhoods, villages, groups or organisations in large and varied countries, and so their coverage of issues, contexts and religious groups is selective, and their conclusions may not be generalisable.

Often the attention of researchers, analysts and political actors is caught by the new, dramatic or alarming. This is reflected in the choice of topics by funders, researchers and publishers (including many referred to in this volume). Although emerging trends and distinctive phenomena may be important and do often influence the quotidian, such a focus has led many researchers and activists to neglect the ordinary, everyday and continuing, which affect the lives of many more people. A focus on the unusual can skew understanding, exaggerate the significance of phenomena such as the emergence of extremist political and religious groups and result in blind spots, because it is taken for granted that we have a good understanding of well-established and ongoing social and religious phenomena and groups.

The analysis in this volume has relied on secondary sources in the English language, including interpretations of the teachings of the religious traditions under consideration by both insiders and outsiders and findings from international sample surveys and ethnographic studies. The synthesis it provides, with its focus on religion and society in Sub-Saharan Africa and southern Asia, is long overdue. Inevitably, however, its coverage is partial, both of the variety within religious traditions and of the societies and cultures of the two sub-continents. In part this is because it seeks to provide an overview, at the risk of superficiality and overgeneralisation. But it is also because of the superficiality of many international (and national) surveys, the lack of methodologically sound comparative research and the patchy subject matter, geographical location and social subjectivity of individual small-scale case studies. Single-case studies can be extremely valuable in generating in-depth understanding, but there is a need for more comparative studies and replication of existing research to assess whether the findings of the case studies available are reinforced, refined or undermined by further research. The coverage, nature and quality of the available evidence therefore varies, restricting the extent to which general conclusions and explanations can be drawn. Instead, based on a consideration of the domains, dimensions and levels of analysis set out in the analytical framework, the analysis has started to weave a picture of the complex links between religions, everyday lives and societies in SSA and southern Asia – a tapestry rather than a grand theory.

It demonstrates, first, as expected, that 'beliefs' are more central to some religious traditions than others, which place more emphasis on states of being or orthopraxis. All are constructed in the interactions between (versions of) the texts, historical experiences, cultural contexts and contemporary social trends. Some adherents hold beliefs because they have been taught them but many because they have grown up in a particular religious habitus:

> Adhering to a religion . . . is not predominantly a matter of personal preference in the minds of individual believers. . . [W]hat people value and find worthwhile doing and being are often unconsciously reproduced and people adopt religion-based values, and behaviours, or follow religious practices simply because this is what people do. . . . even when there are options for reflective choice . . . these are equally permeated and structured by social, political and indeed religious relations.
>
> (Devine and Deneulin, 2011, p. 73)

However, religious beliefs persist because they enable people to make sense of the world in which they live, in which, for many, supernatural forces (God, deities, the ultimate reality, benevolent and malevolent spirits) appear to be present and active and people have what they interpret as religious experiences. As found in research amongst people affiliated with a variety of religious traditions (especially Islam and Hinduism) in rural, urban and semi-urban locations in India and Bangladesh,

> religion is an inherently pragmatic affair permeating everyday concerns and mundane aspirations. . . . The risk is to read into this pragmatism that [adherents] are materialistic or superstitious, or that they are simply instrumental about religion. Nothing could be further from the truth. In our respondents' minds, the everyday and the supernatural, practical needs and transcendental hopes go hand in hand; they are complementary rather than mutually exclusive concerns.
>
> (Devine and Deneulin, 2011, p. 73)

The evidence confirms that the teachings and adherents' ideas about wellbeing and illbeing, evil, death and fate within and between the religious traditions have much in common, but also that there are important differences. For the most part, the traditions teach and adherents believe that, although many aspects of the world and their lives are beyond their control, people have agency – that personal behaviour and actions matter in achieving prosperity, overcoming evil, coming to terms with death and evading fate.

Much scholarship is based on an assumption that religious cosmologies, traditions, sects or denominations offer a coherent discourse and set of practices. However, it is clear from the historical evolution of religious traditions in different cultural contexts and the evidence reviewed that many people have hybrid or syncretic frameworks of thought and understanding and so seek to marshal various powers pragmatically to ensure their own well-being and protect themselves and others from harm. This is achieved in part through engagement in religious practices, the importance of which for many is demonstrated by regular participation in individual and group rituals, the significant resources devoted to constructing and maintaining spaces and buildings perceived as sacred, as well as religiously important occasions such as festivals and pilgrimages. For many, participation in religious practices, thought to be necessary to ensure well-being and counter evil, misfortune and ill health, is also rewarding. While it may demonstrate compliance with the powerful institutions and processes that organise and regulate observance or be instrumental (for example, to achieve social acceptance), it is not merely passive: it also enables participants to achieve a religious subjecthood and identity (Avishai, 2008). Many, especially the religiously observant, report that their religion inspires and motivates them, while living in accordance with religious teachings brings inner feelings of peace and strength. In a number of studies, for example, Muslim informants talked about the sense of beauty or peace they find in reading the Qur'an or doing their prayers (see, for example, White et al., 2012, p. 656), although many also acknowledge that they experience ongoing struggles to temper unruly feelings such as anger and greed. Religious beliefs and practices are not only an important part of people's daily lives, serving to form and reinforce understandings of selfhood and the world, helping to construct individual and social identities and providing ways of responding to rapid social change and catastrophe, even conflict and violence, they are intertwined with social relationships.

In part, these roles and relationships reflect the values and ethics associated with religious teachings. The analysis in this volume focuses on selected values and ethical stances that seem important with respect to religion and day-to-day behaviour, close social relationships and social change: freedom, tolerance, rights and justice; well-being, poverty and equality; altruism, philanthropy and service to others; honesty and integrity; and gender equality, sexuality and the family. It is important to bear in mind that, however pervasive religious adherence appears to be, as shown in the research reviewed here, religion

> is not the only force that inspires what people choose and do with their lives. . . [For example, respondents in research in India and Bangladesh considered that, while] it is important for children to follow the moral code of society and do what is right (honouring parents, respecting people, etc.), [they] judged it as equally important for their children to be educated and choose a life that they wish to have. It is precisely in the complex space that lies between the 'normative' (what God commands) and the 'everyday'

> where actual and real choices are made. Religion inhabits this space; informing both the normative and the everyday, but dominating neither.
>
> (Devine and Deneulin, 2011, p. 74)

As well as religion, the attitudes of religious adherents are shaped by factors such as class, level of education, cultural context and political regime. In addition to differences between religious traditions, there are often marked differences within the traditions, often along a liberal/conservative spectrum. My discussion of the extent to which religious values comply with or depart from internationally accepted benchmarks of rights, equality and integrity demonstrates that, while religious values and moral norms such as freedom, equality, tolerance and well-being may be sophisticated and highly developed, there are many ambivalences and contradictions. These may be between the values and moral norms themselves or between the international emphasis on rights on the one hand and the attitudes and behaviour of religious organisations, leaders and adherents on the other. There are clearly tensions between universal conceptions of individual human rights, some religious beliefs and teachings and more communal conceptions of society.

All societies and religions have means of enforcing compliance with relevant requirements and norms. The disciplinary mechanisms used (e.g. fasting, prayer, penance) vary. For a religion to impose discipline on its adherents, it needs an institutional and organisational apparatus with the power to enforce the prescribed sanctions, in which informal rules and institutions are as if not more important than formal rules and hierarchy, depending on the tradition. In addition, religion's entanglement with everyday life means that rather than relying on religious mechanisms alone, compliance is often ensured through other social institutions, such as the family, school or community, even if not always successfully.

Along with other dimensions of social difference, gender is intertwined with religion – it influences people's interpretation of religious teachings, their experience of religious practices and the ways in which values such as freedom, well-being and altruism are framed, expressed in ethical norms and influence behaviour in practice. Gender considerations have, where possible, been integrated into the analysis throughout, although this is hindered by the gender-blindness of much of the available material or its concentration on women rather than gender and gender relations. Like the discussion of other values, the analysis of gender has been benchmarked against the internationally accepted norms of human rights and equality. Arguably, gender is even more central to people's conception of themselves and their relations with others than dimensions of social difference such as class or ethnicity, because it is inextricably tied up with sexuality, which is vital to social and religious reproduction. The narratives of women (and men) reviewed here reveal diverse social realities, expectations and degrees of ability to exercise agency with respect to their everyday lives, participation in religion and interactions with others in their families and beyond.

Religious teachings and practices are, indeed, patriarchal, not least because they have evolved in social and cultural contexts in which the gender order was and is almost universally hierarchical, based on the idea of male headship and leadership. Like societies at large, religions reproduce these gendered attitudes, ideas and social relationships in ways that appear to leave little scope for change. Many women are pressured into compliance rather than actively choosing to adhere to religious moral and behavioural norms. Nevertheless, and despite the apparent backlash against feminist challenges from outside and within the religious traditions, religions do not operate in a bubble insulated from wider social, cultural, economic and political changes. Like social institutions and legal systems more generally, religious bodies are under pressure to change their views and practices relating to gender equality and sexual rights, whether they like it or not. Thus contextual changes in access to income earning opportunities, information, and education and health care, as well as in political and legal systems, interact with religion, influencing the evolution of values, moral norms and day-to-day behaviour. Although change is slow and the outcomes unpredictable and uneven, dismissal of *all* religion as irredeemably patriarchal and oppressive is oversimplified, because women themselves are increasingly aware of and willing to speak out about the nexus between religion and gender, there are differences between and within religious groups, and constructions of masculinity as well as femininity are being revisited.

What emerges from the review of evidence in this volume is that religions offer complex repertoires of beliefs and values to very many people, although they are clearly not the only source. They can provide not just a cosmological lens through which people believe that they can understand the world and their place in it but also instantiate practices, habits and norms and provide ideas about the right ordering of relationships in families and wider social groups. They thus influence the attitudes of different religious groups to spreading their ideas and recruiting new adherents, values, behaviour and the ways in which groups are organised, which affect the roles a religion plays in wider society and politics, as well as the lives of those affiliated with it. The research reviewed here indicates that religion provides for its adherents in contemporary societies in Sub-Saharan Africa and southern Asia, in no particular order, a valued identity, a sense of dignity and self-worth, a sense of purpose, a source of moral guidance, a means of legitimising the ways in which people live their lives (in families or the pursuit of prosperity), ways of coping with hardship and solving problems (for example, through healing or providing psychological resources) and a rich source of social capital. It provides "a kind of moral compass to help people live their lives in a proper manner" (Devine and Deneulin, 2011, p. 74). This does not mean that adherents, whether leaders or ordinary people, are uncritical of the religious traditions into which (mostly) they are born but that for most, the rewards of religious participation outweigh the disadvantages and burdens.

Religion can and often does establish, reproduce and normalise inequality, exclusion and intolerance through the use of discourses that claim authoritative status and evoke sentiments of affinity and estrangement that influence the construction of social identities and relations (Lincoln, 2014). For example, some religious discourses and practices legitimate discrimination against religious minorities or reinforce the disempowering aspects of culture, especially for women. However, religious discourses are constructed: they change over time in response to both external and internal stimuli, and are often characterised by tensions and contradictions, providing scope for new interpretations. In addition, religious conduct demonstrates not only compliance but also agency. Together, the exercise of agency by adherents, internal challenges and re-interpretation of religious teachings and norms can enable change to occur both within religious traditions and through deconstructing and reconstructing the prevailing social order. First, no religious tradition is homogeneous – all are characterised by internal differences, especially a conservative/liberal spectrum of religious views and social attitudes that reflect internal disputes about beliefs, values and practices. Second, some are quite fluid. For example, Bhatewara and Bradley (2013) show, in their study of Hindus and Buddhists in Pune, India, how the flexibility of Hinduism and Buddhism has offered disadvantaged people opportunities to re-negotiate their identities in a quest for greater dignity and social respect. Finally, contemporary movements within the religious traditions may offer liberatory potential, for example by acting as a source of social capital, enabling women to exercise agency or reinventing ideas about masculinity. The analysis in this volume has thrown light on how the teachings and principles of various religious traditions are interpreted and embodied in social practices, how social and historical processes have led to particular embodiments and how these are manifest in varied and changing contemporary circumstances. It contributes to understanding of how religious beliefs, practices and values shape the lives of adherents in the highly religious societies of Sub-Saharan Africa and southern Asia and reveals some of the ways in which, through their discourses, practices and actions, adherents in turn can shape the traditions with which they are associated.

References

Avishai, O. (2008) 'Doing religion' in a secular world: Women in conservative religions and the question of agency, *Gender and Society*, 22, 4, 409–33.

Bhatewara, Z. and Bradley, T. (2013) 'The people know they need religion in order to develop': Religion's capacity to inspire people in Pune's slums, *European Journal of Development Research*, 25, 288–304.

Devine, J. and Deneulin, S. (2011) Negotiating religion in everyday life: A critical exploration of the relationships between religion, choices and behaviour, *Culture and Religion*, 12, 1, 59–76.

Lincoln, B. (1996) Theses on method, *Method and Theory in the Study of Religion*, 8, 225–7.

Lincoln, B. (2006) *Holy Terrors: Thinking about Religion after September 11*, Chicago: University of Chicago Press, 2nd ed.

Lincoln, B. (2014) *Discourse and the Construction of Society: Comparative Studies of Myth, Ritual and Classification*, New York: Oxford University Press.

White, S.C., Devine, J. and Jha, S. (2012) The life a person lives: Religion, well-being and development in India, *Development in Practice*, 22, 5–6, 651–62.

Index

Note: **Bold** page numbers indicate tables and *italic* page numbers indicate figures.

For Product Safety Concerns and Information please contact our
EU representative GPSR@taylorandfrancis.com Taylor & Francis
Verlag GmbH, Kaufingerstraße 24, 80331 München, Germany